D1711758

Clifford Algebra to Geometric Calculus

Fundamental Theories of Physics

A New International Book Series on The Fundamental Theories of Physics: Their Clarification, Development and Application

Clifford Algebra to Geometric Calculus

A Unified Language for Mathematics and Physics

by

David Hestenes
Department of Physics,
Arizona State University

and

Garret Sobczyk
Institute for Theoretical Physics,
University of Wroclaw

D. Reidel Publishing Company

A MEMBER OF THE KLUWER ACADEMIC PUBLISHERS GROUP

Dordrecht / Boston / Lancaster / Tokyo

Library of Congress Cataloging in Publication Data

CIP

Hestenes, David, 1933–
Clifford algebra to geometric calculus.

(Fundamental theories of physics)
Includes bibliographical references and index.
1. Clifford algebras. 2. Calculus. I. Sobczyk, Garret, 1943–
II. Title. III. Title: Geometric calculus: a unified language for mathematics and physics.
IV. Series.
QA199.H47 1984 512′.57 84-8235
ISBN 90-277-1673-0

Published by D. Reidel Publishing Company,
P.O. Box 17, 3300 AA Dordrecht, Holland.

Sold and distributed in the U.S.A. and Canada
by Kluwer Academic Publishers,
190 Old Derby Street, Hingham, MA 02043, U.S.A.

In all other countries, sold and distributed
by Kluwer Academic Publishers Group,
P.O. Box 322, 3300 AH Dordrecht, Holland.

Reprinted with corrections, 1985

Printed in The Netherlands

Table of Contents

Preface

Matrix algebra has been called "the arithmetic of higher mathematics" [Be]. We think the basis for a better arithmetic has long been available, but its versatility has hardly been appreciated, and it has not yet been integrated into the mainstream of mathematics. We refer to the system commonly called 'Clifford Algebra', though we prefer the name '*Geometric Algebra*' suggested by Clifford himself.

Many distinct algebraic systems have been adapted or developed to express geometric relations and describe geometric structures. Especially notable are those algebras which have been used for this purpose in physics, in particular, the system of complex numbers, the quaternions, matrix algebra, vector, tensor and spinor algebras and the algebra of differential forms. Each of these *geometric algebras* has some significant advantage over the others in certain applications, so no one of them provides an adequate algebraic structure for all purposes of geometry and physics. At the same time, the algebras overlap considerably, so they provide several different mathematical representations for individual geometrical or physical ideas. Consequently, it is not uncommon for mathematicians and physicists to employ two or more different geometric algebras in a single problem, with considerable time required to translate expressions from one representation to another. This state of affairs is hardly satisfactory. A unified theory of the physical world calls for a unified mathematical language to help develop and express it. Geometry too seems to be a unified corpus of ideas, so one can hope to develop a single *Geometric Algebra* capable of expressing the full range of geometrical ideas in all their richness and subtlety. Any system claiming the title *Geometric Algebra* ought to perform at least as well as any one of the special geometric algebras mentioned above in any particular application. The purpose of this book is to show how such a system can be fashioned from Clifford Algebra.

Although Clifford Algebra is known to many mathematicians and physicists, its applications have been limited to a fairly narrow range. Its mathematical range is greatly extended in this book by considering the algebra from a new point of view and developing a system of definitions and algebraic identities to make it an efficient and versatile computational tool. Some of the more well-known mathematical applications of Clifford Algebra which are adequately treated in the literature are not mentioned in this book. Instead, the book is devoted to developing an efficient *Geometric Calculus* which can provide a unified account of the mathematics needed in theoretical physics.

Although every feature of Geometric Calculus has been developed with an eye to the needs of physics, the physical applications are too extensive to be treated in this book. Only one rather esoteric application is considered in detail for its mathematical interest, namely the Petrov classification of curvature tensors (see Section 3-9). However, a broad range of mathematical applications of Geometric Calculus are worked out, from linear and multilinear algebra to differential geometry and Lie groups. Of course, such mathematical applications have, in turn, many applications to physics.

To present the extensive physical applications of Geometric Calculus systematically, a series of books complementing the present volume has been planned. The first two books in the series are to be entitled *New Foundations for Classical Mechanics* (NFI) and *New Foundations for Mathematical Physics* (NFII). The preparation of both these books is well underway. NFI presents a full account of classical mechanics in the language of Geometric Calculus. Since Geometric Algebra smoothly integrates quaternions with the conventional vector algebra, it makes the full power of both systems available together for the first time. Thus, NFI can take full advantage of quaternions in the treatment of rotations and rotational dynamics as well as other topics. Although NFI is designed for use as a textbook at the undergraduate level, it handles a number of topics which are usually regarded as more advanced, because Geometric Algebra makes their treatment feasible at this level. NFI provides a detailed introduction to Geometric Algebra, including a discussion of its historical development and geometric rationale as well as many diagrams and elementary examples. Such considerations are essential for understanding why Geometric Calculus is an ideal language for physics, but they are not broached in the present book, which is devoted to an abstract mathematical treatment of the Calculus.

The book NFII applies Geometric Calculus to the formulation and development of classical field theory, special relativity and non-relativistic quantum mechanics. Most notable is the unified mathematical method it provides for the various branches of physics. Subsequent books will extend the applications of Geometric Calculus to relativistically invariant treatments of electrodynamics, quantum theory and gravitation along the lines already set down in the book *Space-Time Algebra* [H1] and various articles ([H2–12], [S2–5]).

We have no doubt that our treatment of Geometric Calculus can be improved in many ways. We have witnessed significant improvements in repeated rewrites, extending as far back as twenty years for some parts. We are still experimenting with some of the notation and nomenclature. Some of our proofs are admittedly sketchy. And we have left many loose ends of other kinds, some of them beginnings for further research. But on the whole we are quite pleased with the serviceability of the Calculus as it stands. It is our fondest wish that others will find it useful and worthy of refinement.

This book is intended to serve as a reference for mathematicians and physicists who would like to use Geometric Calculus in their research and teaching. No prior knowledge of Clifford Algebra is presumed. However, the reader is presumed to have a good background in linear algebra, and later in differential geometry and then Lie groups when these subjects are taken up. A balanced account of these subjects is not attempted. The main objective is to develop the apparatus and techniques needed to handle these subjects efficiently and completely with Geometric Calculus alone. The effectiveness of the Calculus is illustrated by detailed treatments of a number of special topics.

Our long-range aim is to see Geometric Calculus established as a unified system for handling linear and multilinear algebra, multivariable calculus, complex variable theory, differential geometry and other subjects with geometric content. Mathematicians sympathetic with this aim might like to use this book as text conjointly with a conventional text on any one of the subjects just mentioned. There are no formal exercises in the book, but students will be amply exercised by filling in details of the proofs and examples and by making comparisons with standard texts. We welcome and encourage comparisons of Geometric Calculus with conventional systems in every detail. In our experience, no other activity is quite so efficient for developing insight into the structure of mathematics.

Over the last two decades the senior author (D. Hestenes) has been continuously concerned with the development of Geometric Calculus, both as a mathematical system and as a language for physical science. He has been assisted over the major portion of this book by the second author (G. Sobczyk), and they often worked so closely together that it is difficult to separate the contributions of one from those of the other. In his doctoral dissertation Dr. Sobczyk developed the theory of simplicial derivatives as the basis for new formulations of differential and adjoint outermorphisms. This made it possible for the first time to formulate the theory of induced transformations on manifolds and actually carry out explicit computations without reference to coordinates. As a postdoctoral research fellow supported by the Polish Academy of Science during 1973–74, he made many refinements in the coordinate-free formulation of differential geometry.

Credit should be given to Robert Hecht-Nielsen for contributing to the proof of the inverse function theorem in Section 7-6 and to Alan Jones and Robert Rowley for improving the accuracy of the text. The senior author must accept blame for any deficiencies in the exposition, as he assumed responsibility for writing the entire manuscript.

Our debt is greatest to those unnamed and, in large part, unknown, whose living thoughts have journeyed to inform every page of this book. To them our deepest gratitude and respect.

DAVID HESTENES
Arizona State University

GARRET SOBCZYK
University of Wrocław

Introduction

> Mathematics is an organism for whose vital strength the indissoluble union of its parts is a necessary condition.
>
> HILBERT

In his famous survey of mathematical ideas, *Elementary Mathematics from an Advanced Standpoint*, Felix Klein championed 'the fusion of arithmetic with geometry' as a major unifying principle of mathematics. Promoting the principle with unsurpassed insight, he illuminated themes from the greater part of nineteenth century mathematical thought to reveal them as strands in a single mathematical fabric. So many themes had developed in the hands of independent investigators each with its own vocabulary and symbolism, and concepts are so difficult to separate from their specific representations by symbols, that Klein's monolithic perspective of mathematics was and still is difficult to assimilate from his books despite his inspired presentation.

The picture of mathematics that emerges from Klein's book does not so much resemble the unified perspective that he had achieved in his own mind as a precarious journey from bridge to bridge over cracks and crevasses of mathematical thought. For he was limited in his ability to express his conceptions by the heterogeneous mathematical formalisms of his age. To adequately represent Klein's vision of mathematics as a seamless whole and make his hard-earned insights common property of the mathematical community, the diverse symbolic systems of mathematics must be modified, coordinated and ultimately united in a single mathematical language. Klein organized arithmetic, algebra and analysis with geometrical themes. The fusion to which he aspired can truly be completed only by the development of a universal calculus of geometric thought.

Klein's seminal analysis of the structure and history of mathematics brings to light two major processes by which mathematics grows and becomes organized. They may be aptly referred to as the *algebraic* and the *geometric*. The classification intended here is slightly different from Klein's but quite in accord with the evidence he presents. The one emphasizes algebraic structure while the other emphasizes geometric interpretation. Klein's analysis shows one process alternately dominating the other in the historical development of mathematics. But there is no necessary reason that the two processes should operate in mutual exclusion. Indeed, each process is undoubtedly grounded in one of two great capacities of the human mind: the capacity for language and the capacity for spatial perception. From the psychological point of view, then, the fusion of algebra with geometry is so fundamental that one could well say, 'Geometry without algebra is dumb! Algebra without geometry is blind!'

The interplay between algebraic and geometric processes in the evolution of the real numbers has reached a generally satisfactory climax. It is safe to say that the notation for the real number system and its algebra is one thing which all mathematicians understand and accept. But the coordination of algebraic and geometric processes in the development of 'higher mathematics' has left something to be desired; indeed, it seems that at times the two processes have worked at cross-purposes.

The fact that the real number system was first extended to the system of complex numbers by an algebraic process requiring that every quadratic equation has a root has become an historical commonplace. But as Klein has testified (and nobody was in a better position to observe), the great flourishing of complex analysis in the nineteenth century was fundamentally a geometric process founded on the familiar correspondence of complex numbers with points in a plane. And here lies a subtlety which has been consistently overlooked in the construction of mathematical systems. The geometrical interpretation of complex numbers associates real numbers with a particular direction (the real axis), and this can be distinguished from the interpretation of real numbers as scalars (numbers without direction). This basic distinction is seldom maintained in mathematical work, perhaps because the geometric interpretation is regarded as only incidental to the concept of number. But there is a deeper reason, pregnant with implications, namely, that the geometric concept of direction is not adequately represented by the conventional algebraic concept of number.

The possibility of representing directions by complex numbers is, of course, limited to spaces of two dimensions, and the algebraic representation of direction for spaces of higher dimension requires the formal concept of vector. But the geometric status of complex numbers is confounded by the common practice of working with vector spaces over the complex numbers. It will not be necessary to go into the purely algebraic motivation for this practice. The fact of interest here is that complex vector spaces have not one but several distinct geometrical interpretations, as can be determined by examining how they are used in practice. One can distinguish three fundamentally different geometrical roles tacitly assigned to the unit imaginary $i = \sqrt{-1}$, namely,

(1) the generator of rotations in a plane,
(2) the generator of duality transformations,
(3) the indicator of an indefinite metric.

Confusion is difficult to avoid when i is required to perform more than one of these roles in a single system. Worse yet, in physics all three possibilities are sometimes realized in a single theory confounded with problems of physical interpretation. A specific example is discussed in [H9]. The multiplicity of geometric interpretations for imaginary numbers shows that conventional mathematical formalisms are profoundly deficient in their tacit assumption that there is a unique 'imaginary unit'. Therefore, in the interest of fidelity to geometric interpretation, the convention that complex numbers are scalars should be abandoned in favor of a mathematical system with many roots of minus one, a system in which each basic geometric distinction has a unique algebraic representation. Geometric Algebra has this property.

Geometric Algebra is best regarded as a *geometric extension* of the real number system to provide a *complete* algebraic representation of the geometric notions of direction and magnitude. To be sure, the extension is formally algebraic in that it is completely characterized by a set of algebraic axioms, however, the choice of the axioms is governed by geometric considerations. This may be contrasted with the traditional *algebraic extension* of real to complex numbers made solely for algebraic reasons. For geometric reasons, then, we develop Geometric Algebra over the field of the real numbers. It might be thought that developing the algebra over the complex numbers would produce a more powerful mathematical system. We contend, however, that it would merely cast the system in a different form with the danger of confounding geometric distinctions. In support of our contention, Sections 4-7 and 7-4 of the text show how complex variable theory can be formulated with the algebra over the reals. It will be seen that this formulation generalizes readily to higher dimensions in a way not contemplated in the theory of many complex variables, and it obliterates artificial distinctions between complex variable theory and real variable theory.

Having given our reasons, we shall henceforth identify the scalars with the real numbers without question. Now let us discuss the contents of this book in relation to our objectives.

Our general objective is to develop a comprehensive language for efficiently expressing and exploiting the full range of geometric concepts in mathematics. *The grammar of this language is Clifford Algebra* over the reals. But there is much more to a language than a grammar. To enhance the computational efficiency of the language and prepare it for immediate application to a wide range of problems, we develop an extensive system of secondary concepts, definitions and algebraic identities. We call this purely algebraic part of the language *Geometric Algebra*. We use the term *Geometric Calculus* for the extension of the language to include concepts of analysis, especially differential and integral calculus on manifolds. Again, to make *Geometric Calculus* an efficient and versatile language, we develop an extensive system of concepts and theorems.

Chapter 1 establishes the fundamentals of Geometric Algebra. Section 1-1 provides the backbone of the subject, and its results are used as a matter of course throughout the book. The rest of the chapter is concerned with algebraic applications, primarily to the structure of vector spaces, properties of matrices and the theory of determinants.

Chapter 2 introduces a concept of *vector derivative* unique to Geometric Algebra because of its algebraic properties. This differential operator makes possible a completely *coordinate-free* differential calculus, and the whole subject can be regarded as an elaboration of its properties. Its most important general application is to the invariant definition and analysis of the *differential* and *adjoint* functions, fundamental concepts in the general theory of functions. These concepts are used repeatedly and extended in the rest of the book. The concepts of differential and adjoint are, of course, not new, but Geometric Calculus provides a new way of handling them systematically.

Chapter 3 applies Geometric Calculus to the simplest general class of geometric functions, the linear transformations. The emphasis is on determining canonical forms without resorting to matrices or coordinates. As a general tool Section 3-1 develops the important concept of *outermorphism*, the extension of a linear transformation from a vector space to its entire Geometric Algebra where many of its properties are most readily studied. Section 3-2 gives a new proof of the Cayley–Hamilton theorm, exhibiting it simply as the result of applying a general differential identity to the special class of linear functions. Probably the most useful results in the chapter are the spinor representations of isometries, fully developed in Sections 3-5 and 3-8. These marvelous representations have generated much of the past interest in Clifford Algebra. Here their utility is greatly enhanced by integrating them into a comprehensive approach to linear algebra instead of obtaining them by a specialized technique.

Chapter 4 develops the complete apparatus needed for a coordinate-free treatment of differential calculus on manifolds, including the transformations from one manifold to another. This makes it possible to calculate such things as the Jacobian of a transformation without introducing coordinates, as shown explicitly by the computations in Section 4-6. In Chapter 5 the calculus is applied to a variety of topics in differential geometry. The approach features a new quantity, the shape tensor, from which the curvature tensor can be computed without a connexion. An alternative approach to differential geometry is developed in Chapter 6. Cartan's calculus of differential forms is shown to be a special case of Geometric Calculus.

Chaper 7 develops a theory of integration based on a concept of directed measure to take full advantage of the geometric product. A number of new results are obtained, including a generalization of Cauchy's integral formula to n-dimensional spaces and an explicit integral formula for the inverse of a transformation.

Chapter 8 applies Geometric Calculus to the theory of Lie Groups and Lie Algebras. The theory is not developed very far, but the foundations are established, and the unique perspective which the Calculus brings to the whole subject is made apparent.

Symbols and Notation

Symbol	*Name*	*Page introduced*
AB	geometric product	3
$\langle A\rangle_r = A_{\bar{r}}$	r-vector part of A	4
$\langle A\rangle = \langle A\rangle_0$	scalar part of A	5
$\lvert A\rvert$	magnitude of A	4, 13
$A^\dagger$	reverse of A	5
$A\cdot B$	inner product	6
$A\wedge B$	outer product	6
$A * B \equiv \langle AB\rangle$	scalar product	13
$A\times B$	commutator product	14
$\mathscr{A}_n = \mathscr{G}^1(A)$	n-dimensional vector space	16, 19
$\mathscr{G}(\mathscr{A}_n) = \mathscr{G}(A)$	geometric algebra of $\mathscr{A}_n$, or of A	19
$\mathscr{G}^r(A)$	r-vector subspace of $\mathscr{G}(A)$	19
$P = P_A$	projection operator	18, 20
P	projection operator for a manifold	141
$P^\perp = P_A^\perp$	rejection operator	18
$P_\perp$	rejection operator for a manifold	149
$A\vee B$	meet	25
$\{a_k\}$	frame	27
$\{a^k\}$	reciprocal frame	28
α_r	alternating r-form	33
	differential r-form	241
$\alpha_r\wedge\mu_s$	exterior product of forms	34, 241
$\mu_s\cdot\alpha_r$	interior product of forms	34
$^*\alpha_r$	dual form	241
$\mathrm{d}\alpha_r$	exterior derivative	242
$\delta\alpha_r$	adjoint derivative	243
$\det f = \det b_k^j$	determinant	36, 70
$[a_1, \ldots, a_n]$	determinant	37
$\mathscr{A}_{p,q}$	vector space of signature (p, q)	42

Symbol	*Name*	*Page introduced*
$a \cdot \partial F = F_a$	a-derivative of $F = F(x)$	45, 141
$A * \partial F = F_A$	A-derivative of $F = F(X)$	54
$\underline{F} = \underline{F}(a) = F_a$	(first) differential of $F = F(x)$	45, 141
$\underline{F} = \underline{F}(A) = F_A$	differential of $F = F(X)$	54
$\overline{f} = \overline{f}(a')$	adjoint of $f = f(x)$	50, 166
$\underline{f}(A)$	differential outermorphism of $f = f(x)$	67, 165
$\overline{f}(A)$	adjoint outermorphism of $f = f(x)$	68, 166
F_{ab}	second differential of $F = F(x)$	48, 143
F_{AB}	second differential of $F = F(X)$	54
$\partial F = \partial_x F(x)$	derivative of $F = F(x)$	45, 144
$\partial F = \partial_X F(X)$	derivative of $F = F(X)$	55
$\partial \cdot F$	divergence of F	49
$\dot{g}\,\dot{\partial}\dot{f}$	derivative	50, 254
$\partial \wedge F$	curl of F	49
$[a, b]$	Lie bracket	53, 159
$[A, B]$	Lie bracket	159
$a_{(r)}$	simplicial variable	61
$\partial_{(r)}$	simplicial derivative	61, 166
$x \circ y$	'inner product'	96
	metric tensor	102
$\mathrm{O}(p, q) = \mathrm{O}(\mathscr{A}_{p,q})$	orthogonal group	103
$\mathrm{Vers}\,(p, q)$	versor group	104
$\mathrm{SO}(p, q)$	special orthogonal group	104
$\mathrm{SO}^+(p, q)$		104
$\mathrm{Spin}\,(p, q)$	spin group of $\mathscr{A}_{p,q}$	105
$\mathrm{Spin}^+\,(p, q)$	rotor group of $\mathscr{A}_{p,q}$	105
τ_{ijk}	covariant tensor components	133
τ^{ijk}	contravariant tensor components	133
$I = I(x)$	unit pseudoscalar of a vector manifold	140, 17
$\mathscr{G}(x) = \mathscr{G}(I(x))$	tangent algebra of a vector manifold	141
$T = T(A_1, \ldots, A_k)$	extensor function	142
	extensor field	142
T_a	extensor differential	142
$\mathrm{d}T$	exterior differential	145, 255
P_b	projection differential	147
P_{ab}	projection second differential	164
$S = S(A)$	shape operator	149
S_a	curl tensor	150, 232
S_{ab}	differential of S_a	164
N	spur of a manifold	150, 254
$a \cdot \nabla A$	directional coderivative	155

Symbol	*Name*	*Page introduced*
∇A	coderivative or gradient	155, 156
$\delta_a T$	extensor coderivative	155
δT	exterior codifferential	156
$\nabla \cdot A$	codivergence	156
$\nabla \wedge A$	curl	156
∇^2	colaplacian	158
$\{A, B\}$	dual bracket	161
$R(a \wedge b)$	curvature tensor	191, 232
$R(a)$	Ricci tensor	193
R	scalar curvature	193
$G(a)$	Einstein tensor	194
$S_a \times S_b$	total curvature	195
$H = H(x)$	mean curvature	197
κ	Gaussian curvature	198
$\mathrm{d}a/\mathrm{d}\tau$	derivative	204
$\delta a/\delta\tau$	coderivative	204
Ω_a	total mobile angular velocity	212, 232
$C(a \wedge b)$	projective change in curvature	212, 215
$W(a \wedge b) \cdot c$	Conformal Weyl tensor	219
$a \cdot D$	affine connexion	221
L_a	affine extensor	221
$L(a \wedge b) \cdot c$	affine curvature tensor	223
d_a	affine differential	223
$\{e_k(x)\}$	frame field	228
$\{e^k\}$	reciprocal frame field	228
g_{ij}	metric tensor	226
ω_a	frame 'angular velocity'	230, 233
Ω	mobile angular velocity	237
L_{ij}^k	coefficients of connexion	235, 244
$\mathrm{d}X$	directed measure	250
$\lvert \mathscr{M} \rvert$	volume	251
$\int_{\mathscr{M}} L(\mathrm{d}X)$	directed integral	251
$T(\partial)$	tensor divergence	147, 255
$\delta(x - x')$	delta function	260
$\odot_m$	area of unit $(m - 1)$-sphere	262
J_f	Jacobian of f	166, 266
$\deg(f)$	degree of a mapping	268
χ	Euler characteristic	276
XY	group products	284
$\phi(x, y)$	product function	284
λ_y	left translation	286

Symbol	*Name*	*Page introduced*
$\underline{\lambda}_y$	left differential	286
ρ_y	right translation	286
$\underline{\rho}_y$	right differential	286
$D_\lambda(x)$	Haar measure	289
c_{ij}^k	structure constants	290
$K(A, B)$	Killing form	299

Chapter 1

Geometric Algebra

This chapter defines Geometric Algebra by a set of axioms and develops a system of definitions and identities to make it a versatile and efficient computational tool. These results are used repeatedly in subsequent chapters. Many results are obtained in the form of algebraic identities, but they are seldom presented as theorems, because we wish to emphasize the techniques for generating them, to show how a great variety of useful identities can be generated by a few simple techniques. For example, Section 1-4 shows how easily geometric algebra generates the system of identities making up the theory of determinants. Thus we can see the theory of determinants as only part of a more comprehensive algebraic system.

Clifford Algebra can be introduced in a number of different ways (see [Ch], [Po], [Hu] for standard accounts). We have chosen an unconventional approach for several reasons. Most important, our axioms in Section 1 have been chosen to maximize the system's computational power. It will be noted that they include all the axioms of elementary scalar algebra except commutative multiplication, and it will be found that the breakdown of commutivity has an important geometrical meaning. Only a few additional axioms are needed, including a factoring axiom which generalizes and exploits the factoring concept familiar from elementary algebra.

The axiomatic approach separates algebraic structure from geometric interpretation. A strict separation increases the versatility of the system because a particular algebraic quantity or equation can be given more than one geometric interpretation. Unfortunately, this separation makes it hard for a novice to see any motivation for the algebraic gymnastics in Section 1-1. A few preliminary remarks about geometric interpretation may help the reader negotiate Section 1-1 until the algebra is supplied with a full interpretation in Section 1-2.

The building blocks of geometric algebra are the k-blades, where k is a positive integer called the *grade* of a given blade. The 1-blades are familiar as vectors. Since a nonzero vector determines a unique one-dimensional vector space or line, it may be regarded as representing the line. Similarly, each 2-blade represents a plane, and each k-blade represents a k-dimensional space. The geometric product relates blades of different grade and so describes relations among the spaces they represent. In fact, all directional relations among linear spaces can be represented by this product. For example, if the product of two 1-blades is commutative, then both vectors represent the same line, but if the product is anticommutative, the vectors represent perpendicular lines.

From the geometric product many other kinds of products can be formed. Of these, the inner and outer products are most important geometrically, because they are, respectively, the grade-lowering and the grade-raising operations in geometric algebra. Inner and outer products are the primary operations of multilinear algebra, including the theory of determinants, so we develop their properties at length in Section 1-1.

Since we introduce inner and outer products as secondary concepts defined in terms of the single geometric product, our approach differs significantly from standard expositions of multilinear algebra (for example, [Bo], [Gr] and [Wh]). The main advantage of our approach is that inner and outer products are integrated into a more general and versatile algebraic system from the beginning. But there are computational advantages as well. First, reduction of inner and outer products to a single product reduces the number of axioms needed to establish multilinear algebra. Second, the derivation of identities involving inner and outer products is simplified and systematized by exploiting the associativity of the geometric product.

One other feature of our formulation of geometric algebra deserves some explanation. We define an infinite dimensional algebra at the outset and from it obtain all finite dimensional geometric algebras as subalgebras. We have two reasons for adopting this approach. First, we wish to emphasize that our general results and methods are truly coordinate-free, requiring no reference to a basis or the dimensionality of the algebra. As shown in Section 1-2, this also enables us to define any finite dimensional algebra in the same simple way that we define its subalgebras. Section 1-2 shows that in a finite dimensional algebra the inner product is related to the outer product by duality. Conventional approaches to multilinear algebra (such as [Bo]) use this relation to define the inner product in terms of the outer product, but this restricts them to finite dimensional spaces because only then is duality well defined. The procedure becomes especially awkward when spaces of different dimension are compared. Therefore, we think it is important to define the inner product without reference to duality, as we have done.

Our second reason for defining an infinite dimensional geometric algebra is that it is essential to our treatment of manifolds in Chapter 4. Actually, we are not so much concerned with an algebra of infinite dimension as with algebras of unspecified dimension. The dimension of an algebra is important chiefly in connexion with the question of closure arising when elements are to be expanded in a basis. Since our method avoids such expansions, questions of closure and dimension are generally irrelevant to our considerations. The question of closure for finite dimensional algebras is treated in Section 1-2 separate from the general axioms in Section 1-1. For the purposes of this book, it was not necessary to examine the interesting question of closure for the infinite dimensional algebra. Some recent articles on infinite dimensional Clifford Algebras are listed among the References.

1-1. Axioms, Definitions and Identities

This section presents a convenient set of axioms for geometric algebra. In terms of the fundamental geometric product, several other products of special geometric and algebraic significance are defined, namely, the inner and outer products, the scalar product, and the commutator. A comprehensive set of basic identities is established, identities which are needed for efficient application of the algebra to almost any problem. The axioms, definitions and results of this section will be used repeatedly throughout the rest of this book.

We define geometric algebra by a set of axioms which enable us to carry out proofs and computations without reference to basis in the algebra. For the sake of simplicity and ease of application, we have not attempted to eliminate all redundancy from the axioms.

An element of *the Geometric Algebra* $\mathcal{G}$ will be called a multivector. We assume that $\mathcal{G}$ is algebraically closed, that is, that the sum or product of any pair of multivectors is a unique multivector. The *geometric sum* and *product* of multivectors $A, B, C, \ldots$ have the following properties:

Addition is commutative;

$$A + B = B + A. \tag{1.1}$$

Addition and multiplication are associative;

$$(A + B) + C = A + (B + C); \tag{1.2}$$

$$(AB)C = A(BC). \tag{1.3}$$

Multiplication is distributive with respect to addition;

$$A(B + C) = AB + AC, \tag{1.4}$$

$$(B + C)A = BA + CA. \tag{1.5}$$

There exist unique additive and multiplicative identities 0 and 1;

$$A + 0 = A, \tag{1.6}$$

$$1A = A. \tag{1.7}$$

Every multivector A has a unique additive inverse $-A$;

$$A + (-A) = 0. \tag{1.8}$$

Geometric algebra is set apart from other associatve algebras by a few additional axioms which classify multivectors into different types or, as we shall say, *grades*.*
We assume that any multivector A can be written as the sum

$$A = \langle A\rangle_0 + \langle A\rangle_1 + \langle A\rangle_2 + \cdots = \sum_r \langle A\rangle_r. \tag{1.9}$$

* We have chosen the word 'grade' over the terms 'degree' and 'dimension', which have frequently been used in the literature (e.g. in [H1] and [Wh]), to avoid possible confusion with other widely understood meanings of those terms.

The quantity $\langle A \rangle_r$ is called the *r-vector part* of A. If $A = \langle A \rangle_r$ for some positive integer r, then A is said to be *homogeneous* of grade r and will be called an *r-vector*. The terms *scalar, vector, bivector, trivector*, . . . are often used as alternatives to the terms 0-vector, 1-vector, 2-vector, 3-vector, . . . respectively.

The *grade operator* $\langle \ldots \rangle_r$ enjoys the properties

$$\langle A + B \rangle_r = \langle A \rangle_r + \langle B \rangle_r, \tag{1.10}$$

$$\langle \lambda A \rangle_r = \lambda \langle A \rangle_r = \langle A \rangle_r \lambda, \quad \text{if } \lambda = \langle \lambda \rangle_0, \tag{1.11}$$

$$\langle \langle A \rangle_r \rangle_r = \langle A \rangle_r. \tag{1.12}$$

Axioms (1.10) and (1.11) imply that the space $\mathscr{G}^r$ of all r-vectors is a linear subspace of $\mathscr{G}$, and, indeed, that $\mathscr{G}$ itself is a linear space. Axiom (1.11) also implies that the scalars compose a commutative subalgebra of $\mathscr{G}$. Without further ado, we assume that the space $\mathscr{G}^0$ of all scalars is identical with the set of *real numbers*. As argued elsewhere in this book, we regard any wider definition of the scalars (for example as the complex numbers) to be entirely unnecessary and, indeed, inimical to the purposes of geometric algebra.

Equation (1.12) exhibits the characteristic property of a projection operator, so $\langle A \rangle_r$ can be regarded as the projection of A into the space $\mathscr{G}^r$. Actually, (1.12) need not be regarded as an axiom, because it can be derived with the help of our remaining axioms which fix the relations among multivectors of different grade.

Multiplication of vectors is related to scalars by the assumption that the 'square' of any nonzero vector a is equal to the square of a unique *positive* scalar $|a|$ called the *magnitude* of a, that is

$$aa = a^2 = \langle a^2 \rangle_0 = |a|^2 > 0. \tag{1.13}$$

The multiplicative relation of vectors to r-vectors is specified by assuming that, for any integer $r > 0$, an r-vector can be expressed as a sum of r-blades. A multivector A_r is called an *r-blade* or a *simple r-vector* if and only if it can be *factored* into a product of r anticommuting vectors $a_1, a_2, \ldots, a_r$, that is

$$A_r = a_1 a_2 \ldots a_r, \tag{1.14a}$$

where

$$a_j a_k = -a_k a_j \tag{1.14b}$$

for $j, k = 1, 2, \ldots, r$, and $j \neq k$.

As a final axiom, we assume that for every nonzero r-blade A_r, there exists a nonzero vector a in $\mathscr{G}$ such that $A_r a$ is an $(r + 1)$-blade. This guarantees the existence of nontrivial blades of every finite grade. In fact, it implies that each $\mathscr{G}^r$ and, of course, all of $\mathscr{G}$ is a linear space of infinite dimension. This leads ultimately to delicate questions of convergence for homogeneous multivectors with infinite grade. But we will not be concerned with such questions in this book.

Many consequences of the above axioms can be obtained by well-known arguments, so we need not dwell on them here. Instead, we emphasize arguments and results which are especially significant in geometric algebra.

As a rule, we use Greek letters $\alpha, \beta, \lambda, \mu, \ldots$ to denote scalars, lower case Latin letters $a, b, c, u, v, x, \ldots$ to denote vectors and capital letters $A, B, C, U, V, X, \ldots$ to denote other multivectors.

To avoid needless repetition, we specify at once that in this section we use A_r, B_s and C_t to denote multivectors of homogeneous grades r, s and t respectively. On the other hand, we write a_k for the kth member of the set of vectors $a_1, a_2, \ldots, a_r$. In this latter case, the subscripts are indices with no relation to grade. If we wish to indicate unambiguously that a subscript specifies grade, we mark it with an overbar. Thus, we have the two notations

$$A_{\bar{r}} \equiv \langle A \rangle_r \tag{1.15}$$

for the projection of A into $\mathscr{G}^r$. Each notation has its own merits, and we have not succeeded in devising a single notation that is satisfactory in every way. The adoption of two notations seems to be justified, because the grade operator is used so frequently in geometric calculus. Also the scalar grade operator is sufficiently distinctive to merit the special notation

$$\langle A \rangle \equiv \langle A \rangle_0 \equiv A_{\bar{0}}. \tag{1.16}$$

In algebraic computations it is often desirable to reorder the factors in a product. As a general aid to this activity it is convenient to introduce the operation of *reversion* defined by the equations

$$(AB)^\dagger = B^\dagger A^\dagger, \tag{1.17a}$$

$$(A + B)^\dagger = A^\dagger + B^\dagger, \tag{1.17b}$$

$$\langle A^\dagger \rangle = \langle A \rangle, \tag{1.17c}$$

$$a^\dagger = a \quad \text{where } a = \langle a \rangle_1. \tag{1.17d}$$

It follows immediately that the *reverse* of a product of vectors is

$$(a_1 a_2 \ldots a_r)^\dagger = a_r \ldots a_2 a_1. \tag{1.18}$$

This justifies the name 'reverse'.

If the vectors in (1.18) anticommute, then, using (1.14b) to reorder the vectors on the right side of (1.18), we easily prove that

$$\langle A^\dagger \rangle_r = \langle A \rangle_r^\dagger = (-1)^{r(r-1)/2} \langle A \rangle_r. \tag{1.19}$$

Although we have only indicated the proof for simple r-vectors, the result (1.19) obtains for arbitrary multivectors by virtue of the linearity of the grade operator.

Of course, the associative axiom (1.3) was also used in the proof. Hereafter, we will usually omit mention of such familiar uses of the axioms.

Using (1.17a), we get as immediate corollaries of (1.19), the relations

$$\langle AB\rangle_r = (-1)^{r(r-1)/2}\langle B^\dagger A^\dagger\rangle_r, \tag{1.20a}$$

$$\langle A_r B_s\rangle_r = \langle B_s^\dagger A_r\rangle_r = (-1)^{s(s-1)/2}\langle B_s A_r\rangle_r, \tag{1.20b}$$

$$\langle AB_r C\rangle_r = \langle C^\dagger B_r A^\dagger\rangle_r, \tag{1.20c}$$

$$\langle A_r B_s C_t\rangle_q = (-1)^\epsilon\langle C_t B_s A_r\rangle_q, \tag{1.20d}$$

where $\epsilon = \frac{1}{2}(q^2 + r^2 + s^2 + t^2 - q - r - s - t)$. These relations suffice to show how reordering of r-vector parts is most easily accomplished.

We define the *inner product* of homogeneous multivectors by

$$A_r \cdot B_s \equiv \langle A_r B_s\rangle_{|r-s|}, \quad \text{if } r, s > 0, \tag{1.21a}$$

$$A_r \cdot B_s \equiv 0, \quad \text{if } r = 0 \text{ or } s = 0. \tag{1.21b}$$

The inner product of arbitrary multivectors is then defined by

$$A \cdot B \equiv \sum_r A_{\bar{r}} \cdot B = \sum_s A \cdot B_{\bar{s}} = \sum_r \sum_s A_{\bar{r}} \cdot B_{\bar{s}}. \tag{1.21c}$$

The equivalence of the three expressions on the right side of (1.21c) is an obvious consequence of (1.10) and the distributivity of the geometric product.

In a similar way, we define the *outer product* of homogeneous multivectors by

$$A_r \wedge B_s \equiv \langle A_r B_s\rangle_{r+s}. \tag{1.22a}$$

Note that, in contrast to the special treatment given to scalars in (1.21b), we allow (1.22a) to yield

$$A_r \wedge \lambda = \lambda \wedge A_r = \lambda A_r \quad \text{if } \lambda = \langle\lambda\rangle. \tag{1.22b}$$

The outer product of arbitrary multivectors is defined by

$$A \wedge B \equiv \sum_r A_{\bar{r}} \wedge B = \sum_s A \wedge B_{\bar{s}} = \sum_r \sum_s A_{\bar{r}} \wedge B_{\bar{s}}. \tag{1.22c}$$

According to (1.21a), if $0 < r \leqslant s$, then $A_r \cdot B_s$ is an $(s-r)$-vector, so inner multiplication of B_s by A_r 'lowers the grade' of B_s by r units. In order to be consistent with (1.27a) and other equations below, multiplication by a scalar is treated in (1.21b) as a trivial exception to this rule. However, (1.22a) says that, without exception, outer multiplication by an r-vector 'raises grade' by r units. The outer and inner products are the grade raising and lowering operations in Geometric Algebra. For this reason, we must be thoroughly familiar with their properties if we wish to apply Geometric Algebra effectively.

The general algebraic properties of inner and outer products are straightforward consequences of our definitions and the axioms of geometric algebra. Like the geometric product, the inner and outer products are not commutative. However, by using (1.19) on the right side of (1.20a), we find, for homogeneous multivectors, the reordering rules

$$A_r \cdot B_s = (-1)^{r(s-1)} B_s \cdot A_r \quad \text{for } r \leqslant s, \tag{1.23a}$$

$$A_r \wedge B_s = (-1)^{rs} B_s \wedge A_r. \tag{1.23b}$$

Distributivity of the geometric product implies the distributive rules

$$A \cdot (B + C) = A \cdot B + A \cdot C, \tag{1.24a}$$

$$A \wedge (B + C) = A \wedge B + A \wedge C. \tag{1.24b}$$

The outer product is associative;

$$A \wedge (B \wedge C) = (A \wedge B) \wedge C. \tag{1.25a}$$

The inner product is not associative, but homogeneous multivectors obey the rules

$$A_r \cdot (B_s \cdot C_t) = (A_r \wedge B_s) \cdot C_t \quad \text{for } r + s \leqslant t \text{ and } r, s > 0, \tag{1.25b}$$

$$A_r \cdot (B_s \cdot C_t) = (A_r \cdot B_s) \cdot C_t \quad \text{for } r + t \leqslant s. \tag{1.25c}$$

Equations (1.25a, b, c) are consequences of the associativity of the geometric product, but we delay the proof until we have established another general property of the product.

Using (1.23b) and (1.25a) it is easy to establish the useful fact that the outer product is antisymmetric under an interchange of any pair of vectors, that is, for vectors a and b,

$$a \wedge A \wedge b \wedge B = -b \wedge A \wedge a \wedge B. \tag{1.26a}$$

This immediately gives the useful result

$$a \wedge A \wedge a \wedge B = 0 \tag{1.26b}$$

for any A and B.

At this point it is convenient to introduce explicitly the convention that, if there is ambiguity, indicated inner and outer products should be performed before an adjacent geometric product. Thus

$$(A \wedge B)C = A \wedge BC \neq A \wedge (BC),$$

$$(A \cdot B)C = A \cdot BC \neq A \cdot (BC).$$

This convention eliminates an appreciable number of parentheses, especially in complicated expressions. Other parentheses can be eliminated by the convention that outer products have 'preference' over inner products, so

$$A \cdot (B \wedge C) = A \cdot B \wedge C \neq (A \cdot B) \wedge C,$$

but we use this convection much less often than the previous one.

The inner and outer products are united by the geometric product. This is evident in the fundamental formulas for the product of a vector with an r-vector:

$$a \cdot A_r = \langle aA_r \rangle_{r-1} = \tfrac{1}{2}(aA_r - (-1)^r A_r a), \tag{1.27a}$$

$$a \wedge A_r = \langle aA_r \rangle_{r+1} = \tfrac{1}{2}(aA_r + (-1)^r A_r a). \tag{1.27b}$$

$$aA_r = a \cdot A_r + a \wedge A_r = \langle aA_r \rangle_{r-1} + \langle aA_r \rangle_{r+1}. \tag{1.28}$$

It will be understood that the grade operator results in zero when its index is negative. This convention makes (1.27a) meaningful when $r = 0$.

Clearly any two of the three equations (1.27a, b), (1.28) imply the third. Equations (1.27a, b) reduce inner and outer products to the geometric product, and they will shortly be established. But first, we use the fact that the sign of $(-1)^r$ depends only on the evenness or oddness of r to get somewhat more general formulas.

We say that a multivector A_+ is *even* if $\langle A_+ \rangle_r = 0$ for all odd values of r. Similarly, A_- is said to be *odd* if $\langle A_- \rangle_r = 0$ for even r. Clearly any multivector A can be written as the sum of an even part A_+ and an odd part A_-;

$$A = A_+ + A_-. \tag{1.29}$$

By virtue of the distributive rule, (1.27a, b) implies

$$a \cdot A_+ = \tfrac{1}{2}(aA_+ - A_+ a), \tag{1.30a}$$

$$a \wedge A_+ = \tfrac{1}{2}(aA_+ + A_+ a), \tag{1.30b}$$

$$a \cdot A_- = \tfrac{1}{2}(aA_- + A_- a), \tag{1.30c}$$

$$a \wedge A_- = \tfrac{1}{2}(aA_- - A_- a). \tag{1.30d}$$

The sum of these expressions gives the generalization of (1.28); for any A,

$$aA = a \cdot A + a \wedge A. \tag{1.31}$$

Turning now to a proof of (1.27a), we first consider the case $r = 1$. For vectors a and b, the distributive rule gives

$$(a + b)^2 = a^2 + ab + ba + b^2.$$

Since $a + b$ is a vector, we have, from (1.13),

$$ab + ba = |a + b|^2 - |a|^2 - |b|^2,$$

which is a clearly a scalar. Since also $\langle ab \rangle = \langle ba \rangle$, we have established

$$a \cdot b = \langle ab \rangle = \tfrac{1}{2}(ab + ba), \tag{1.32}$$

which is identical to (1.27a) when $A_r = b = \langle b \rangle_1$. Note that (1.32) implies that $a \cdot b = 0$ if and only if $ab = -ba$. If $a \cdot b = 0$, we say that the two vectors a and b are *orthogonal*.

We can proceed now to a proof of (1.27a, b) in general by establishing the valuable vector identity

$$a \cdot (a_1 a_2 \ldots a_r) = \sum_{k=1}^{r} (-1)^{k+1} a \cdot a_k (a_1 \ldots \check{a}_k \ldots a_r), \tag{1.33}$$

where the inverted circumflex means the kth vector is omitted from the product. Identity (1.33) holds for any choice of vectors. One instructive way to prove (1.33) is to notice that (1.32) can be written in the form $ab = 2a \cdot b - ba$ and used to reverse the order of factors in a vector product. Applying this trick r successive times, we move the vector a from left to right in the product

$$\begin{aligned} aa_1 a_2 \ldots a_r &= 2a \cdot a_1 a_2 \ldots a_r - a_1 a a_2 \ldots a_r \\ &= 2a \cdot a_1 a_2 \ldots a_r - 2a \cdot a_2 a_1 a_3 \ldots a_r + a_1 a_2 a a_3 \ldots a_r \\ &= \ldots \\ &= 2 \sum_{k=1}^{r} (-1)^{k+1} a \cdot a_k a_1 \ldots \check{a}_k \ldots a_r + (-1)^r a_1 a_2 \ldots a_r a. \end{aligned}$$

This gives (1.33) if

$$a \cdot (a_1 a_2 \ldots a_r) = \tfrac{1}{2}(aa_1 \ldots a_r - (-1)^r a_1 \ldots a_r a). \tag{1.34}$$

This is true by virtue of (1.30) since, as we shall shortly see, the product $a_1 a_2 \ldots a_r$ is an even (odd) multivector if r is even (odd). But for the moment we simply notice that, for a simple r-vector $A_r = a_1 a_2 \ldots a_r$ factored in accordance with axiom (1.14), (1.34) is equivalent to (1.27a). Moreover, in this case (1.14) also implies that each term on the right side of (1.33) is an $(r-1)$-vector. Thus we have proved (1.27a) for simple r-vectors. The generalization to arbitrary r-vectors is trivial.

To prove (1.27b), we continue to use the factorization $A_r = a_1 a_2 \ldots a_r$. Because of (1.13),

$$a_k^{-1} a_k = 1 \quad \text{where } a_k^{-1} = |a_k|^{-2} a_k.$$

So (1.33) can now be written

$$a \cdot A_r = \sum_{k=1}^{r} (-1)^{k+1} a \cdot a_k a_k^{-1} a_k a_1 \ldots \check{a}_k \ldots a_n$$

$$= \sum_{k=1}^{r} a \cdot a_k a_k^{-1} A_r.$$

Substituting this in (1.28) we get

$$a \wedge A_r = aA_r - a \cdot A_r = bA_r, \tag{1.35}$$

where $b = a - \Sigma_{k=1}^{r} a \cdot a_k a_k^{-1}$. This expression for b implies that $b \cdot a_k = 0$ so, by (1.32), $ba_k = -a_k b$. Thus, the factorability axiom (1.14) has been satisfied by the last term in (1.35), so that term is an $(r+1)$-vector. This completes our proof of (1.27b) for simple r-vectors.

We are now in a position to prove the fundamental formula

$$\begin{aligned} A_r B_s &= \langle A_r B_s \rangle_{|r-s|} + \langle A_r B_s \rangle_{|r-s|+2} + \cdots + \langle A_r B_s \rangle_{r+s} \\ &= \sum_{k=0}^{m} \langle A_r B_s \rangle_{|r-s|+2k}, \end{aligned} \tag{1.36}$$

where $m = \frac{1}{2}(r+s-|r-s|)$. We can prove (1.36) for simple A_r by using the factorization (1.14) of A_r into anticommuting vectors and applying (1.28) r times. The idea should be clear from examining the simplest case. For $A_2 = a_1 \wedge a_2 = a_1 a_2$ and $s \geqslant 2$,

$$\begin{aligned} A_2 B_s &= a_1 a_2 B_s = a_1 (a_2 \cdot B_s + a_2 \wedge B_s) \\ &= a_1 \cdot (a_2 \cdot B_s) + a_1 \wedge (a_2 \cdot B_s) + a_1 \cdot (a_2 \wedge B_s) + a_1 \wedge a_2 \wedge B_s \\ &= A_2 \cdot B_s + \langle A_2 B_s \rangle_s + A_2 \wedge B_s, \end{aligned} \tag{1.37a}$$

where

$$A_2 \cdot B_s = \langle A_2 B_s \rangle_{|s-2|} = a_1 \cdot (a_2 \cdot B_s), \tag{1.37b}$$

$$\langle A_2 B_s \rangle_s = a_1 \wedge (a_2 \cdot B_s) + a_1 \cdot (a_2 \wedge B_s), \tag{1.37c}$$

$$A_2 \wedge B_s = \langle A_2 B_s \rangle_{s+2} = a_1 \wedge a_2 \wedge B_s. \tag{1.37d}$$

Equation (1.36) shows that, in general, the product of homogeneous multivectors is not homogeneous. The first and last terms on the right side of (1.36) will be identified as $A_r \cdot B_s$ and $A_r \wedge B_s$ respectively. The other terms are of intermediate grade differing from $|r-s|$ and $r+s$ by some multiple of two. It follows that the product of even multivectors is always even. Therefore the set of all even multivectors is a subalgebra of $\mathcal{G}$, naturally called *the even subalgebra* of $\mathcal{G}$. On the other hand, it is obvious that the set of odd multivectors is not closed under multiplication.

The preceding discussion makes it clear that the product $a_1 a_2 \ldots a_r$ is even if r is even, because each successive pair of vectors $a_1 a_2, a_3 a_4, \ldots, a_{r-1} a_r$ is even and their product is even. One more vector in the product will then make the result odd. This is the argument we promised to complete the proof of (1.33).

We are now justified in pointing out that a very useful formula can be obtained from the $(r-1)$-vector part of (1.33), which, by virtue of the distributive rule, can be written

$$a \cdot \langle a_1 a_2 \ldots a_r \rangle_r = \sum_{k=1}^{r} (-1)^{k+1} a \cdot a_k \langle a_1 \ldots \check{a}_k \ldots a_r \rangle_{r-1}.$$

But $\langle a_1 a_2 \ldots a_r \rangle_r = a_1 \wedge a_2 \wedge \ldots \wedge a_r$, so

$$a \cdot (a_1 \wedge a_2 \wedge \ldots \wedge a_r) = \sum_{k=1}^{r} (-1)^{k+1} a \cdot a_k a_1 \wedge \ldots \wedge \check{a}_k \wedge \ldots \wedge a_r. \quad (1.38)$$

The special case

$$a \cdot (a_1 \wedge a_2) = a \cdot a_1 a_2 - a \cdot a_2 a_1 \quad (1.39)$$

generalizes a well-known formula; in fact the formulas for inner and outer products given here include generalizations of all the formulas in the vector algebra of Gibbs. All that will be developed in detail in the sequel to this book entitled *New Foundations for Classical Mechanics*.

A valuable generalization of (1.38) is the formula

$$\begin{aligned} B_r \cdot (a_1 \wedge \ldots \wedge a_n) &= B_r \cdot (a_1 \wedge \ldots \wedge a_r) a_{r+1} \wedge \ldots \wedge a_n - \\ &\quad - B_r \cdot (a_2 \wedge \ldots \wedge a_{r+1}) a_1 \wedge a_{r+2} \wedge \ldots \wedge a_n + \cdots \\ &= \sum_{j_1 < \ldots < j_r} \epsilon(j_1 \ldots j_n) B_r \cdot (a_{j_1} \wedge \ldots \wedge a_{j_r}) \\ &\quad a_{j_{r+1}} \wedge \ldots \wedge a_{j_n}, \end{aligned} \quad (1.40)$$

where B_r is an r-vector of grade $r \leqslant n$, each j_k is a distinct positive integer not greater than n, $j_{r+1} < \ldots < j_n$ as well as $j_1 < \ldots < j_r$, and the 'permutation symbol' $\epsilon(j_1 \ldots j_n)$ has the value 1 (or -1) if $(j_1 \ldots j_n)$ is an even (or odd) permutation of $(1, 2, \ldots, n)$. The number of terms in the expansion is $\binom{n}{r}$, which is the number of ways B_r can be 'dotted' with r vectors from a collection of n vectors. Note that if $a_k \cdot B_r = 0$, any coefficient 'containing' the kth vector a_k will vanish; there can be $n-r$ such vectors if B_r is simple, in which case the expansion has at most one nonvanishing term. Equation (1.40) can be proved by using (1.25b) and iterating with (1.38), but it will be established later as a by-product of our systematic study of bases in Section 3.

Returning to some unfinished business, we easily supply the promised proofs of (1.25a) and (1.25b). Using (1.36) to evaluate each side of

$$\langle A_r(B_sC_t)\rangle_{r+s+t} = \langle (A_rB_s)C_t\rangle_{r+s+t},$$

we have a proof of (1.25a) for homogeneous multivectors. Similarly, we prove (1.25b) by using (1.36) to evaluate

$$\langle A_r(B_sC_t)\rangle_{(t-s)-r} = \langle (A_rB_s)C_t\rangle_{t-(r+s)}.$$

This wraps up the loose ends we left dangling earlier.

To our arsenal of formulas, we can add one more group of identities of general applicability:

$$a\cdot(A_rB) = a\cdot A_rB + (-1)^r A_ra\cdot B \tag{1.41a}$$

$$= a\wedge A_rB - (-1)^r A_ra\wedge B, \tag{1.41b}$$

$$a\wedge(A_rB) = a\wedge A_rB - (-1)^r A_ra\cdot B \tag{1.41c}$$

$$= a\cdot A_rB + (-1)^r A_ra\wedge B. \tag{1.41d}$$

We can prove any of these identities, for example (1.41a), by using (1.27) and (1.36) as follows:

$$\begin{aligned} a\cdot(A_rB_s) &= \tfrac{1}{2}(aA_rB_s - (-1)^{r+s}A_rB_sa)\\ &= \tfrac{1}{2}(aA_rB_s - (-1)^r A_raB_s) + (-1)^r\tfrac{1}{2}(A_raB_s - (-1)^s A_rB_sa)\\ &= a\cdot A_rB_s + (-1)^r A_ra\cdot B_s. \end{aligned}$$

The identities (1.41a, b, c, d) are useful as they stand, but we can project out of them a number of convenient identities involving inner and outer products alone. For example, from (1.41a) we get

$$a\cdot(A_r\wedge B_s) = (a\cdot A_r)\wedge B_s + (-1)^r A_r\wedge(a\cdot B_s), \tag{1.42}$$

and from (1.41d) we get

$$a\wedge(A_r\cdot B_s) = (a\cdot A_r)\cdot B_s + (-1)^r A_r\cdot(a\wedge B_s), \quad \text{for } s\geqslant r>1. \tag{1.43}$$

The identity (1.42) should be compared with (1.33).

The inner and outer products along with their identities suffice to express and exploit all possible relations among vectors of a given set. But to deal with more general multivectors, it is convenient to have the *scalar product* defined by

$$A * B \equiv \langle AB \rangle. \tag{1.44}$$

From (1.36) and (1.21a) we see that

$$A_{\bar{r}} * B_{\bar{s}} = 0 \quad \text{if } r \neq s, \tag{1.45a}$$

$$A_{\bar{r}} * B_{\bar{r}} = A_{\bar{r}} \cdot B_{\bar{r}} = B_{\bar{r}} * A_{\bar{r}} \quad \text{if } r \neq 0. \tag{1.45b}$$

The restriction $r \neq 0$ in (1.45b) is necessary only because the inner product of scalars was assumed to vanish in (1.21b). Of course,

$$A_{\bar{0}} * B_{\bar{0}} = A_{\bar{0}} B_{\bar{0}} = \langle A \rangle \langle B \rangle.$$

From (1.45), the axioms of geometric algebra lead immediately to the expansion of any scalar product in terms of homogeneous parts, that is

$$A * B = \sum_r A_{\bar{r}} * B = \sum_r A_{\bar{r}} * B_{\bar{r}} = \langle A \rangle \langle B \rangle + \sum_r A_{\bar{r}} \cdot B_{\bar{r}}. \tag{1.46}$$

Moreover, the scalar product is seen to be symmetric and linear, that is

$$A * B = \langle AB \rangle = \langle BA \rangle = B * A, \tag{1.47a}$$

$$A * (\alpha B + \beta C) = \alpha A * B + \beta A * C, \tag{1.47b}$$

where α and β are scalars. And, referring to (1.20a), we see that

$$A * B = A^\dagger * B^\dagger. \tag{1.48}$$

Our scalar product is 'positive definite', that is, we can associate with any multivector A a unique positive scalar *magnitude* $|A|$ defined by

$$|A|^2 = A^\dagger * A = \sum_r |A_{\bar{r}}|^2 \geqslant 0, \tag{1.49}$$

where $|A| = 0$ if and only if $A = 0$. It suffices to establish (1.49) for simple r-vectors. Using (1.18) and the axiom (1.13) we get the general formula

$$(a_1 \ldots a_r)^\dagger (a_1 \ldots a_r) = |a_1 \ldots a_r|^2 = |a_1|^2 \ldots |a_r|^2 \geqslant 0. \tag{1.50}$$

From this we see immediately that

$$A_r^\dagger A_r = |A_r|^2 \geqslant 0 \quad \text{if } A_r \text{ is simple}. \tag{1.51}$$

We note that, in general,

$$A * B \leqslant |A|\,|B|, \tag{1.52a}$$

and

$$|AB| = |A|\,|B| \quad \text{if } A^\dagger A = |A|^2. \tag{1.52b}$$

In addition, the distributive axiom leads immediately to the 'law of cosines'

$$|A + B|^2 = |A|^2 + 2A^\dagger * B + |B|^2, \tag{1.53a}$$

where an 'angle' between multivectors (denoted by $\angle AB$) can be defined by writing

$$\cos \angle AB = \frac{A^\dagger * B}{|A|\,|B|}. \tag{1.53b}$$

This angle has a simple geometric interpretation if A and B are k-blades; it reduces to the so-called 'dihedral angle' between intersecting planes if A and B are noncommuting 2-blades. Unfortunately, we will not have the occasion to examine the sundry geometrical interpretations of (1.53a, b) in more detail.

An important consequence of (1.51) is the fact that every nonzero r-blade A_r has an inverse

$$A_r^{-1} = |A_r|^{-2} A_r^\dagger. \tag{1.54}$$

Division by an r-blade is, of course, equivalent to successive division by r vectors. The possibility of division by vectors greatly expedites algebraic manipulation. Division cannot be defined for either the inner or outer product alone. It requires the full geometric product. This is one of the reasons for regarding the geometric product as more fundamental than either the inner or outer product.

Sometimes it is convenient to employ the *commutator* product, which we define by

$$A \times B = \tfrac{1}{2}(AB - BA). \tag{1.55}$$

It is easily established that this product is anticommutative

$$A \times B = -B \times A, \tag{1.56a}$$

linear

$$(\alpha A + \beta B) \times C = \alpha A \times C + \beta B \times C, \tag{1.56b}$$

and, instead of being associative, satisfies the so-called 'Jacobi identity'

$$A \times (B \times C) + B \times (C \times A) + C \times (A \times B) = 0. \tag{1.56c}$$

In addition, the commutator is related to the geometric product by

$$A \times (BC) = (A \times B)C + B(A \times C). \tag{1.57}$$

The commutator of a vector a with any multivector A can be expressed as inner and outer products with the even and odd parts of A; from (1.30) we get

$$a \times A = a \cdot A_+ + a \wedge A_- . \tag{1.58}$$

The commutator of this with another vector b is then

$$b \times (a \times A) = b \wedge (a \cdot A_+) + b \cdot (a \wedge A_-). \tag{1.59}$$

With the help of (1.57) or the Jacobi identity we find the expressions

$$(b \wedge a) \times A = b \cdot (a \wedge A) - a \cdot (b \wedge A) \tag{1.60a}$$

$$= b \wedge (a \cdot A) - a \wedge (b \cdot A) \tag{1.60b}$$

$$= ba \cdot A - A \cdot ba \tag{1.60c}$$

$$= ba \wedge A - A \wedge ba = (ba) \times A. \tag{1.60d}$$

To prove the equivalence of (1.60c) to (1.60d), it is easiest to begin with

$$bAa = b(A \cdot a + A \wedge a) = (b \cdot A + b \wedge A)a.$$

Taking $A = c$ to be a vector in (1.60a) we get the Jacobi identity for vectors in terms of inner and outer products;

$$a \cdot (b \wedge c) + b \cdot (c \wedge a) + c \cdot (a \wedge b) = 0. \tag{1.61}$$

We seldom use the commutator product unless one of the factors is a bivector. To show that the commutator with a bivector is especially useful, we note that (1.37) along with (1.23) imply for a *bivector* B and $r \neq 1$,

$$BA_r = B \cdot A_r + \langle BA_r \rangle_r + B \wedge A_r,$$

$$A_r B = B \cdot A_r - \langle BA_r \rangle_r + B \wedge A_r.$$

So the symmetric part of the product BA_r is

$$\tfrac{1}{2}(BA_r + A_r B) = B \cdot A_r + B \wedge A_r \quad \text{if } r \neq 1, \tag{1.62a}$$

while the antisymmetric part is

$$B \times A_r = \langle BA_r \rangle_r . \tag{1.62b}$$

Thus we have quite generally

$$BA = B \cdot A + B \times A + B \wedge A, \tag{1.63}$$

if $B = \langle B \rangle_2$ and $\langle A \rangle_1 = 0$. The exclusion of vectors from (1.62a) and (1.63) is necessary because, by (1.58),

$$B \times a \equiv \tfrac{1}{2}(Ba - aB) = B \cdot a. \tag{1.64}$$

Equations (1.64) and (1.63) should be compared with (1.27a) and (1.31).

It is important to note that (1.62b) says that the commutator with a bivector preserves grade. So we immediately get from (1.57), for $B = \langle B\rangle_2$,

$$B \times (A \cdot C) = (B \times A) \cdot C + A \cdot (B \times C), \tag{1.65}$$

$$B \times (A \wedge C) = (B \times A) \wedge C + A \wedge (B \times C). \tag{1.66}$$

Iteration of (1.66) leads to the expansion formula for vectors a_k,

$$\begin{aligned} B \times (a_1 \wedge \ldots \wedge a_r) &= (B \times a_1) \wedge a_2 \wedge \ldots \wedge a_r + \\ &\quad + a_1 \wedge (B \times a_2) \wedge a_3 \wedge \ldots \wedge a_r + \cdots + a_1 \wedge \ldots \wedge a_{r-1} \wedge (B \times a_r) \\ &= \sum_{k=1}^{r} (-1)^{k+1} (B \cdot a_k) \wedge a_1 \wedge \ldots \check{a}_k \ldots \wedge a_r. \end{aligned} \tag{1.67}$$

Equation (1.62b) implies that the space of bivectors is closed under the commutator product. It follows that under the commutator product the bivectors make up a Lie algebra, which is, as is well-known and easy to show with geometric algebra, the Lie algebra of rotations in Euclidean space. The so-called structure equations for this Lie algebra can be written in the form

$$\begin{aligned} (a \wedge b) \times (c \wedge d) &= b \cdot c a \wedge d - b \cdot d a \wedge c + a \cdot d b \wedge c - a \cdot c b \wedge d \\ &= a \wedge (b \cdot c \wedge d) - b \wedge (a \cdot c \wedge d), \end{aligned} \tag{1.68}$$

where a, b, c, d are any vectors. Equation (1.68) is easily derived by using (1.39) in (1.60a). Putting $d = b$ in (1.68), we find that

$$(a \wedge b) \times (c \wedge b) = (a \wedge b \wedge c) \cdot b. \tag{1.69}$$

This relation has many applications; for example, the 'law of sines' in spherical trigonometry follows from it almost trivially.

1.2. Vector Spaces, Pseudoscalars and Projections

In this section we show that every n-dimensional vector space $\mathcal{A}_n$ determines a unique geometric algebra $\mathcal{G}(\mathcal{A}_n)$ which can be interpreted as an algebra of subspaces of $\mathcal{A}_n$. Projection operators are defined in terms of the geometric product and their properties are derived without resorting to a basis. Standard concepts of linear algebra such as 'direct sum', 'intersection' and 'factor space' are characterized by relations in geometric algebra, making it easier to describe relations among subspaces and work out their implications. The result is an efficient calculus for, among other things, formulating and proving the theorems of projective geometry.

Every nonzero n-blade A determines a unique n-dimensional vector space $\mathscr{A}_n$ consisting of all vectors a which satisfy the equation

$$a \wedge A = 0. \tag{2.1}$$

Hence it is appropriate to call $\mathscr{A}_n$ the *vector space* of A.

Conversely, every n-dimensional vector space $\mathscr{A}_n$ uniquely determines two unit n-blades $\pm I$. The n-blade formed by outer multiplication of any set of n vectors $a_1, \ldots, a_n$ in $\mathscr{A}_n$ is proportional to I, that is

$$a_1 \wedge a_2 \wedge \ldots \wedge a_n = \lambda I, \tag{2.2}$$

and the scalar λ vanishes if and only if the a_k are linearly dependent. Any nonzero scalar multiple of I is called *a pseudoscalar of* $\mathscr{A}_n$. Assignment of an orientation to $\mathscr{A}_n$ is equivalent to associating a unique unit n-blade I with $\mathscr{A}_n$; in that case, we say that I is *the tangent* or *the direction* of $\mathscr{A}_n$. Thus $\mathscr{A}_n$ and I are 'equivalent' in the sense that specification of either uniquely determines the other.

The asserted equivalence of $\mathscr{A}_n$ and I will now be established. Since $A = |A|I$ is simple, it can be factored into an outer product of n vectors:

$$A = a_1 \wedge a_2 \wedge \ldots \wedge a_n. \tag{2.3}$$

It follows by (1.26b) that

$$a_k \wedge A = 0, \tag{2.4}$$

so each of the a_k is in $\mathscr{A}_n$. Furthermore, the a_k are linearly independent; for supposing one of them, say a_n, can be expressed as a linear combination of the others, we substitute

$$a_n = \sum_{k=1}^{n-1} \alpha_k a_k$$

into (2.3), whence

$$A = \sum_{k=1}^{n-1} \alpha_k a_1 \wedge \ldots \wedge a_{n-1} \wedge a_k = 0,$$

because, by (1.26b), each term in the sum vanishes. But this contradicts the assumption that $A \neq 0$. Hence the a_k compose a basis for $\mathscr{A}_n$.

To show that every a satisfying (2.1) is in $\mathscr{A}_n$, we use (1.28) to get

$$aA = a \cdot A. \tag{2.5}$$

Since A has a multiplicative inverse,

$$a = a \cdot AA^{-1} = a \cdot A^{-1}A. \tag{2.6}$$

The proof is completed by showing that the right side of (2.6) can be expressed as an expansion of a in terms of the a_k. If the a_k are orthogonal then $A = a_1 a_2 \ldots a_n$, $A^{-1} = a_n^{-1} \ldots a_2^{-1} a_1^{-1}$, so using (1.33) to expand $a \cdot A^{-1}$, and anticommutativity to reorder the products, we obtain

$$a = \sum_{k=1}^{n} a \cdot a_k^{-1} a_k. \tag{2.7}$$

Though (2.7) suffices for present purposes, it is worth mentioning that the expansion with nonorthogonal a_k is obtained in the next section.)

The proof that A determines $\mathscr{A}_n$ uniquely is now complete. In essence, by reversing the above argument it can be proved that $\mathscr{A}_n$ determines I, the main point being to show that all n-vectors generated by multiplication of vectors in $\mathscr{A}_n$ are scalar proportional. This last fact is established in the course of the general considerations below.

Incidentally, we have proved in the course of our discussion that n vectors $a_1, a_2, \ldots, a_n$ are linearly independent if and only if $a_1 \wedge a_2 \wedge \ldots \wedge a_n \neq 0$.

The pseudoscalar A of $\mathscr{A}_n$ facilitates the decomposition of any vector b into components in and orthogonal to $\mathscr{A}_n$; thus,

$$b = bAA^{-1} = (b \cdot A + b \wedge A)A^{-1},$$

so

$$b = b_\parallel + b_\perp, \tag{2.8a}$$

where

$$b_\parallel = P_A(b) \equiv b \cdot AA^{-1} \tag{2.8b}$$

and

$$b_\perp = P_A^\perp(b) \equiv b \wedge AA^{-1}. \tag{2.8c}$$

It is readily verified that $b_\parallel A = b_\parallel \cdot A$ or $b_\parallel \wedge A = 0$ so $b_\parallel$ is in $\mathscr{A}_n$, and $b_\perp A = b_\perp \wedge A$ or $b_\perp \cdot A = 0$ so b is orthogonal to every vector in $\mathscr{A}_n$.

The important thing about (2.8) is that it shows how orthogonal projections can be expressed in terms of geometric multiplication and addition; this makes it possible to derive the properties of projections by elementary algebraic computation. The representation (2.8b) for the orthogonal *projection* $P_A(b)$ of b into $\mathscr{A}_n$ should be compared with the matrix representation of a projection which is usually employed for computations. The matrix representation requires a choice of some basis in $\mathscr{A}_n$, while (2.8b) is independent of any such choice. The matrix representation is more unwieldy because the simple concept of an r-vector is absent from matrix algebra. It should be noted further that the function $P_A^\perp(b)$, called the orthogonal *rejection* of b *from* $\mathscr{A}_n$, is defined algebraically by (2.8c) without reference to any 'enveloping' vector space in which $\mathscr{A}_n$ is embedded. Such an enveloping space is required for the matrix representation of $P_A^{\perp}(b)$ as the projection of b into the 'orthogonal complement' of $\mathscr{A}_n$, though for many purposes it is as irrelevant as the choice of vector basis.

By multiplication and addition the vectors of $\mathscr{A}_n$ generate a subalgebra $\mathscr{G}(\mathscr{A}_n)$ of the complete Geometric Algebra called the *Geometric Algebra of* $\mathscr{A}_n$. In view of the 'equivalence' of $\mathscr{A}_n$ with A, $\mathscr{G}(\mathscr{A}_n)$ can with equal justice be called the geometric algebra of A and denoted by $\mathscr{G}(A)$. It is not difficult to show that $\mathscr{G}(A)$ is a 2^n-dimensional linear space which is closed under the geometric product. From any set of n linearly independent vectors spanning $\mathscr{A}_n = \mathscr{G}^1(A)$ we can form $\binom{n}{r}$ linearly independent r-blades by taking the outer product of each combination of r different vectors selected from the given set. It is not difficult to prove that these r-blades compose a basis for the space $\mathscr{G}^r(A)$ of all r-vectors in $\mathscr{G}(A)$. The properties of a basis for $\mathscr{G}^r(A)$ are developed in detail in the next section. Thus $\mathscr{G}^r(A)$ is a linear subspace of $\mathscr{G}(A)$ with dimension $\binom{n}{r}$. The dimension of $\mathscr{G}(A)$ is therefore given by

$$\dim \mathscr{G}(A) = \sum_{r=0}^{n} \dim \mathscr{G}^r(A) = \sum_{r=0}^{n} \binom{n}{r} = 2^n.$$

We can give each blade in $\mathscr{G}(A)$ a geometric interpretation, and this determines a geometric interpretation for the whole of $\mathscr{G}(A)$. Just as the n-blade A determines the vector space $\mathscr{A}_n = \mathscr{G}^1(A)$, so each nonzero r-blade A_r in $\mathscr{G}^r(A)$ determines a unique oriented vector space $\mathscr{A}_r = \mathscr{G}^1(A_r)$, consisting of all vectors a which satisfy the equation $a \wedge A_r = 0$. Indeed, $\mathscr{A}_r$ is a subspace of $\mathscr{A}_n$. Thus, *every nonzero r-blade in* $\mathscr{G}^r(A)$ *determines a unique oriented r-dimensional subspace of* $\mathscr{A}_n = \mathscr{G}^1(A)$. Conversely, each oriented r-dimensional subspace $\mathscr{A}_r$ of $\mathscr{A}_n$ determines a unique unit r-blade I_r, 'the direction' of $\mathscr{A}_r$. This one-to-one correspondence between oriented subspaces and unit blades makes it possible to describe every relation between subspaces by an algebraic relation or equation for the corresponding blades. For example, the fact that $\mathscr{A}_r = \mathscr{G}^1(I_r)$ is a subspace of $\mathscr{A}_n = \mathscr{G}^1(I)$ is expressed algebraically by the fact that the direction I_r is a factor of the unit n-blade I, that is, there exists a unique unit $(n-r)$-blade I_{n-r} such that

$$I_r I_{n-r} = I.$$

This can be solved for $I_{n-r} = I_r^{\dagger} I$. The subspace $\mathscr{A}_{n-r} = \mathscr{G}^1(I_{n-r})$ is (disregarding orientation) often called the *orthogonal complement* of $\mathscr{A}_r$ (with respect to $\mathscr{A}_n$). Thus, orthogonal complements are easily represented or determined by a factorization of the pseudoscalar. With this example as an illustration, we assert that $\mathscr{G}(\mathscr{A}_n)$ *can be interpreted as an algebra of directions in* $\mathscr{A}_n$, because it completely characterizes the so-called 'lattice of subspaces of $\mathscr{A}_n$'. We elaborate on this theme throughout the rest of the chapter.

A word about nomenclature is in order. Because we always use the word 'vector' to refer to a one-vector in geometric algebra, we find it convenient to distinguish between the terms 'vector space' and 'linear space'. We use the word 'linear space' in the usual sense, and, as usual, we take a vector space always to be a linear space, but more, we always assume that the elements of a vector space have the multiplicative properties of a one-vector in Geometric Algebra.

The expression (2.8b) for the projection of a vector into $\mathscr{A}_n$ can be generalized to an algebraic expression for the orthogonal projection $P_A(B)$ of any multivector B into $\mathscr{G}(\mathscr{A}_n)$; specifically,

$$\begin{aligned} P_A(B) &= (B \cdot A) \cdot A^{-1} = A^{-1} \cdot (A \cdot B) \\ &= P_I(B) = (B \cdot I) \cdot I^\dagger = I^\dagger \cdot (I \cdot B) = (-1)^{n(n-1)/2} (B \cdot I) \cdot I, \end{aligned} \tag{2.9a}$$

where $A^{-1} = |A|^{-1} I^\dagger = (-1)^{n(n-1)/2} |A|^{-1} I$. Unfortunately, Eqn. (2.9a) gives $P(\langle B \rangle) = 0$ and $P(\langle B \rangle_n) = 0$ because of the convention (1.21b), but of course the projection operator should have the properties

$$P(\langle B \rangle) = \langle B \rangle, \tag{2.9b}$$

$$P(\langle B \rangle_n) = \langle B \rangle_n \cdot AA^{-1}. \tag{2.9c}$$

To ensure this, we can stipulate that (2.9a) does not apply to scalars or pseudoscalars and postulate (2.9b, c) separately. Or, whenever we use (2.9a), we can disregard the exceptional convention (1.21b) and require that (1.21a) define the inner product even for zero grades, in which case (2.9b, c) follow from (2.9a). Desirable as this convention is when we work with (2.9a), it cannot be adopted universally without complicating many other formulas such as (1-1.28) and (1-1.41).

Now let us consider the properties of (2.9a) in more detail. If B is a simple s-vector, then $B \cdot A$ is simple, so it is factorable into a product of $|n - s|$ vectors when $s \neq n$, that is $B \cdot A = c_1 c_2 \ldots c_{|s-n|}$; moreover, if $s > n$, then $c_k \cdot A = 0$ for each vector factor c_k; but if $s < n$, then $c_k \wedge A = 0$ for each c_k. Hence

$$P_A(B_{\bar{s}}) = 0 \quad \text{if } s > n, \tag{2.10a}$$

and

$$P_A(B_{\bar{s}}) = B_{\bar{s}} \cdot AA^{-1} \quad \text{if } s \leqslant n. \tag{2.10b}$$

Equation (2.10b) differs from (2.9a) only in the absence of a 'dot', but that difference makes it much easier to use. For example, the right side of (2.10b) makes it obvious that

$$P_A(B_{\bar{s}}) = B_{\bar{s}} \quad \text{if } B_{\bar{s}} A = B_{\bar{s}} \cdot A. \tag{2.11}$$

It is usually convenient to take (2.11) as the 'defining property' of $\mathscr{G}(A)$. To show that it is equivalent to the definition of $\mathscr{G}(A)$ as the algebra generated by vectors in $\mathscr{A}_n$, it is necessary to prove the assertion that

$$P_A(B) = B \quad \text{iff } B \text{ is in } \mathscr{G}(A)^*. \tag{2.12}$$

* 'iff' means 'if and only if'.

This assertion has already been proved for one-vectors in connection with (2.8). The proof for a simple s-vector can be accomplished by factoring it into a product of vectors as shown in the argument following Eqn. (2.9). Then only linearity is needed to complete the proof for an arbitrary multivector.

The projection operator defined by (2.9) has the following general properties in addition to those already mentioned.

$$P(\alpha B + \beta C) = \alpha P(B) + \beta P(C), \tag{2.13a}$$

$$P(\langle B\rangle_r) = \langle P(B)\rangle_r, \tag{2.13b}$$

$$P(P(B)) = P(B), \tag{2.13c}$$

$$P(C \wedge B) = P(C) \wedge P(B), \tag{2.13d}$$

$$P(BC) = BP(C) \quad \text{if } P(B) = B. \tag{2.13e}$$

The abbreviation $P = P_A$ is convenient when a single projection is of interest. It will be noted that, because of (2.13b), (2.13e) includes the important special case

$$P(B \cdot C) = B \cdot P(C) \quad \text{if } P(B) = B, \tag{2.13f}$$

but, in general, $P(B \cdot C) \neq P(B) \cdot P(C)$.

In general, the relation between two different projection operators, say P_A and P_B, is quite complicated, though it is algebraically determined by (2.9). However, the relation has a simple form in two important special cases. If

$$BA = B \wedge A, \tag{2.14a}$$

then we have the operator equations*

$$P_{BA} = P_B + P_A, \tag{2.14b}$$

$$P_B P_A = P_A P_B = 0. \tag{2.14c}$$

* In (2.14c) and (2.15c) we have followed the common convention of indicating a composite linear function by juxtaposition; thus (2.14c) is equivalent to

$$P_B P_A(a) \equiv P_B(P_A(a)) = P_A(P_B(a)) = 0 \quad \text{for all } a.$$

Unfortunately this convention increases the risk of confusing symbols which denote functions with symbols which denote multivectors, because the geometric product is also indicated by juxtaposition. The problem is even more serious when the value of a function is represented by the same symbol as the function itself, a practice which is common and convenient if not absolutely necessary in application to physics. However the resulting simplification of equations seems to be worth the risk, and we will often rely on the context in which an equation is presented to remove any ambiguity in its meaning. Thus, only the context tells us that (2.14c) is not to be interpreted as the equation

$$P_B(a)P_A(a) = 0.$$

But if

$$BA = B \cdot A \quad \text{and} \quad \text{grade}\, B \leqslant \text{grade}\, A, \tag{2.15a}$$

then

$$P_{BA} = P_A - P_B, \tag{2.15b}$$

$$P_B P_A = P_A P_B = P_B. \tag{2.15c}$$

It will be noted that $\mathscr{G}(B)$ is a subalgebra of $\mathscr{G}(A)$ if and only if (2.15a) is satisfied. Also it should be understood that the projection operators P_A, P_B are well defined only for simple multivectors A, B.

Now let us say a few words about the proofs of the projection properties. Properties (2.13a, b) are obvious consequences of the linearity of the inner product and, in fact, were used in the argument leading from (2.9) to (2.10). A simple variation of that argument also proves (2.13c). Let us prove (2.13d) for $C_r = \langle C\rangle_r$ and $B_s = \langle B\rangle_s$. If $r + s > n$, then the left side of (2.13d) vanishes by (2.10a), while the right side vanishes because of (2.12) and (2.10a). If $r + s \leqslant n$, then, with the help of (2.10b), (1.25b) and (1.23b), we have

$$\begin{aligned} P(C_r \wedge B_s)A &= (C_r \wedge B_s) \cdot A \\ &= C_r \cdot (B_s \cdot A) = C_r \cdot (P(B_s) \cdot A) \\ &= [C_r \wedge P(B_s)] \cdot A = (-1)^{rs} [P(B_s) \wedge C_r] \cdot A \\ &= (-1)^{rs} [P(B_s) \wedge P(C_r)] \cdot A = P(C_r) \wedge P(B_s)A. \end{aligned}$$

The proof of (2.13e) will be given presently, after the duality of inner and outer products has been explained. The simple proofs of (2.14) and (2.15) will be omitted, as they involve no new ideas.

We call the $|n - s|$-vector $B_{\bar{s}} \cdot A$ the *dual of* $B_{\bar{s}}$ *by the n-blade* A, because it can be regarded as an algebraic formulation of the duality concept in geometry. Often the word 'dual' is used when the direction but not the magnitude and orientation of A is specified. Ordinarily, the notion of duality by A is applied only to elements of $\mathscr{G}(A)$, but the more general notion used here ties up duality with orthogonal projections. Thus, the right side of (2.9a) says that, except for a possible difference in sign, the projection into $\mathscr{G}(A)$ is equal to the 'double dual' by the pseudoscalar A. If B is any multivector in $\mathscr{G}(A)$, then (2.12) guarantees that the dual of B is simply BA.

From (1.41b, d) we get the valuable relations

$$(a \cdot B)A = a \wedge (BA) \quad \text{if}\ a \wedge A = 0, \tag{2.16a}$$

$$(a \wedge B)A = a \cdot (BA) \quad \text{if}\ a \wedge A = 0. \tag{2.16b}$$

These relations hold even if B is not in $\mathcal{G}(A)$. As a special case of (2.16), we have

$$(a \cdot B) \cdot A = a \wedge (B \cdot A) \quad \text{if } a \wedge A = 0, \tag{2.17a}$$

$$(a \wedge B) \cdot A = a \cdot (B \cdot A) \quad \text{if } a \wedge A = 0. \tag{2.17b}$$

In accordance with the definition of dual in the last paragraph, (2.17) may be interpreted as expressing the *duality of inner and outer products* with respect to A. In (2.17a), the dual of the inner product is equal to the outer product of the dual. Of course, by virtue of (2.11), Eqns. (2.16) and (2.17) are identical if B is in $\mathcal{G}(A)$, in which case (2.16) is the better expression of duality. It should be noted that the duality of inner and outer products is independent of the scale and orientation of A.

By iterating (2.16) it is easy to establish the more general 'duality relations'

$$C_{\bar{r}} \cdot B_{\bar{s}}A = C_{\bar{r}} \wedge (B_{\bar{s}}A) \quad \text{for } r \leqslant s, \tag{2.18a}$$

$$C_{\bar{r}} \wedge B_{\bar{s}}A = C_{\bar{r}} \cdot (B_{\bar{s}}A) \quad \text{for } r \leqslant |s - n|, \tag{2.18b}$$

where $C_{\bar{r}}A = C_{\bar{r}} \cdot A$, $r \leqslant n$ and $A = A_{\bar{n}}$. The restrictions on grade in (2.18) arise from using (1.25b) in the proof.

We now prove that

$$P(aB) = aP(B) \quad \text{if } P(a) = a. \tag{2.19}$$

The more general result (2.13e) is easily obtained from this by iteration and linearity. Since, by (2.10a), $\langle P(B)\rangle_r = P(\langle B\rangle_r)$ vanishes if $r > n$, it suffices to prove (2.19) under the assumption that $\langle B\rangle_r = 0$ for $r > n$, in which case we can use (2.10b) along with (2.16). Thus,

$$P(a \cdot B)A = (a \cdot B) \cdot A = a \wedge (B \cdot A) = a \wedge [P(B)A] = a \cdot P(B)A.$$

From (2.13d) we see immediately that $P(a \wedge B) = a \wedge P(B)$, so (2.19) can be established by using

$$P(aB) = P(a \cdot B + a \wedge B) = P(a \cdot B) + P(a \wedge B).$$

To describe properties of vector spaces it is common to introduce set relations and operations such as inclusion, intersection and addition. However, since every vector space is determined by a pseudoscalar, the properties of vector spaces can alternatively be described as algebraic properties of their pseudoscalars. This approach admits a more detailed description of vector spaces, because the Geometric Algebra has more structure than the set language. Furthermore, it has considerable computational advantages, because, as will be shown in chapter 3, all linear vector functions can be expressed in terms of geometric algebra. To translate the 'set language' into Geometric Algebra, we give a small dictionary.

A vector space $\mathscr{B} = \mathscr{G}^1(B)$ is a subspace of $\mathscr{A} = \mathscr{G}^1(A)$ if and only if its pseudoscalar B is a *factor* of A; that is,

$$\mathscr{B} \subset \mathscr{A} \text{ iff } BA = B \cdot A, \text{grade } B < \text{grade } A. \tag{2.20}$$

This is a fairly obvious consequence of our previous results, particularly (2.11). Of course, we assume that A and B are simple here. That the equation $BA = B \cdot A$ says that B is a factor of A is clear by solving the equation; thus,

$$A = BB^{-1} \cdot A = BA' = B \wedge A'.$$

The factor $A' = B^{-1} \cdot A$ orthogonal to B is obviously unique. However, there are other factorizations of A of the form $A = B \wedge C \neq BC$, where C is not uniquely determined by B and A, though C is simple and grade C = grade A' = grade A − grade B. Such a factorization is equivalent to the decomposition of a vector space into a *direct sum* of subspaces, as expressed by

$$\mathscr{G}^1(B \wedge C) = \mathscr{G}^1(B) \cup \mathscr{G}^1(C). \tag{2.21}$$

The outer product describes more than the direct sum, because it relates orientations of the spaces. So we can interpret (2.21) as specifying an oriented direct sum of oriented vector spaces.

If C is a *common factor of maximum grade* of blades A and B, then

$$\langle AB \rangle_{\max} = (A \cdot C) \wedge (C^{-1} \cdot B) = A' \wedge B', \tag{2.22}$$

where $\langle AB \rangle_{\max}$ denotes the part of AB with maximum grade, and A', B' are blades satisfying $A = A' \wedge C$, $B = C^{-1} \wedge B'$. Furthermore, the vector space $\mathscr{G}^1(C)$ is the *intersection* of vector spaces $\mathscr{A} = \mathscr{G}^1(A)$ and $\mathscr{B} = \mathscr{G}^1(B)$;

$$\mathscr{G}^1(C) = \mathscr{G}^1(A) \cap G^1(B) = \mathscr{A} \cap \mathscr{B}. \tag{2.23}$$

The *sum* $\mathscr{A} \cup \mathscr{B}$ of vector spaces is related to the product AB by

$$\mathscr{A} \cup \mathscr{B} = \mathscr{G}^1(C \wedge \langle AB \rangle_{\max}) = \mathscr{A} \cap \mathscr{B} + \mathscr{G}^1(\langle AB \rangle_{\max}). \tag{2.24}$$

If A and B have no common factor, then $C = 0$ and $\langle AB \rangle_{\max} = A \wedge B$, so (2.24) reduces to

$$\mathscr{A} \cup \mathscr{B} = \mathscr{G}^1(A \wedge B). \tag{2.25}$$

Thus, a nonvanishing outer product of blades describes the direct sum of non-intersecting vector spaces. Unfortunately, when blades have a nonvanishing common factor C, that factor cannot be computed from products of A and B alone, so the intersection of vector spaces (2.23) cannot be determined without additional information needed to compute the common factor. We shall see what information is needed below.

The correspondence between operations of set theory and geometric algebra can be made more direct by introducing a new operation defined in terms of inner and outer products. It will be convenient to introduce the notation

$$\widetilde{A} \equiv AI^{\dagger} = A \cdot I^{\dagger} \tag{2.26}$$

for the dual of a multivector A in $\mathscr{G}(I)$, where I is a unit n-vector. The double dual is then

$$(\widetilde{A})^{\sim} = (-1)^{n(n-1)/2}A. \tag{2.27}$$

Now we can define the *meet* $A \vee B$ of multivectors A and B in $\mathscr{G}(I)$ by

$$A \vee B \equiv (\widetilde{A} \wedge \widetilde{B}) \cdot I = (-1)^{n(n-1)/2}(\widetilde{A} \wedge \widetilde{B})^{\sim}. \tag{2.28}$$

Multiplication by $I^{\dagger}$ gives

$$(A \vee B)^{\sim} = \widetilde{A} \wedge \widetilde{B}. \tag{2.29}$$

This exhibits the meet as the dual of the outer product. It is, of course, a disguised form for the duality between inner and outer products which we noted earlier. Indeed, if grade $(\widetilde{A} \wedge \widetilde{B}) \leqslant n$, then (2.18b) implies

$$(\widetilde{A} \wedge \widetilde{B}) \cdot I = \widetilde{A} \cdot (\widetilde{B}I) = \widetilde{A} \cdot (BI^{\dagger}I) = \widetilde{A} \cdot B.$$

Hence (2.29) reduces to

$$A \vee B = \widetilde{A} \cdot B = A \cdot I^{\dagger}B \quad \text{if grade } (\widetilde{A} \wedge \widetilde{B}) \leqslant n. \tag{2.30}$$

Of course, $\widetilde{A} \wedge \widetilde{B} = 0$ if grade $(\widetilde{A} \wedge \widetilde{B}) > n$, so the meet vanishes if (2.30) is not satisfied. With this proviso, we could define the meet by (2.30) instead of (2.28).

From the associativity and the anticommutivity (1.23b) of the outer product it follows that the meet is associative and anticommutative with

$$A \vee B = (-1)^{(n-r)(n-s)}B \vee A \tag{2.31}$$

if r and s are the grades of blades A and B. Also, from (2.26) we have

$$\widetilde{A} \cdot I = (AI^{\dagger})I = A = II^{\dagger}A = \widetilde{I}A$$

so from (2.30) we obtain

$$A \vee I = I \vee A = A. \tag{2.32}$$

Now, a geometrical interpretation for the meet is determined by choosing I to be the pseudoscalar for

$$\mathscr{A} \cup \mathscr{B} = \mathscr{G}^1(I). \tag{2.33}$$

So from (2.24) we have

$$C \wedge \langle AB \rangle_{\max} = C \langle AB \rangle_{\max} = \lambda I,$$

where λ is a scale factor to be chosen for our convenience. With this choice for I, if A and B have a common factor, then their duals do not; hence,

$$\langle AB \rangle_{\max} = \langle AI^{\dagger} IB \rangle_{\max} = (-1)^k \langle \tilde{A}\tilde{B} \rangle_{\max} = (-1)^k \tilde{A} \wedge \tilde{B},$$

where k is an integer. Now we choose λ so that

$$C^{-1} \wedge \tilde{A} \wedge \tilde{B} = C^{-1} \tilde{A} \wedge \tilde{B} = I^{\dagger},$$

which can be solved for

$$C = \tilde{A} \wedge \tilde{B} I = A \vee B.$$

Therefore,

$$\mathscr{A} \cap \mathscr{B} = \mathscr{G}^1(A \vee B) \tag{2.34}$$

when (2.33) is satisfied. It should be noted that $A \vee B$ specifies an orientation for $\mathscr{A} \cap \mathscr{B}$ which depends on the orientation of $\mathscr{B} \cup \mathscr{A}$ as well as the orientations of $\mathscr{A}$ and $\mathscr{B}$. The relation among these orientations is completely determined by (2.30).

The meet was introduced originally by Doublet *et al*. [Do] as part of a formalism that is completely equivalent to geometric algebra as we have used it here. Rota and Stein [Ro] have shown that the formalism is 'ideally suited to proving theorems in projective geometry', and the reader is referred to their article for an account of this application.

For the last entry in our dictionary, let us see how to formulate the notion of a factor space in terms of Geometric Algebra. First, note that if a k-blade B is a factor of a $(k+1)$-blade C, then the set of all vectors a satisfying the equation $a \wedge B = C$ is a *k-plane* with tangent (or pseudoscalar) B; the k-plane is located a distance $d = a \wedge BB^{-1} = CB^{-1}$ from the origin. If $\mathscr{B} = \mathscr{G}^1(B)$ is a subspace of $\mathscr{A} = \mathscr{G}^1(A)$, then the set of all such k-planes in $\mathscr{A}$ is the factor space $\mathscr{A}/\mathscr{B}$. Clearly $\mathscr{A}/\mathscr{B}$ is isomorphic to the set of all $(k+1)$-vectors in $\mathscr{G}^{k+1}(A)$ with a common k-factor B; that is,

$$\mathscr{A}/\mathscr{B} = \mathscr{G}^1(A)/\mathscr{G}^1(B) \longleftrightarrow \{a \wedge B\}. \tag{2.35}$$

The right side of the isomorphism (2.35) is not only easier to handle algebraically, but it is often more *apropos* in applications than the factor space.

1-3. Frames and Matrices

It is common practice to define a vector as an ordered set of scalars and develop matrix algebra as a system for manipulating such sets. In contrast, geometric algebra puts vectors and scalars on an equal footing and makes a definite algebraic distinction between vectors and ordered sets of scalars. With the geometric product it is easier to base matrix algebra on manipulations with ordered sets of vectors instead of scalars.

This section shows how geometric algebra simplifies and systemizes the fundamental procedures for finding orthogonal and reciprocal sets of vectors, as well as solving vector equations for scalar coefficients, of which Cramer's rule is a special case. The construction of a basis for finite dimensional geometric algebras is also carried out.

The last topic of the section is the construction of an inverse for a nonsingular matrix. The construction requires that a matrix be regarded as a set of inner products of vectors, which is always possible. This is one of many examples which show, we think, that matrix algebra is best developed *ab initio* from a multilinear algebra of vectors.

We say that a set of vectors $a_1, a_2, \ldots, a_n$ is a *frame* if and only if $A_n = a_1 \wedge a_2 \wedge \ldots \wedge a_n \neq 0$. With the complete Geometric Algebra and the identities of Section 1 at our disposal, we find that the condition $A_n \neq 0$ is usually easier to use than the equivalent condition that the a_k be linearly independent. Manipulations with a frame can be simplified by orthogonalization or by introducing a reciprocal frame, so we discuss these procedures first.

A given frame $a_1, a_2, \ldots, a_n$ can be systematically orthogonalized by constructing the 'graded sequence' of multivectors

$$A_0 = 1, \qquad A_1 = a_1, \qquad A_2 = a_1 \wedge a_2, \ldots, \qquad A_n = a_1 \wedge \ldots \wedge a_n, \tag{3.1}$$

from which we obtain the frame of vectors

$$c_k = A_{k-1}^{\dagger} A_k, \quad k = 1, \ldots, n. \tag{3.2}$$

The c_k are obviously orthogonal, because (3.2) expresses each c_k as the dual by A_k of the pseudoscalar A_{k-1} of the vector space spanned by the preceding $k-1$ vectors. Using (1.51), we find that

$$c_1 c_2 \ldots c_k = A_k |A_{k-1}|^2 \ldots |A_2|^2 |A_1|^2, \tag{3.3}$$

which gives a factorization of A_k into a product of orthogonal vectors. The c_k can be used as they are in (3.2) or, by iterating with (1.25b) and (1.38) or using (1.40), they can be expressed as a linear combination of the a_k. For $k > 1$, we find

$$\begin{aligned} c_k &= A_{k-1}^{\dagger} \cdot A_k = (a_{k-1} \wedge \ldots \wedge a_1) \cdot (a_1 \wedge \ldots \wedge a_k) \\ &= \sum_{i=1}^{k} (-1)^{k-i} (a_{k-1} \wedge \ldots \wedge a_1) \cdot (a_1 \wedge \ldots \check{a}_i \ldots \wedge a_k) a_i. \end{aligned} \tag{3.4}$$

The coefficients on the right side of (3.4) are determinants and their properties will be discussed in more detail in the next section. In particular, we will show how to express the determinant in terms of the inner products $a_j \cdot a_k$, which is especially desirable when the inner products are given at the beginning of a problem.

The result (3.4) is identical to that obtained by the so-called 'Gram–Schmidt orthogonalization process', but it is in a more compact and perspicuous form.

When one desires to work directly with a given frame $a_1, a_2, \ldots, a_n$ of non-orthogonal vectors, it is convenient to introduce the *reciprocal frame* $a^1, \ldots, a^n$, specified by the equations

$$a^k \cdot a_j = \delta_j^k, \tag{3.5}$$

where $j, k = 1, \ldots, n$, and δ_j^k, the *Kronecker delta*, has the value 1 if $j = k$ and the value 0 if $j \neq k$. Geometric Algebra makes it especially easy to solve equation (3.5) for the a^k in terms of the a_j. Note that

$$A_n = a_1 \wedge \ldots \wedge a_n = (-1)^{k-1} a_k \wedge (a_1 \wedge \ldots \check{a}_k \ldots \wedge a_n). \tag{3.6}$$

So, because of (1.26b) and (2.16b)

$$\begin{aligned} \delta_j^k &= a_j \wedge [(-1)^{k-1} a_1 \wedge \ldots \check{a}_k \ldots \wedge a_n] A_n^{-1} \\ &= a_j \cdot [(-1)^{k-1} (a_1 \wedge \ldots \check{a}_k \ldots \wedge a_n) A_n^{-1}]. \end{aligned}$$

Hence

$$a^k = (-1)^{k-1} a_1 \wedge \ldots \check{a}_k \ldots \wedge a_n A_n^{-1} \tag{3.7}$$

satisfies (3.5). The frames $\{a^k\}$ and $\{a_k\}$ are sometimes said to be *dual* to one another, because, as (3.7) shows, a^k is the dual of an $(n-1)$-vector composed of the a_j. Equation (3.7) is the most useful expression for the a^k, but if desired we can get explicit expressions for the a^k as linear combinations of the a_k in the same way that we arrived at (3.4). Thus, according to (1.54) and (1.51),

$$A_n^{-1} = \frac{A_n^\dagger}{A_n A_n^\dagger} = \frac{a_n \wedge \ldots \wedge a_1}{(a_1 \wedge \ldots \wedge a_n)(a_n \wedge \ldots \wedge a_1)},$$

so we get from (3.7),

$$\begin{aligned} a^k &= (-1)^{k-1} \frac{(a_1 \wedge \ldots \check{a}_k \ldots \wedge a_n) \cdot (a_n \wedge \ldots \wedge a_1)}{(a_1 \wedge \ldots \wedge a_n) \cdot (a_n \wedge \ldots \wedge a_1)} \\ &= \sum_{i=1}^{n} (-1)^{k+i} \frac{(a_1 \wedge \ldots \check{a}_k \ldots \wedge a_n) \cdot (a_n \wedge \ldots \check{a}_i \ldots \wedge a_1) a_i}{(a_1 \wedge \ldots \wedge a_n) \cdot (a_n \wedge \ldots \wedge a_1)}. \end{aligned} \tag{3.8}$$

While we are on the subject of frames, we may as well show how to construct a basis and its reciprocal for the complete Geometric Algebra $\mathcal{G}(A_n)$ from a basis for $\mathcal{G}^1(A_n)$. But the mere existence of a basis for $\mathcal{G}(A_n)$ is all that we shall ever appeal to in the rest of the book, so the details of the following construction may be passed over without loss.

From a frame $\{a_k\}$ with pseudoscalar $A_n = a_1 \wedge \ldots \wedge a_n$, a basis for the space of r-vectors $\mathcal{G}^r(A_n)$ can be constructed by outer multiplication, namely, the simple r-blades

$$a_{k_1} \wedge a_{k_2} \wedge \ldots \wedge a_{k_r}, \quad 0 < k_1 < \ldots < k_r \leqslant n. \tag{3.9}$$

A dual basis can be constructed from the dual frame $\{a^k\}$, namely

$$a^{j_1} \wedge a^{j_2} \wedge \ldots \wedge a^{j_r}, \quad 0 < j_1 < \ldots < j_r \leqslant n. \tag{3.10}$$

Each such r-vector is the dual of an $(n - r)$-vector, specifically

$$a^{j_1} \wedge \ldots \wedge a^{j_r} = (-1)^{\Sigma_{i=1}^{r} (j_i - 1)} a_{j_{r+1}} \wedge \ldots \wedge a_{j_n} A_n^{-1}, \tag{3.11}$$

where $0 < j_{r+1} < \ldots < j_n \leqslant n$, and all the j's are distinct integers. Equation (3.11) can be proved by using (2.11), (1.25b) and (1.38) as follows:

$$a^{j_1} A_n = a^{j_1} (a_1 \wedge \ldots \wedge a_n) = (-1)^{j_1 - 1} a_1 \wedge \ldots \check{a}_{j_1} \ldots \wedge a_{j_n}$$

$$a^{j_2} \wedge a^{j_1} A_n = a^{j_2} \cdot [a^{j_1} \cdot A_n] = (-1)^{j_1 - 1} (-1)^{j_2 - 2} a_1 \wedge \ldots \check{a}_{j_1} \ldots \check{a}_{j_2} \ldots \wedge a_{j_n}$$

$$\cdot \quad \cdot \quad \cdot$$

$$a^{j_r} \wedge \ldots \wedge a^{j_1} A_n = (-1)^{\Sigma_{i=1}^{r} (j_i - i)} a_{j_{r+1}} \wedge \ldots \wedge a_{j_n}. \tag{3.12}$$

Division of (3.12) by A_n gives (3.11), since

$$a^{j_r} \wedge \ldots \wedge a^{j_1} = (-1)^{r(r-1)/2} a^{j_1} \wedge \ldots \wedge a^{j_r} = (-1)^r (-1)^{\Sigma_{i=1}^{r} i} a^{j_1} \wedge \ldots \wedge a^{j_r}. \tag{3.13}$$

The inner product of (3.9) with (3.10) can be evaluated with the help of (3.11); thus

$$(a_{k_r} \wedge \ldots \wedge a_{k_1}) \cdot (a^{j_1} \wedge \ldots \wedge a^{j_r})$$

$$= (-1)^{\Sigma_{i=1}^{r} (j_i - 1)} a_{k_r} \wedge \ldots \wedge a_{k_1} \wedge a_{j_{r+1}} \wedge \ldots \wedge a_{j_n} A_n^{-1}$$

$$= \delta_{k_1}^{j_1} \delta_{k_2}^{j_2} \ldots \delta_{k_r}^{j_r}. \tag{3.14}$$

If $r = n$, (3.14) gives $(a^n \wedge \ldots \wedge a^1)(a_1 \wedge \ldots \wedge a_n) = 1$, so

$$A^n \equiv a^n \wedge \ldots \wedge a^1 = A_n^{-1}. \tag{3.15}$$

The projection of an r-vector B_r into $\mathscr{G}(A_n)$ can be expanded in a basis in $\mathscr{G}^r(A_n)$. For $1 \leqslant r \leqslant n$,

$$P(B_r) = (B_r \cdot A_n)A_n^{-1} = \sum_{j_1 < \ldots < j_r} B_r \cdot (a_{j_r} \wedge \ldots \wedge a_{j_1})a^{j_1} \wedge \ldots \wedge a^{j_r}. \tag{3.16}$$

This can be used to supply the promised proof of (1.40). Multiplying (3.16) by A_n to get

$$B_r \cdot A_n = \sum_{j_1 < \ldots < j_r} B_r \cdot (a_{j_r} \wedge \ldots \wedge a_{j_1})(a^{j_1} \wedge \ldots \wedge a^{j_r}) \cdot A_n,$$

and substituting (3.12), we get (1.40) with the following expression for the permutation symbol:

$$\epsilon(j_1 \ldots j_n) = (-1)^{\Sigma_{i=1}^{r} (j_i - i)}, \tag{3.17}$$

which holds for $j_1 < \ldots < j_r$ and $j_{r+1} < \ldots < j_n$.

As an application of (3.16), recall that we have shown that any r-dimensional subspace $\mathscr{B}_r$ of the vector space $\mathscr{A}_n = \mathscr{G}^1(A_n)$ is uniquely determined by specifying an r-blade B_r. Expanding B_r in terms of a basis as in (3.16), we have

$$B_r = \sum_{j_1 < \ldots < j_r} \beta_{j_r \ldots j_1} a^{j_1} \wedge \ldots \wedge a^{j_r}.$$

The ratios of the coefficients $\beta_{j_r \ldots j_1} = B_r \cdot (a_{j_r} \wedge \ldots \wedge a_{j_1})$ are independent of the magnitude of B_r; they are called the *Plücker coordinates* of the space $\mathscr{B}_r = \mathscr{G}^1(B_r)$ in the mathematical literature. Since B_r is simple, the $\beta_{j_r \ldots j_1}$ cannot be chosen freely, being constrained by the condition that $B_r^\dagger B_r = |B_r|^2$. For example, for $n = 4$, $r = 2$, we have $B_r^\dagger \wedge B_r = 0$ which, when expressed as a condition on the coefficients, becomes

$$\beta_{12}\beta_{34} + \beta_{31}\beta_{24} + \beta_{23}\beta_{14} = 0.$$

We have no need for Plücker coordinates, because geometric algebra enables us to characterize the r-blade B_r directly by its algebraic properties without reference to coordinates.

To get a compact notation for a basis in $\mathscr{G}(A_n)$, we represent sets of indicies by a single capital letter. We write

$$J = (j_1, \ldots, j_n), \tag{3.18a}$$

where $j_k = k$ or 0. If we delete elements for which $j_k = 0$, any one of the base blades (3.9) can be written in the form

$$a_J \equiv a_{j_1} \wedge a_{j_2} \wedge \ldots \wedge a_{j_n}. \tag{3.18b}$$

For the case when all the j_k vanish, we take $a_J = a_{(0, \ldots, 0)} = 1$. For the corresponding reciprocal blades we have the indices in opposite order;

$$a^J \equiv a^{j_n} \wedge \ldots \wedge a^{j_2} \wedge a^{j_1}. \tag{3.18c}$$

Evidently (3.14) can now be written in the form

$$a_K * a^J = \delta_K^J. \tag{3.19}$$

The projection (2.9) of any multivector B into $\mathcal{G}(A_n)$ can now be written

$$P(B) = \sum_J B * a^J a_J = \sum_J B * a_J a^J. \tag{3.20}$$

Use of the scalar product in (3.20) (instead of the inner product as in (3.16)) is essential to pick out $\langle B \rangle_r$ when a_J is an r-vector.

Perhaps the most basic problem in linear algebra is to solve a vector equation of the form

$$a = \beta^1 b_1 + \beta^2 b_2 + \cdots + \beta^n b_n \tag{3.21}$$

for the scalar coefficients β^k. A unique solution exists if $B_n = b_1 \wedge \ldots \wedge b_n \neq 0$, which, we repeat, is the most felicitous way of expressing that the b_k are linearly independent. The solution is easily effected by outer-multiplying equation (3.21) on the left by $b_1 \wedge \ldots \wedge b_{k-1}$ and on the right by $b_{k+1} \wedge \ldots \wedge b_n$ and noting that, by (1.26b), all terms on the right but one vanish leaving

$$b_1 \wedge \ldots \wedge b_{k-1} \wedge a \wedge b_{k+1} \wedge \ldots \wedge b_n = \beta^k b_1 \wedge \ldots \wedge b_n.$$

Dividing by $b_1 \wedge \ldots \wedge b_n \neq 0$, we get β^k as the ratio of proportional or, better, *codirectional* n-blades:

$$\beta^k = \frac{b_1 \wedge \ldots \wedge b_{k-1} \wedge a \wedge b_{k+1} \wedge \ldots \wedge b_n}{b_1 \wedge \ldots \wedge b_n}. \tag{3.22}$$

Division is well-defined in (3.22) since numerator and denominator commute. The appropriate way to get a definite real number from the right side of (3.22) is determined by the form in which our information is given. We consider one common way below.

Instead of the vector Eqn. (3.21), we might have been given the system of n linear scalar equations in n unknowns β^k:

$$\alpha^j = b^j_k \beta^k, \quad j, k = 1, \ldots, n. \tag{3.23}$$

In (3.23) and often hereafter we use the convention that summation over all values of repeated pairs of upper and lower indices is understood.

We can reformulate (3.23) as a vector equation of the form (3.21) by introducing any frame of vectors $a_1, \ldots, a_n$ and writing

$$a = \alpha^j a_j, \qquad b_k = b^j_k a_j. \tag{3.24}$$

The solution of (3.23) is then given by (3.22), but we would like the answer in terms of the scalars α^j and b^j_k given in (3.23). To accomplish this, we note that, because of (3.5), multiplication of (3.24) gives us back the scalars; thus,

$$\alpha^j = a^j \cdot a, \tag{3.25a}$$

$$b^j_k = a^j \cdot b_k. \tag{3.25b}$$

Now, in the same way that we rationalize the denominator of a scalar fraction, we can 'scalarize' the denominator of (3.22) by multiplying numerator and denominator by the n-vector $a^n \wedge \ldots \wedge a^1$. Noting that the product of any pair of the n-vectors is a scalar, we get from (3.22),

$$\beta^k = \frac{(b_1 \wedge \ldots \wedge b_{k-1} \wedge a \wedge b_{k+1} \wedge \ldots \wedge b_n) \cdot (a^n \wedge \ldots \wedge a^1)}{(b_1 \wedge \ldots \wedge b_n) \cdot (a^n \wedge \ldots \wedge a^1)}. \tag{3.26}$$

This result is the well-known *Cramer's Rule* for the solution of a set of simultaneous linear equations. Equation (3.26) expresses β^k as a ratio of determinants, and we shall see in the next section how, using (3.25), these determinants can be evaluated in terms of the original scalars.

An $n \times n$ matrix b^j_k together with a frame of vectors $\{a_k\}$ with pseudoscalar $A_n = a_1 \wedge \ldots \wedge a_n$ determines a linear transformation f of $\mathscr{A}_n = \mathscr{G}^1(A_n)$ into $\mathscr{A}_n$;

$$f: a_k \to f(a_k) = b_k = b^j_k a_j. \tag{3.27a}$$

Conversely, the linear transformation f of a frame $\{a_k\}$ with dual $\{a^k\}$ determines the matrix

$$b^j_k = a^j \cdot b_k = a^j \cdot f(a_k). \tag{3.27b}$$

If m is the largest grade of nonvanishing multivectors formed by outer multiplication of the b_k, then f (or b^j_k) is said to have *rank m*. If the b_k are in $\mathscr{G}^1(A_n)$, then b^j_k is said to be a *matrix representation* of f. It follows that f is a linear transformation of $\mathscr{A}_n$ to an m-dimensional subspace $\mathscr{B}_m = \mathscr{G}^1(B_m)$. A frame $\{b^{i_1}, b^{i_2}, \ldots, b^{i_m}\}$ dual to $\{b_{i_1}, \ldots, b_{i_m}\}$ can be defined as before.

We say that f (or b^j_k) is *nonsingular* if and only if $B_n \equiv b_1 \wedge \ldots \wedge b_n = f(a_1) \wedge \ldots \wedge f(a_n) \neq 0$. If f is nonsingular, it has an inverse

$$f^{-1}: b_i \to f^{-1}(b_i) = a_i = a^j_i b_j \tag{3.28a}$$

with associated matrix

$$a_j^i = b^i \cdot a_j = b^i \cdot f^{-1}(b_j). \tag{3.28b}$$

The problem of finding f^{-1} from f is equivalent to the problem of finding the frame $\{b^i\}$ dual to given $\{b_k\}$. Equation (3.7) gives the solution immediately.

$$\begin{aligned} b^i &= (-1)^{i-1}(b_1 \wedge \ldots \check{b}_i \ldots \wedge b_n)A^n(A^n)^{-1}B_n^{-1} \\ &= (-1)^{i-1}\frac{(b_1 \wedge \ldots \check{b}_i \ldots \wedge b_n) \cdot A^n}{A^n \cdot B_n}, \end{aligned} \tag{3.29}$$

where $A^n = A_n^{-1}$. Substitution of (3.29) into (3.28b) gives

$$a_j^i = \frac{(b_1 \wedge \ldots \wedge b_{i-1} \wedge a_j \wedge b_{i+1} \wedge \ldots \wedge b_n) \cdot A^n}{A^n \cdot B_n}. \tag{3.30}$$

From this one can identify the 'classical adjugate matrix'

$$ad(b_j^i) = (b_1 \wedge \ldots \wedge b_{i-1} \wedge a_j \wedge b_{i+1} \wedge \ldots \wedge b_n) \cdot A^n. \tag{3.31}$$

The quantity $A^n B_n$ is the *determinant* of f (or b_j^i); it will be discussed in the next section. Of course (3.30) is just a form of Cramer's rule (3.26).

1-4. Alternating Forms and Determinants

Alternating forms play a prominent role in many modern mathematical works. But when Geometric Algebra is used, alternating forms are much less important, because all their properties are merely simple special cases of the results in Section 1-1. Thus, many computations with forms are more easily carried out by using Geometric Algebra directly. In this section we briefly discuss a typical version of the 'algebra of forms' primarily to show how it can be translated into Geometric Algebra.

A determinant can be regarded as a special kind of alternating form. Indeed, a determinant is nothing more nor less than the scalar product of two blades. So we get even the more complicated expansion theorems for determinants immediately from our previous work.

A scalar-valued function of r vector variables is said to be an *alternating r-form* (or just an 'r-form', for short) if it is skew-symmetric and linear in each argument. Section 2-3 will provide a simple proof that any r-form $\alpha_r = \alpha_r(a_1, a_2, \ldots, a_r)$ can be written in the form

$$\alpha_r = A_r^\dagger \cdot (a_1 \wedge a_2 \wedge \ldots \wedge a_r) \tag{4.1}$$

where $A_r^\dagger$ is a unique r-vector.

From (4.1) all the properties of r-forms are obvious consequences of the properties of inner and outer products already established. So it should be quite sufficient to list some of them:

$$\alpha_r \text{ is a linear function of each vector argument } a_k. \tag{4.2a}$$

$$\alpha_r \text{ vanishes if two arguments are equal.} \tag{4.2b}$$

$$\alpha_r \text{ changes sign if any two arguments are interchanged.} \tag{4.2c}$$

$$\alpha_r \text{ vanishes if any argument is zero.} \tag{4.2d}$$

$$\alpha_r \text{ vanishes if the arguments are linearly dependent.} \tag{4.2e}$$

We now explain how fundamentals of the 'algebra of forms' which has recently become popular with mathematicians can easily be formulated in terms of geometric algebra.* By virtue of (4.1), the sum of r-forms α_r and β_r is an r-form

$$\begin{aligned} \alpha_r + \beta_r &= A_r^\dagger \cdot (a_1 \wedge \ldots \wedge a_r) + B_r^\dagger \cdot (a_1 \wedge \ldots \wedge a_r) \\ &= (A_r^\dagger + B_r^\dagger) \cdot (a_1 \wedge \ldots \wedge a_r). \end{aligned} \tag{4.3}$$

The 'exterior product' of α_r with an s-form $\beta_s = B_s^\dagger \cdot (b_1 \wedge \ldots \wedge b_s)$ is defined to be the $(r + s)$-form

$$\alpha_r \wedge \beta_s \equiv (A_r \wedge B_s)^\dagger \cdot (a_1 \wedge \ldots \wedge a_r \wedge b_1 \wedge \ldots \wedge b_s). \tag{4.4}$$

It follows from our results in Section 1 that the exterior product of forms is associative, distributive and satisfies

$$\alpha_r \wedge \beta_s = (-1)^{rs} \beta_s \wedge \alpha_r. \tag{4.5}$$

An $(r-1)$-form α_{r-1}, called the 'contraction' of α_r by the vector b, is obtained by fixing the first argument of α_r at the value b; thus

$$\begin{aligned} \alpha_{r-1} = \alpha_{r-1}(a_2, \ldots, a_r) &\equiv A_r^\dagger \cdot (b \wedge a_2 \wedge \ldots \wedge a_r) \\ &= (A_r^\dagger \cdot b) \cdot (a_2 \wedge \ldots \wedge a_r) = (b \cdot A_r)^\dagger \cdot (a_2 \wedge \ldots \wedge a_r). \end{aligned} \tag{4.6}$$

This shows that α_{r-1} is uniquely determined by the $(r-1)$-vector $b \cdot A_r$. Alternatively, α_{r-1} can be regarded as the 'interior product' of α_r with the 1-form $\beta_1 = b \cdot a_1$. The 'interior product' of an r-form with an s-form can be defined (for $r > s$) to be the $(r - s)$-form

$$\beta_s \cdot \alpha_r \equiv (B_s \cdot A_r)^\dagger \cdot (a_{s+1} \wedge \ldots \wedge a_r). \tag{4.7}$$

* In Section 6-4 we discuss differential forms.

This much should suffice to show how to make translations from the 'language of forms' to Geometric Algebra. It should also be clear that within the language of geometric algebra the notions of interior and exterior products of forms are quite superfluous; they are supplanted by the more versatile inner and outer products of multivectors.

An r-form $\alpha_r = A_r^\dagger \cdot (b_1 \wedge \ldots \wedge b_r)$ is said to be simple if A_r is simple. A *determinant* of rank r is a simple r-form. Since A_r is simple, it can be factored into a product of vectors, $A_r = a_1 \wedge a_2 \wedge \ldots \wedge a_r$, and the determinant α_r takes the form

$$\alpha_r = (a_r \wedge \ldots \wedge a_1) \cdot (b_1 \wedge \ldots \wedge b_r). \tag{4.8}$$

The a_j are called 'row vectors' and the b_k 'column vectors' of the determinant α_r.

Of course a determinant has all the properties (4.2a) to (4.2e) enumerated above for a general r-form, but it has additional properties because it is simple. Thus, by (1.45b) and (1.19),

$$(a_r \wedge \ldots \wedge a_1) \cdot (b_1 \wedge \ldots \wedge b_r) = (b_r \wedge \ldots \wedge b_1) \cdot (a_1 \wedge \ldots \wedge a_r), \tag{4.9}$$

which shows that the value of a determinant is unaffected by an interchange of rows and columns. Moreover, using (1.23b), (1.25b) and (1.38), we obtain

$$\begin{aligned} \alpha_r &= (-1)^{j+1} (a_r \wedge \ldots \check{a}_j \ldots \wedge a_1) \cdot [a_j \cdot (b_1 \wedge \ldots \wedge b_r)] \\ &= \sum_{k=1}^{r} (-1)^{j+k} a_j \cdot b_k (a_r \wedge \ldots \check{a}_j \ldots \wedge a_1) \cdot (b_1 \wedge \ldots \check{b}_k \ldots \wedge b_r). \end{aligned} \tag{4.10}$$

This is the familiar expansion of a determinant by the jth row. To get a more general formula, use (1.23b) repeatedly to write A_r in the form

$$A_r = (-1)^{k_1 - 1 + k_2 - 2 + \cdots + k_s - s} A_s \wedge A_{r-s}, \tag{4.11a}$$

where

$$A_s = a_{k_1} \wedge \ldots \wedge a_{k_s}, \qquad A_{r-s} = a_{k_{s+1}} \wedge \ldots \wedge a_{k_r}, \tag{4.11b}$$

and all the k_i are distinct positive integers not exceeding n such that $k_1 < \ldots < k_s$, $k_{s+1} < \ldots < k_r$. Using (1.25b) and then (1.40) and (3.17), we obtain

$$\alpha_r = \sum_{j_1 < \ldots < j_s} (-1)^{\Sigma_{i=1}^{s} (k_i + j_i)} A_s^\dagger \cdot (b_{j_1} \wedge \ldots \wedge b_{j_s}) A_{r-s}^\dagger \cdot (b_{j_{s+1}} \wedge \ldots \wedge b_{j_r}). \tag{4.11c}$$

This is the so-called 'Laplace expansion' of a determinant by the rows k_1, k_2, $\ldots$, k_s.

Successive application of the expansions (4.10) or (4.11) shows that $\alpha_r = (a_r \wedge \ldots \wedge a_1) \cdot (b_1 \wedge \ldots \wedge b_r)$ is completely determined by the matrix of inner products $a_i \cdot b_j$. Accordingly, α_r is called the *determinant of the matrix* $a_i \cdot b_j$.

The *determinant* of the linear transformation f defined by (3.27) is defined to be the same as the *determinant* of its associated matrix $b^j_k = a^j \cdot b_k$; thus,

$$\begin{aligned} \det f = \det b^j_k &= (a^n \wedge \ldots \wedge a^1) \cdot (b_1 \wedge \ldots \wedge b_n) \\ &= A^n \cdot B_n = A^n B_n = \frac{B_n}{A_n}. \end{aligned} \tag{4.12}$$

The important thing to note here is that $A_n = a_1 \wedge \ldots \wedge a_n$, $A^n = A_n^{-1}$, and B_n are all pseudoscalars of the same vector space $\mathscr{A}_n$: hence they are scalar multiples of one another, so it is permissible to remove the dot denoting inner product in (4.12); also, A_n commutes with B_n, so division is unambiguously expressed by the usual notation used in scalar algebra. The value of $\det b^j_k$ is completely independent of the choice of frame $\{a^j\}$ in $\mathscr{A}_n$. We prove this implicitly in Chapter 3 by giving a manifestly frame-free definition of $\det f$ and showing that it is equivalent to (4.12).

Multiplying (4.12) by A_n, we get

$$B_n = (\det f) A_n. \tag{4.13}$$

If f is followed by a linear transformation g, the pseudoscalar B_n is mapped into

$$C_n = (\det g) B_n. \tag{4.14}$$

Substituting (4.13) into (4.14), we obtain, for the composite transformation gf the well-known result

$$\det gf = (\det g)(\det f). \tag{4.15}$$

The derivation of (4.15) from (4.12) depends crucially on the relation $A^n \cdot B_n = A^n B_n$. This relation holds only because A^n and B_n are pseudoscalars of the same space, and so does not apply to all determinants. However, we can derive a generalization of (4.12) which has universal validity. To any pair of r-vectors $A_r = a_1 \wedge \ldots \wedge a_r$, $B_r = b_1 \wedge \ldots \wedge b_r$, we can associate a nonvanishing n-vector $E_n = e_1 \wedge \ldots \wedge e_n$ with the properties

$$A_r E_n = A_r \cdot E_n, \qquad B_r E_n = B_r \cdot E_n. \tag{4.16}$$

This is equivalent to assuming that the a_j and b_k are elements of the vector space $\mathscr{E}_n = \mathscr{G}^1(E_n)$ spanned by the frame $\{e_k\}$ and its dual $\{e^k\}$. It follows immediately from (2.12) and (3.16) that

$$\begin{aligned} A_r^\dagger \cdot B_r &= (a_r \wedge \ldots \wedge a_1) \cdot (b_1 \wedge \ldots \wedge b_r) \\ &= \sum_{k_1 < \ldots < k_r} A_r^\dagger \cdot (e^{k_r} \wedge \ldots \wedge e^{k_1})(e_{k_1} \wedge \ldots \wedge e_{k_r}) \cdot B_r. \end{aligned} \tag{4.17}$$

This expansion of a determinant is sometimes called *Lagrange's Identity*. It is, of course, just the familiar expansion of a scalar product in terms of coordinates.

The conventional notion of a determinant refers implicitly to a unit n-vector I. In this case, a convenient notation for the determinant formed by vectors $a_1, \ldots, a_n$ is

$$[a_1 \ldots a_n] \equiv a_1 \wedge a_2 \wedge \ldots \wedge a_n I^\dagger = (a_1 \wedge \ldots \wedge a_n) \cdot I^\dagger, \tag{4.18}$$

where it is understood that all vectors are elements of the vector space $\mathcal{G}^1(I)$. A *unimodular basis* for $\mathcal{G}^1(I)$ is defined by the property $[e_1 \ldots e_n] = 1$, or equivalently.

$$e_1 \wedge \ldots \wedge e_n = I. \tag{4.19}$$

Vectors $\bar{a}_k$ *adjugate* to the vectors a_k are defined by

$$\bar{a}_k = (-1)^{k+1} a_1 \wedge \ldots \check{a}_k \ldots \wedge a_n I^\dagger, \tag{4.20}$$

so

$$a_j \cdot \bar{a}_k = [a_1 \ldots a_n]\, \delta_{jk}. \tag{4.21}$$

For a unimodular basis, then, the adjugate vectors make up the dual basis as it was defined earlier, and

$$\bar{e}_1 \wedge \ldots \wedge \bar{e}_n = I. \tag{4.22}$$

The determinant of a set of vectors is equivalent to the determinant of a matrix, that is,

$$[a_1 \ldots a_n] = (a_1 \wedge \ldots \wedge a_n) \cdot (\bar{e}_n \wedge \ldots \wedge \bar{e}_1) = \det (a_j \cdot \bar{e}_k). \tag{4.23}$$

The determinant

$$[\bar{a}_1 \ldots \bar{a}_n] = \det (\bar{a}_j \cdot e_k) \tag{4.24}$$

is called the *adjugate* of $[a_1 \ldots a_n]$; it is the determinant of the $(n-1) \times (n-1)$ minors of the matrix $a_j \cdot \bar{e}_k$, as shown by

$$\begin{aligned} \bar{a}_j \cdot e_k = e_k \cdot \bar{a}_j &= [a_1 \ldots a_{j-1} e_k a_{j+1} \ldots a_n] \\ &= (-1)^{j+1} (a_1 \wedge \ldots \check{a}_j \ldots \wedge a_n) \cdot (I^\dagger \cdot e_k) \\ &= (-1)^{j+k} (a_1 \wedge \ldots \check{a}_j \ldots \wedge a_n) \cdot (\bar{e}_n \wedge \ldots \check{\bar{e}}_k \ldots \wedge \bar{e}_1). \end{aligned} \tag{4.25}$$

With this preparation, we can follow Rota and Stein [Ro] to derive with great ease the classical identities for compound determinants collected by Turnbull [Tu].

A *compound determinant* is a determinant whose elements are minors of another determinant. The simplest compound determinant is the adjugate, whose value can be related to the original determinant in the following way:

$$
\begin{aligned}
[a_1 \ldots a_n]\ [\bar{a}_1 \ldots \bar{a}_n] &= a_1 \wedge \ldots \wedge a_n I^\dagger I \bar{a}_n \wedge \ldots \wedge \bar{a}_1 \\
&= (a_1 \wedge \ldots \wedge a_{n-1}) \cdot (a_n \cdot (\bar{a}_n \wedge \ldots \wedge \bar{a}_1)) \\
&= [a_1 \ldots a_n]\ (a_1 \wedge \ldots \wedge a_{n-1}) \cdot (\bar{a}_{n-1} \wedge \ldots \wedge \bar{a}_1) \\
&= [a_1 \ldots a_n]^n . \qquad (4.26)
\end{aligned}
$$

Hence *Cauchy's identity*

$$[\bar{a}_1 \ldots \bar{a}_n] = [a_1 \ldots a_n]^{n-1}. \qquad (4.27)$$

For further calculations, it is more useful in the form

$$\bar{a}_1 \wedge \ldots \wedge \bar{a}_n = [a_1 \ldots a_n]^{n-1} I. \qquad (4.28)$$

Then

$$
\begin{aligned}
(\bar{a}_1 \wedge \ldots \wedge \bar{a}_n) a_n \wedge \ldots \wedge a_{s+1} &= \bar{a}_1 \wedge \ldots \wedge \bar{a}_s [a_1 \ldots a_n]^{n-s} \\
&= [a_1 \ldots a_n]^{n-1} I a_n \wedge \ldots \wedge a_{s+1},
\end{aligned}
$$

so

$$\bar{a}_1 \wedge \ldots \wedge \bar{a}_s = [a_1 \ldots a_n]^{s-1} I a_n \wedge \ldots \wedge a_{s+1}. \qquad (4.29)$$

Dotting this with an arbitrary s-blade $b_1 \wedge \ldots \wedge b_s$, we obtain *Bazin's identity*

$$(b_1 \wedge \ldots \wedge b_s) \cdot (\bar{a}_s \wedge \ldots \wedge \bar{a}_1) = [a_1 \ldots a_n]^{s-1} [b_1 \ldots b_s a_{s+1} \ldots a_n]. \qquad (4.30)$$

To describe $r \times r$ minors, we introduce r-blades A_J defined by

$$A_J \equiv a_{j_1} \wedge a_{j_2} \wedge \ldots \wedge a_{j_r}, \qquad (4.31)$$

where $j_1 < j_2 < \ldots < j_r$. The set of indices can be put in *lexical order* so J has the integer values $J = 1, 2, \ldots, N = \binom{n}{r}$. A set of adjugate r-blades is defined by

$$\bar{A}_J = (-1)^{j_1 + \cdots + j_r + r} I a_n \wedge \ldots \check{a}_{j_r} \ldots \check{a}_{j_1} \ldots \wedge a_1, \qquad (4.32)$$

so that

$$A_J \cdot \bar{A}_K = [a_1 \ldots a_n]\ \delta_{JK}. \qquad (4.33)$$

From (4.29) applied to (4.32) we obtain

$$\bar{a}_{j_r} \wedge \ldots \wedge \bar{a}_{j_1} = [a_1 \ldots a_n]^{r-1} \bar{A}_J. \tag{4.34}$$

The matrix of $r \times r$ determinants of the matrix $a_j \cdot \bar{e}_k$ is, by (4.32) and (4.34),

$$\begin{aligned} A_J \cdot \bar{E}_K &= (a_{j_1} \wedge \ldots \wedge a_{j_r}) \cdot (\bar{e}_{k_r} \wedge \ldots \wedge \bar{e}_{k_1}) \\ &= [e_1 \ldots e_{k_1-1} a_{j_1} e_{k_1+1} \ldots e_{k_r-1} a_{j_r} e_{k_r+1} \ldots e_n]. \end{aligned} \tag{4.35}$$

Again using (4.32) and (4.34), we obtain *Jacobi's identity*

$$\begin{aligned} (e_{k_1} \wedge \ldots \wedge e_{k_r}) \cdot (\bar{a}_{j_r} \wedge \ldots \wedge \bar{a}_{j_1}) &= [a_1 \ldots a_n]^{r-1} \bar{A}_J \cdot E_K \\ &= [a_1 \ldots a_n]^{r-1} [a_1 \ldots a_{j_1-1} e_{k_1} a_{j_1+1} \ldots a_{j_r-1} e_{k_r} a_{j_r+1} \ldots a_n] \\ &= [a_1 \ldots a_n]^{r-1} (\bar{e}_1 \wedge \ldots \check{\bar{e}}_{k_1} \ldots \check{\bar{e}}_{k_r} \ldots \wedge \bar{e}_n) \cdot \\ &\quad \cdot (a_n \wedge \ldots \check{a}_{j_r} \ldots \check{a}_{j_1} \ldots \wedge a_1)(-1)^{\Sigma_i (j_i + k_i)}. \end{aligned} \tag{4.36}$$

This relates minors of a determinant to *complementary minors* of its adjugate.

The generalization of Cauchy's identity to $r \times r$ minors is most easily accomplished by regarding the A_J as vectors in a new vector space of dimension $N = \binom{n}{r}$ with determinant $[A_1 \ldots A_N]$. Then, with this ambiguity in notation, from (4.33) we derive

$$[A_1 \ldots A_N]\, [\bar{A}_1 \ldots \bar{A}_N] = (A_1 \ldots A_N) \cdot (\bar{A}_N \wedge \ldots \wedge \bar{A}_1) = [a_1 \ldots a_n]^N \tag{4.37}$$

in the same way that we derived (4.26). Now, from (4.31) and (4.32) we see that there are $n - r$ of the vectors a_k in $\bar{A}_J$ for r of them in A_J, hence

$$[\bar{A}_1 \ldots \bar{A}_N]^r = [A_1 \ldots A_N]^{n-r}. \tag{4.38}$$

Combining this with (4.37), we obtain *Sylvester's identity*

$$[A_1 \ldots A_N] = [a_1 \ldots a_n]^{\binom{n-1}{r-1}}, \tag{4.39}$$

and

$$[\bar{A}_1 \ldots \bar{A}_N] = [a_1 \ldots a_n]^{\binom{n-1}{r}}. \tag{4.40}$$

For $r = n - 1$, Sylvester's identity reduces to Cauchy's identity.

Now using Lagrange's identity we obtain

$$\begin{aligned} [a_1 \ldots a_n]^s &= (A_1 \wedge \ldots \wedge A_s) \cdot (\bar{A}_s \wedge \ldots \wedge \bar{A}_1) \\ &= \sum_{i_1 < i_2 < \ldots < i_s} (A_1 \wedge \ldots \wedge A_s) \cdot (\bar{E}_{i_s} \wedge \ldots \wedge \bar{E}_{i_1}) \cdot \\ &\quad \cdot (\bar{E}_{i_1} \wedge \ldots \wedge \bar{E}_{i_s}) \cdot (\bar{A}_s \wedge \ldots \wedge \bar{A}_1). \end{aligned} \tag{4.41}$$

On the other hand, using Sylvesters identity and the Laplace expansion, we obtain

$$[a_1 \dots a_n]^{\binom{n-1}{r-1}} = (A_1 \wedge \dots \wedge A_N) \cdot (\bar{E}_N \wedge \dots \wedge \bar{E}_1)$$
$$= \sum (A_1 \wedge \dots \wedge A_s) \cdot (E_{i_s} \wedge \dots \wedge E_{i_1}) \cdot$$
$$(E_{i_{s+1}} \wedge \dots \wedge E_{i_N}) \cdot (A_N \wedge \dots \wedge A_{s+1}).$$

Comparing this with (4.41) and equating corresponding coefficients of $(A_1 \wedge \dots \wedge A_s) \cdot (\bar{E}_{i_s} \wedge \dots \wedge \bar{E}_{i_1})$, we obtain *Franke's identity*

$$[a_1 \dots a_n]^{\binom{n-1}{r-1}-s} (\bar{A}_1 \wedge \dots \wedge \bar{A}_s) \cdot (E_{i_s} \wedge \dots \wedge E_{i_1})$$
$$= (A_{s+1} \wedge \dots \wedge A_N) \cdot (\bar{E}_{i_N} \wedge \dots \wedge \bar{E}_{i_{s+1}}) \tag{4.42}$$

relating complementary minors of the compound determinants.

If we repeat the steps in the derivation of Franke's identity with the E_J and $\bar{E}_J$ replaced by r-blades B_J and $\bar{B}_J$ formed in the same way from a set of vectors $b_1, \dots, b_n$, we obtain *Picquet's identity*

$$[a_1 \dots a_n]^{\binom{n-1}{r-1}-s} [b_1 \dots b_n]^{\binom{n-1}{r}-s} (\bar{A}_1 \wedge \dots \wedge \bar{A}_s) \cdot (B_{i_s} \wedge \dots \wedge B_{i_1})$$
$$= (A_{s+1} \wedge \dots \wedge A_N) \cdot (\bar{B}_{i_N} \wedge \dots \wedge \bar{B}_{i_{s+1}}). \tag{4.43}$$

Finally, from (4.39) and (4.40), we obtain *Reiss's identity*

$$[a_1 \dots a_n]^{\binom{n-1}{r-1}} [b_1 \dots b_n]^{\binom{n-1}{r}} = (A_1 \wedge \dots \wedge A_N) \cdot (\bar{B}_N \wedge \dots \wedge \bar{B}_1). \tag{4.44}$$

The theory of permanents can be incorporated into the present formalism in a simple and elegant way. Suppose that I is a blade of grade $2n$, and let $\{B_1, \dots, B_n\}$ be a set of n bivectors in $\mathcal{G}^2(I)$; we define the *permanent* of this set by

$$[B_1 \dots B_n] \equiv B_1 \wedge \dots \wedge B_n I^\dagger. \tag{4.45}$$

According to this definition, a permanent is just a determinant built out of bivectors instead of vectors. To relate it to the conventional definition, we choose a set of n orthogonal 2-blades E_k so that

$$E_1 \wedge E_2 \wedge \dots \wedge E_n = E_1 E_2 \dots E_n = I. \tag{4.46}$$

Furthermore, we assume that each of the B_k lies in the n-dimensional subspace of $\mathcal{G}^2(I)$ spanned by the E_k. Then

$$[B_1 \dots B_n] = (E_n^\dagger \wedge \dots \wedge E_1^\dagger) \cdot (B_1 \wedge \dots \wedge B_n) = \text{perm}\,(E_j^\dagger \cdot B_k). \tag{4.47}$$

Thus $[B_1 \ldots B_n]$ is the permanent of the $n \times n$ matrix of scalars $E_j^\dagger \cdot B_k$. The properties of permanents now follow from the properties of inner and outer products as did the properties of determinants. Note, for example, from (4.47) that the permanent is symmetric under interchange of rows and columns, because the outer product of bivectors is symmetric. For this reason, the terms in the 'Laplace expansion'

$$[B_1 \ldots B_n] = ((E_n^\dagger \wedge \ldots \wedge E_1^\dagger) \cdot B_1) \cdot (B_2 \wedge \ldots \wedge B_n)$$

$$= \sum_{k=1}^{n} E_k^\dagger \cdot B_1 (E_n^\dagger \wedge \ldots E_k \ldots \wedge E_1^\dagger) \cdot (B_2 \wedge \ldots \wedge B_n) \quad (4.48)$$

do not alternate in sign as in the corresponding expansion with vectors.

1-5. Geometric Algebras of Pseudo-Euclidean Spaces

Recall axiom (1.13) which requires that the square of any nonzero vector be a positive scalar. Let us refer to it as the 'Euclidean axiom'. The Euclidean axiom is independent of other axioms of Geometric Algebra, and for some purposes, especially for applications to space-time physics, it is more convenient to adopt the axiom (let us call it the 'pseudo-Euclidean axiom') which specifies merely that the square of a vector is a scalar.

The pseudo-Euclidean axiom allows vectors to have positive, negative or zero square, so at first sight it appears to admit a more general mathematical system than the Euclidean axiom. But actually it is merely different. Both axioms require that the square of a vector be a definite scalar. In Section 3-7 we show that any metrical properties represented by using the pseudo-Euclidean axiom can be represented with the Euclidean axiom as well. So the difference is primarily a matter of convenience and this depends on which applications are of interest.

Our above formulation of the pseudo-Euclidean axiom is incomplete. It must be supplemented by axioms asserting the existence of vectors with positive and negative square and specifying their relation to vectors with zero square. We do this below for finite dimensional spaces.

This section discusses certain consequences of the pseudo-Euclidean axiom necessary for applications of geometric algebra to spacetime physics. Except where otherwise indicated, we assume the Euclidean axiom in the rest of this book for the sake of simplicity. But those few results which depend on this assumption can easily be recognized and, in the light of this section, altered to fit the pseudo-Euclidean case.

Many of the important formulas for inner and outer products established in Section 1 depend on the fact that the square of a vector is a scalar, but none require that the scalar be positive, so they remain valid in the pseudo-Euclidean case. Only our proof of (1.27b) would need to be revised if the pseudo-Euclidean axiom were adopted at the beginning.

The only results in Section 1-1 which are peculiar to the Euclidean axiom are consequences of the theorem that $\langle A^\dagger A\rangle \equiv A^\dagger * A$ is a positive scalar if $A \neq 0$. In the pseudo-Euclidean case we define the magnitude of A to be the positive scalar

$$|A| = |A^\dagger * A|^{1/2}, \tag{5.1}$$

where $|A^\dagger * A|$ is the absolute value of the scalar $A^\dagger * A$. We say that A is *singular* or *null* if $|A| = 0$ when $A \neq 0$.

An n-blade $A = \langle A\rangle_n$ can be factored into a product of vectors, $A = a_1 a_2 \ldots a_n$, so that

$$A^\dagger * A = A^\dagger A = a_1^2 a_2^2 \ldots a_n^2. \tag{5.2}$$

Therefore, a blade is null if and only if it has a null vector for a factor. If p of the vectors in (5.2) have positive square and $q = n - p$ of them have negative square, then A is said to be a blade with *signature* (p, q). In this case, A obviously has an inverse

$$A^{-1} = (-1)^q \frac{A^\dagger}{|A|^2} = \frac{A}{A^2}. \tag{5.3}$$

Of course a null blade has no inverse.

The Geometric Algebra of a nonsingular blade can be defined and analyzed by the method of Section 1-2, so we can run through the highlights quickly here, to bring out special features of the pseudo-Euclidean case.

If A is a blade with signature (p, q), then the set of all vectors a satisfying the equation

$$a \wedge A = 0 \tag{5.4}$$

constitute a $(p + q)$-dimensional *vector space* $\mathscr{A}_{p,q}$ *with signature* (p, q) and *pseudoscalar* A. Any basis of $\mathscr{A}_{p,q}$ composed of factors of A will include p vectors with positive square and q vectors with negative square. If p and q are not zero, then $\mathscr{A}_{p,q}$ contains null vectors, but the null vectors cannot be factors of A.

The geometric algebra of A or of $\mathscr{A}_{p,q}$ is denoted by

$$\mathscr{G}(A) = \mathscr{G}(\mathscr{A}_{p,q}), \tag{5.5}$$

and called a *Clifford Algebra* or a *Geometric Algebra of signature* (p, q). The elements of $\mathscr{G}(\mathscr{A}_{p,q})$ are commonly called *Clifford numbers*, but we will stick with the geometrical term multivector. As in the Euclidean case, the projection of any multivector B into $\mathscr{G}(A)$ is given by

$$P_A(B) = (B \cdot A) \cdot A^{-1}. \tag{5.6}$$

The development of bases and reciprocal bases for $\mathscr{A}_{p,q} = \mathscr{G}^1(A)$ and $\mathscr{G}(A)$ and the computation of a matrix inverse carried out in Section 1-3 apply to the pseudo-Euclidean as well as the Euclidean cases, because they require only that the pseudoscalar be nonsingular.

Chapter 2

Differentiation

Geometric Calculus is primarily concerned with the theory and techniques for differentiating and integrating *geometric functions*, that is, functions whose domain and range are subsets of the Universal Geometric Algebra $\mathscr{G}$. This chapter deals with the differentiation of functions defined on linear subspaces of $\mathscr{G}$. Differentiation can be defined on more general subsets of $\mathscr{G}$ called manifolds, and this will be considered in Chapter 4. But in the interest of simplicity, it is best to study calculus on linear spaces first. Calculus on more general manifolds involves differential geometry.

This chapter develops the general properties of differentiation by multivectors and enough special properties to make applications routine. The results of this chapter are used extensively throughout the rest of the book. Section 2-1 has particularly wide application to physics. As shown in [H1], it includes the results of standard vector analysis and generalizes them to apply to spacetime. The results of Sections 2-2 and 2-3 will be applied to the study of linear transformations in Chapter 3.

2-1. Differentiation by Vectors

We saw in Chapter 1 that Geometric Algebra makes it possible to divide by vectors. Besides its obvious algebraic convenience, this feature provides a boon to analysis – it permits a unique theory of differentiation by vectors. Concepts of gradient, divergence and curl, which are developed separately in vector analysis, are reduced to a single concept of vector derivative in Geometric Calculus. As we shall see, the vector derivative can be decomposed into divergence and curl, but each of these is, so to speak, only 'half' the derivative. The geometric product alone unites divergence and curl, indeed, reduces all differential operators to a single concept of differentiation.

This section defines the vector derivative and develops its basic properties. Chapter 4 extends the theory to vector manifolds, and Chapter 7 determines the relation of differentiation to integration.

We begin by noting that standard definitions of continuity and scalar differentiation apply to geometric functions, because the scalar product defined in Section

1-1 determines a unique 'distance' $|A - B|$ between any two elements A and B in $\mathcal{G}$. The derivative of a (multivector valued) function $F = F(\tau)$ of a scalar parameter τ is expressed by

$$\partial_\tau F(\tau) = \frac{\partial F(\tau)}{\partial \tau} = \lim_{\Delta\tau \to 0} \frac{F(\tau + \Delta\tau) - F(\tau)}{\Delta\tau} \tag{1.1}$$

or, suppressing the argument, by $\partial_\tau F = \partial F/\partial\tau$. Leibnitz's notation $\mathrm{d}F/\mathrm{d}\tau$ or $\partial F/\partial\tau$ emphasizes the definition of derivative as the limit of a difference quotient. It will be seen that differentiation by a general multivector cannot be defined by a difference quotient, so Leibnitz's notation is appropriate only for scalar variables.

Now let $F = F(x)$ be a (multivector valued) function defined on an n-dimensional vector space $\mathcal{A}_n = \mathcal{G}^1(I)$ with unit pseudoscalar I. For a vector a in $\mathcal{A}_n$, the *derivative* of F in the *direction* a, or briefly, the *a-derivative* of F at x is given by

$$a \cdot \partial F(x) \equiv \left.\frac{\partial F(x + \tau a)}{\partial \tau}\right|_{\tau=0} = \lim_{\tau \to 0} \frac{F(x + \tau a) - F(x)}{\tau}, \tag{1.2}$$

provided this limit exists. The function F is said to be *continuously differentiable* at x if for each fixed a, $a \cdot \partial F(y)$ exists and is a continuous function of y for each y in a neighborhood of x. It can be proved that if F is defined and continuously differentiable at x, then, for fixed x, the function $a \cdot \partial F(x)$ is a linear function of a. In this case we write

$$\underline{F}(x, a) = F_a(x) \equiv a \cdot \partial F(x), \tag{1.3a}$$

or, suppressing the argument,

$$\underline{F} = \underline{F}(a) = F_a \equiv a \cdot \partial F. \tag{1.3b}$$

This function $\underline{F}$ of two variables x and a is called the *first differential* or just the *differential* of F.

Geometric Algebra provides the apparatus necessary to define the derivative with respect to a vector. The *derivative* (by x) of the function $F = F(x)$ will be denoted by $\partial_x F(x)$ or, more simply, by ∂F, where differentiation of F by its argument is understood. The differential operator ∂_x can be defined by assuming that it possesses the algebraic properties of a vector in $\mathcal{A}_n = \mathcal{G}^1(I)$ and that the inner product $a \cdot \partial_x$ of ∂_x with a vector a in $\mathcal{A}_n$ is equal to the a-derivative operator defined by (1.2).

Let us formulate the basic algebraic properties of ∂_x explicitly. Recall that the condition that a vector x be an element of $\mathcal{G}^1(I)$ can be expressed by the equations $I \wedge x = 0$ and $Ix = I \cdot x$. Similarly, irrespective of the functions on which it operates, the derivative by x has the properties

$$I \wedge \partial_x = 0\,, \tag{1.4a}$$

$$I \partial_x = I \cdot \partial_x\,. \tag{1.4b}$$

Equivalently, like any other vector in $\mathscr{G}^1(I)$, the derivative has the properties

$$\partial_x = P(\partial_x) = \sum_k a^k a_k \cdot \partial_x \tag{1.5}$$

where, in accordance with (1-3.16) and (1-3.20), $P = P_I$ is the projection operator for $\mathscr{G}(I)$, and the right side is an expansion of ∂_x in terms of a basis $\{a^k\}$ spanning $\mathscr{G}^1(I)$. The expansion (1.5) in a basis separates the algebraic properties of ∂_x from its differential properties; the algebraic properties reside in the vectors a^k, while the $a_k \cdot \partial_x$ are scalar differential operators. This separation has the defect of depending on the choice of basis $\{a^k\}$, whereas the combination of algebraic and differential properties possessed by ∂_x is independent of any such choice.

It should be noted that, though each term on the right of (1.5) can be defined as the limit of a difference quotient, the sum of terms composing ∂_x cannot be so expressed. A definition of ∂_x directly in terms of a limit which makes no reference to the directional derivative will be given in Chapter 7, but that definition employs an integral and so might be regarded as less elementary than the one presented here.

We can define the differential of $F = F(x)$ in terms of $\partial = \partial_x$ by using (1-1.32) to write

$$\underline{F}(a) \equiv a \cdot \partial F = \tfrac{1}{2}(a\,\partial F + \dot{\partial} a \dot{F}), \tag{1.6}$$

where the overdots are meant to indicate that only F is to be differentiated when it is desired to regard a also as a function of x. Equation (1.6) explains why we used the notation $a \cdot \partial$ in the definition (1.2) for the directional derivative, for once the derivative ∂_x by a vector has been defined, $a \cdot \partial_x$ can indeed be regarded as the inner product of the vector a with ∂_x. It is important to note that (1.6) defines the differential $\underline{F}(b)$ for all vectors b, and not just elements of $\mathscr{G}^1(I)$ as presumed in (1.2). Of course, from (1.5) it follows that

$$a \cdot \partial_x = P(a) \cdot \partial_x, \tag{1.7}$$

and $P(a)$ is indeed in $\mathscr{G}^1(I)$. Inserting this in (1.6), we have

$$\underline{F}(a) = \underline{F}(P(a)), \tag{1.8}$$

which shows that, though $\underline{F}(a)$ is defined for all vectors a, it vanishes if $P(a) = 0$.

It should be mentioned that the operator $a \cdot \partial$ preserves grade, that is

$$a \cdot \partial \langle F \rangle_r = \langle a \cdot \partial F \rangle_r. \tag{1.9}$$

This is a consequence of axioms (1-1.10) and (1-1.11) applied to the definition (1.2). Since (1.9) is similar to the grade-preserving property of scalar multiplication expressed by (1-1.11), we say that *$a \cdot \partial$ is a scalar differential operator*.

The differential has four other general properties, besides (1.8) and (1.9), which can be derived from its definition (1.2) in terms of a limit. First we have linearity:

$$\underline{F}(a + b) = \underline{F}(a) + \underline{F}(b), \tag{1.10a}$$

$$\underline{F}(\lambda a) = \lambda \underline{F}(a). \tag{1.10b}$$

Of course, the linearity property (1.10) is a trivial consequence of (1.6) and the linearity of the inner product. But (1.6) requires that ∂F be well defined, which is equivalent to the requirement that F be continuously differentiable. Later on it will be convenient to allow ∂F to be singular but well defined in the sense of distribution theory [GS]. But we do not wish to consider exceptions to the linearity of the differential at this time. Henceforth, unless otherwise stated, we tacitly assume that any required differentiations are well defined.

Two other general properties of the differential are the familiar rules for the differentials of the sum and product of functions $F = F(x)$ and $G = G(x)$; we have

$$a \cdot \partial(F + G) = a \cdot \partial F + a \cdot \partial G, \tag{1.11a}$$

$$a \cdot \partial(FG) = (a \cdot \partial F)G + F(a \cdot \partial G), \tag{1.12a}$$

or, using the underbar notation for differentials,

$$\underline{F + G} = \underline{F} + \underline{G}, \tag{1.11b}$$

$$\underline{FG} = \underline{F}G + F\underline{G}. \tag{1.12b}$$

The proofs of (1.11) and (1.12) are standard, so they need not be reproduced here. Of course, the noncommutativity of the geometric product plays no essential role in the proof since $a \cdot \partial$ is by (1.9) a scalar operator.

The remaining general property of differentials concerns composite functions. Let $x' = f(x)$ be a function defined on $\mathscr{A}_n = \mathscr{G}^1(I)$ with values in some vector space $\mathscr{A}'_n = \mathscr{G}^1(I')$. To determine the differential of the composite function

$$F(x) = G(f(x)), \tag{1.13}$$

we use (1.2) and the usual 'Taylor expansion'

$$f(x + \tau a) = f(x) + \tau a \cdot \partial f(x) + \frac{\tau^2}{2!}(a \cdot \partial)^2 f(x) + \cdots$$

Thus,

$$\begin{aligned} a \cdot \partial_x G(f(x)) &= \partial_\tau G(f(x + \tau a))\big|_{\tau=0} \\ &= \partial_\tau G(f(x) + \tau \underline{f}(a))\big|_{\tau=0} = \underline{f}(a) \cdot \partial_{x'} G(x')\big|_{x'=f(x)}. \end{aligned}$$

Since composite functions are evaluated at corresponding points, it is safe to suppress the point of evaluation. So, we write

$$a \cdot \partial F = a \cdot \partial_x G(f(x)) = \underline{f}(a) \cdot \partial G, \tag{1.14a}$$

or, equivalently,

$$\underline{F}(a) = \underline{G}(\underline{f}(a)), \tag{1.14b}$$

which, in completely unambiguous notation, is written

$$\underline{F}(x, a) = \underline{G}(f(x), \underline{f}(x, a)). \tag{1.14c}$$

The fundamental result (1.14) is called the *chain rule*. In words, it says that *the differential of a composite function is the composite of differentials*. The most significant feature of (1.14) is that, in contrast to other formulations of the chain rule, no coordinates or matrices were used in its derivation. Nor will they be needed for its application.

There is another version of the chain rule for scalar functions. If $\tau = \tau(x)$ is a scalar-valued function, then

$$a \cdot \partial_x F(\tau(x)) = (a \cdot \partial\tau)\, \partial_\tau F. \tag{1.15}$$

The proof of (1.15) differs from that of (1.14) only in the use of (1.1) instead of (1.2) in the first step of the argument. In the next section we shall see (1.14) and (1.15) as special cases of a more general chain rule.

In consonance with (1.3), the *second differential* of the function $F = F(x)$ is defined by

$$F_{ab}(x) \equiv b \cdot \dot{\partial} a \cdot \partial \dot{F}(x), \tag{1.16a}$$

or, suppressing the argument,

$$F_{ab} \equiv b \cdot \dot{\partial} a \cdot \partial \dot{F}. \tag{1.16b}$$

From the definition (1.2) of the differential as a limit, we can derive the so-called *integrability condition*:

$$F_{ab} = F_{ba}. \tag{1.17}$$

Other properties of the second differential obviously obtain from our discussion of the first differential. To sum up, the second differential is a symmetric bilinear function of its differential arguments.

As in the scalar differential calculus, it is necessary to appeal to the definition of the differential in terms of a limit to evaluate only a few elementary functions. Differentials of more complex functions can then be determined from the simpler ones by using the linearity of the differential, the product rule and the chain rule. The one elementary result we need is the *differential of the identity function* $F(x) = x$.

$$a \cdot \partial_x x = P(a) = \partial_x (x \cdot a). \tag{1.18}$$

The left equality of (1.18) follows immediately from (1.8) and (1.2). To get the right equality, note that for base vectors a_k, we have

$$a_k \cdot \partial x = P(a_k) = a_k.$$

So, using (1.5), we have

$$\partial_x (x \cdot a) = \sum_k a^k a_k \cdot \partial_x (x \cdot a) = \sum_k a^k a_k \cdot a = P(a),$$

as desired.

From (1.18) and (1.5), we get the operator identity

$$\partial_x = P(\partial_x) = \partial_a a \cdot \partial_x. \tag{1.19}$$

We can use this to obtain the derivative of a function immediately from its differential; thus,

$$\partial_x F(x) = \partial_a a \cdot \partial_x F(x) = \partial_a \underline{F}(x, a). \tag{1.20a}$$

Introducing the notation $\underline{\partial}$ for the derivative with respect to the differential argument of $\underline{F}$, we can write (1.20a) in the useful compact form

$$\partial F = \underline{\partial}\underline{F}. \tag{1.20b}$$

The general properties of the derivative can now be obtained easily from the properties of the differential, we have only to take due account of the vector character of the derivative.

The derivative does not have the grade preserving property (1.9) of the differential. But, according to (1-1.31), the vector property of $\partial = \partial_x$ allows us to write

$$\partial F = \partial \cdot F + \partial \wedge F. \tag{1.21a}$$

If $F = \phi(x)$ is scalar valued, then, according to (1-1.27a), $\partial \cdot \phi = 0$, and (1.21a) reduces to

$$\partial \phi = \partial \wedge \phi, \tag{1.21b}$$

which will be recognized as the vector valued function usually called the *gradient* of ϕ. We call $\partial \cdot F$ the *divergence* of F and $\partial \wedge F$ the *curl* of F. This terminology is fully in accord with standard nomenclature in vector and tensor analysis.

It is of the utmost significance that (1.21) displays divergence and curl as distinct parts of a single more general operator, which we have identified as THE VECTOR DERIVATIVE. The fundamental significance of $\partial = \partial_x$ is conclusively established in Chapter 7, where it is shown that, in contrast to other kinds of derivative, ∂_x has a definite inverse, and this leads to many results, of which the generalized Cauchy's integral formula is perhaps the most important.

Using (1.20b) and the distributive rule, we easily establish from (1.11b) that the derivative of a sum is a sum of derivatives:

$$\partial(F + G) = \partial F + \partial G. \tag{1.22}$$

From (1.12b) we get the product rule in the form

$$\partial(FG) = \underline{\partial}\underline{F}G + \underline{\partial}F\underline{G}. \tag{1.23}$$

The underbars on the right side of (1.23) can be regarded as instructions as to which function is to be differentiated, quite apart from their connexion with differentials. In the first term, the instructions can be given in the usual way by using parentheses, thus, $\underline{\partial}\underline{F}G = (\partial F)G$. But parentheses cannot be used in the second term by writing, for example, $F(\partial G)$, because the product of F with ∂ is noncommutative. The quantity $\underline{\partial F G}$ is just the derivative of the function defined by the product FG with F regarded as a constant; explicitly,

$$\underline{\partial} F \underline{G} = \partial_y (F(x)G(y)) \big|_{y=x}.$$

We sometimes use overdots instead of underbars to designate quantities to be differentiated, especially when we are using differentials for something else. Thus, instead of (1.23), we may write

$$\partial(FG) = \dot{\partial}\dot{F}G + \dot{\partial}F\dot{G}. \tag{1.24a}$$

As a rule, we use the usual convention that ∂ differentiates only quantities to its right; thus,

$$F\,\partial G = F\,\underline{\partial}\underline{G} = F\,\dot{\partial}\dot{G}.$$

When it is necessary to indicate that ∂ differentiates quantities to the left or both to the left and the right, we use overdots or underbars. For example,

$$\dot{F}\,\dot{\partial}\dot{G} = \dot{F}\,\dot{\partial}G + F\,\dot{\partial}\dot{G} \tag{1.24b}$$

is another form of the product rule.

To get the most perspicuous form of the chain rule we define the adjoint of a function. We have seen that the differential $\underline{f}(a)$ of a vector valued function $x' = f(x)$ is a vector valued linear function of its differential argument a. We define the *adjoint* of f by the equation

$$\overline{f}(a') \equiv \underline{\partial}\underline{f} \cdot a' = \partial f \cdot a'. \tag{1.25a}$$

More explicitly, this is written

$$\overline{f}(x, a') \equiv \partial_a \underline{f}(x, a) \cdot a' = \partial_x f(x) \cdot a'. \tag{1.25b}$$

To establish the equality on the right of (1.25), we applied (1.19) to the definition of the differential.

Like the differential, the adjoint $\overline{f}(a')$ is obviously a vector-valued linear function of a'. Thus, because of the linearity of the inner product, we have from the definition (1.25),

$$\overline{f}(a' + b') = \overline{f}(a') + \overline{f}(b'), \tag{1.26a}$$

$$\overline{f}(\alpha a') = \alpha\overline{f}(a'). \tag{1.26b}$$

Application of (1.19) to (1.25) yields

$$P(\overline{f}(a')) = \overline{f}(a'), \tag{1.27a}$$

so the range of $\overline{f}$ is $\mathscr{A}_n = \mathscr{G}^1(I)$. From (1.25) it is also clear that

$$\overline{f}(a') = \partial_a \underline{f}(a) \cdot a' = \overline{f}(P'(a')), \tag{1.27b}$$

where P' is the projection into the range of $\underline{f}$, namely, $\mathscr{A}'_n = \mathscr{G}^1(I')$. Combining (1.27a, b), we get

$$P\overline{f}P'(a') = \overline{f}(a'). \tag{1.27c}$$

for all a'. This should be compared with

$$\underline{P}'\underline{f}P(a) = \underline{f}(a), \tag{1.27d}$$

the corresponding relation satisfied by the differential $\underline{f}$.

Applying (1.19) to (1.14a) and using (1.25), we now get the chain rule for the derivative: If $F(x) = G(f(x))$, then

$$\partial_x F(x) = \bar{f}(\partial_{x'})G(x'). \tag{1.28a}$$

Thus, the 'change of variables' from x to $x' = f(x)$ induces the adjoint linear transformation of the derivative $\partial_{x'}$ to

$$\partial_x = \bar{f}(\partial_{x'}). \tag{1.28b}$$

If we differentiate the function $F = F(x)$ twice, we have, using (1.19) and the definition (1.16) of the second differential,

$$\partial_x^2 F(x) = \partial_b \partial_a F_{ab} = (\partial_b \cdot \partial_a + \partial_b \wedge \partial_a)F_{ab}.$$

Because of the integrability condition (1.17), we have

$$\partial_x \wedge \partial_x F(x) = \partial_b \wedge \partial_a F_{ab} = -\partial_b \wedge \partial_a F_{ab} = 0.$$

Hence, the operator identity

$$\partial_x \wedge \partial_x = 0 \tag{1.29}$$

expresses the *integrability condition for the vector derivative*. Conversely, from (1.29) we can derive the differential form of the integrability condition (1.17); using (1-4.10) and (1.16), we have

$$(a \wedge b) \cdot (\partial \wedge \partial)F = F_{ba} - F_{ab} = 0. \tag{1.30}$$

Because of integrability, the second derivative is a scalar differential operator,

$$\partial_x^2 = \partial_x \cdot \partial_x. \tag{1.31}$$

This operator is called the *Laplacian*.

In computations, of course, we will frequently need the derivatives of elementary functions. The most commonly used derivatives besides (1.18) are assembled in the following table, where $\partial = \partial_x$ and obvious singularities at $x = 0$ are understood to be excluded.

$$\partial |x|^2 = \partial x^2 = 2P(x) = 2x, \tag{1.32}$$

$$\partial \wedge x = 0, \tag{1.33}$$

$$\partial x = \partial \cdot x = n, \tag{1.34}$$

$$\partial |x|^k = k|x|^{k-2}x, \tag{1.35}$$

$$\partial\left(\frac{x}{|x|^k}\right) = \frac{n-k}{|x|^k}, \tag{1.36}$$

$$\partial \log |x| = \frac{x}{|x|^2} = x^{-1}. \tag{1.37}$$

If $A = P(A) = \langle A \rangle_r$,

$$\dot{\partial}(\dot{x} \cdot A) = A \cdot \partial x = rA, \tag{1.38}$$

$$\dot{\partial}(\dot{x} \wedge A) = A \wedge \partial x = (n - r)A, \tag{1.39}$$

$$\dot{\partial} A \dot{x} = \sum_k a^k A a_k = (-1)^r (n - 2r)A. \tag{1.40}$$

Let us briefly comment on how the results in this table can be established. Equation (1.32) follows from (1.18) by the product rule. Equation (1.33) follows from (1.32) by the integrability condition (1.29). Equation (1.34) follows from (1.33), (1.5), (1.18) and (1-3.5); thus,

$$\partial \cdot x = \sum_k a^k \cdot (a_k \cdot \partial x) = \sum_k a^k \cdot a_k = n.$$

Equations (1.35), (1.36) and (1.37) are obtained by using the 'scalar chain rule' (1.15).

An instructive way to establish (1.38) is to first reformulate it by supposing that A is a r-blade, so that division by A introduces, according to (1-2.10b), the projection operator as follows:

$$A^{-1}A \cdot \partial x = P_A(\partial)x = \partial P_A(x) = r. \tag{1.41a}$$

But the difference between (1.41a) and (1.34) is no more than a matter of notation, for, by virtue of (1.4) and (1.5), (1.34) can be written

$$\partial x = I^{-1}I\, \partial x = P_I(\partial)x = \partial P_I(x) = n.$$

The projection operator P_A in (1.41a) serves merely to limit the domain of x to an r-dimensional subspace of $\mathscr{A}_n = \mathscr{G}^1(I)$. The validity of (1.38) for non-simple A follows from the simple case by the distributive rule as usual. Equation (1.39) can be proved in a similar way, by noticing that for simple A, Eqn. (1-2.16b) helps us put it in the form

$$A^{-1}A \wedge \partial x = (IA)^{-1}(IA) \cdot \partial x = P_{IA}(\partial)x = \partial P_{IA}(x) = n - r. \tag{1.41b}$$

The projection operator P_{IA} 'confines' the variable x to the $(n - r)$-dimensional subspace of $\mathscr{A}_n = \mathscr{G}^1(I)$ with pseudoscalar $IA = I \cdot A$. Finally, (1.40) can be obtained by combining (1.38) and (1.39), or by differentiating the expression $Ax = (-1)^r(xA - 2x \cdot A)$ obtained from (1-1.27a).

Since the derivative has the algebraic properties of a vector, we get a large assortment of 'differential identities' by replacing some vector by ∂ in any of the algebraic identities in Section 1-1 and taking due account of the rule for differentiating a product. For example, let $a = a(x)$, $b = b(x)$, $c = c(x)$ be vector-valued functions. From the identity (1-1.39), we get

$$b \cdot (\partial \wedge a) = b \cdot \partial\, a - \dot{\partial}\, \dot{a} \cdot b, \tag{1.42}$$

whence

$$\partial(a \cdot b) = a \cdot \partial b + b \cdot \partial a - a \cdot (\partial \wedge b) - b \cdot (\partial \wedge a). \tag{1.43}$$

Using (1-1.39) differently, we get

$$[a, b] = \partial \cdot (a \wedge b) - b\,\partial \cdot a + a\,\partial \cdot b, \tag{1.44}$$

where

$$[a, b] \equiv a \cdot \partial b - b \cdot \partial a \tag{1.45}$$

is a conventional notation for the so-called *Lie bracket*. Dotting (1.42) with c, we get

$$\begin{aligned}(c \wedge b) \cdot (\partial \wedge a) &= b \cdot \dot{\partial}\dot{a} \cdot c - c \cdot \dot{\partial}\dot{a} \cdot b \\ &= b \cdot \partial(a \cdot c) - c \cdot \partial(a \cdot b) + [c, b] \cdot a.\end{aligned} \tag{1.46}$$

Similarly, from the Jacobi identity (1-1.61), we get

$$\begin{aligned}a \cdot (\partial \wedge b) &= \dot{b} \cdot (\dot{\partial} \wedge a) + \dot{\partial} \cdot (a \wedge \dot{b}) \\ &= (a \wedge \partial) \cdot b + a \cdot \partial b - a\,\partial \cdot b\end{aligned} \tag{1.47}$$

and

$$\begin{aligned}\partial \cdot (a \wedge b) &= \dot{a} \cdot (\dot{\partial} \wedge \dot{b}) - \dot{b} \cdot (\dot{\partial} \wedge \dot{a}) \\ &= (b \wedge \partial) \cdot a + a \cdot (\partial \wedge b) - (a \wedge \partial) \cdot b - b \cdot (\partial \wedge a).\end{aligned} \tag{1.48}$$

From (1-1.60c, d) we get generalizations of (1.42) and (1.47):

$$\begin{aligned}A \times (\partial \wedge b) &= A \cdot \partial b - \dot{\partial}\dot{b} \cdot A \\ &= A \wedge \partial b - \dot{\partial}\dot{b} \wedge A.\end{aligned} \tag{1.49}$$

This is especially useful when $\partial \wedge b = 0$, and shows that $\partial \wedge x = 0$ was the essential condition leading to the result in (1.38) and (1.39).

As our last example of a differential identity, we differentiate the outer product of homogeneous multivector functions $A = \langle A_r \rangle = A(x)$ and $B = \langle B \rangle_s = B(x)$. Using (1-1.23b) to re-order factors, we have

$$\begin{aligned}\dot{A} \wedge \dot{\partial} \wedge \dot{B} &= (-1)^r\, \partial \wedge (A \wedge B) \\ &= A \wedge \partial \wedge B + (-1)^r (\partial \wedge A) \wedge B \\ &= A \wedge \partial \wedge B + (-1)^{r+s(r+1)} B \wedge \partial \wedge A.\end{aligned} \tag{1.50}$$

2-2. Multivector Derivatives, Differentials and Adjoints

Section 2-1 was concerned with differentiating functions of scalar and vector variables. This section deals with the more general theory of differentiating functions of any multivector variable. We will run through the theory quickly, merely stating the results, and omitting many details which were covered in the preceding section.

The multivector derivative emerges as the central concept of geometric calculus. It characterizes the local properties of geometric functions. The derivative associates with every continuously differentiable function two auxiliary functions, the differential and the adjoint, which can be interpreted as linear approximations to the function at each 'point'. In Section 2-1, the adjoint was defined only for vector-valued functions, whereas the differential was not so restricted. In this section, the differential and adjoint are given a universal significance. The multivector derivative makes the adjoint easy to define and use. Consequently, the adjoint assumes a conceptual and computational status comparable to that of the differential.

Let $F = F(X)$ be a function defined on the Geometric Algebra $\mathcal{G}(I)$ with unit pseudoscalar $I = \langle I\rangle_n$ and let $P(A) = (A \cdot I) \cdot I^\dagger$ be the projection of a given multivector A into $\mathcal{G}(I)$. The *A-derivative* of F at X is defined by

$$A * \partial_X F(X) \equiv \partial_\tau F(X + \tau P(A))\big|_{\tau=0}. \tag{2.1}$$

When this quantity is to be regarded as a linear function of A, we call it the *differential* of F (at X) and employ any of the notations

$$\underline{F} = \underline{F}(A) = \underline{F}(X, A) = F_A(X) = A * \partial F. \tag{2.2}$$

From the definition (2.1) it follows that the differential has the following general properties:

$$\underline{F}(A) = \underline{F}(P(A)), \tag{2.3}$$

$$A * \partial\langle F\rangle_r = \langle A * \partial F\rangle_r = \langle \underline{F}(A)\rangle_r, \tag{2.4}$$

$$\underline{F}(A + B) = \underline{F}(A) + \underline{F}(B), \tag{2.5}$$

$$\underline{F}(\lambda A) = \lambda \underline{F}(A) \quad \text{if } \lambda = \langle\lambda\rangle. \tag{2.6}$$

If $G = G(X)$ like $F = F(X)$ is defined on $\mathcal{G}(I)$, we have the *sum and product rules*

$$A * \partial(F + G) = A * \partial F + A * \partial G, \tag{2.7}$$

$$A * \partial(FG) = (A * \partial F)G + F(A * \partial G). \tag{2.8}$$

If $X' = f(X)$ is a function defined on $\mathcal{G}(I)$ with values in $\mathcal{G}(I')$, where $I' = \langle I'\rangle_n$, then the differential of the composite function $F(X) = G(f(X))$ is given by the *chain rule*

$$A * \partial F = A * \partial_X G(f(X)) = \underline{f}(A) * \partial G, \tag{2.9}$$

or, equivalently

$$\underline{F}(A) = \underline{G}(\underline{f}(A)). \tag{2.10}$$

The *second differential* of $F = F(X)$, defined by

$$F_{AB} = F_{AB}(X) \equiv B * \dot{\partial} A * \partial \dot{F}, \tag{2.11}$$

satisfies the integrability condition

$$F_{AB}(X) = F_{BA}(X). \tag{2.12}$$

The *derivative* $\partial = \partial_X$ with respect to a variable X in $\mathcal{G}(I)$ has the general algebraic properties of a multivector in $\mathcal{G}(I)$; specifically

$$\partial X = P(\partial_X) = \sum_J a^J a_J * \partial_X, \tag{2.13}$$

where $\{a_J\}$ is a basis for $\mathcal{G}(I)$ as defined by (1-3.18). Moreover,

$$\partial_X = \sum_r \partial_{\langle X\rangle_r}, \tag{2.14a}$$

where

$$\partial_{\langle X\rangle_r} = \langle\partial_X\rangle_r = \sum_J \langle a^J\rangle_r \langle a_J\rangle_r * \partial_X \tag{2.14b}$$

is the derivative with respect to a variable $X_{\bar{r}} = \langle X\rangle_r$ in $\mathcal{G}^r(I)$. Often we will write (2.14) in the more succinct form

$$\partial = \sum_r \partial_{\bar{r}}. \tag{2.15}$$

The A-derivative can be obtained from the derivative by taking the scalar product of A with ∂; thus,

$$A * \partial = \sum_r A_{\bar{r}} * \partial = \sum_r A * \partial_{\bar{r}} = \sum_r A_{\bar{r}} * \partial_{\bar{r}}. \tag{2.16}$$

Conversely, the derivative can be obtained from the differential by using the identity

$$\partial_X = \partial_A A * \partial_X. \tag{2.17}$$

This enables us to write

$$\partial F \equiv \partial_X F(X) = \partial_A \underline{F}(X, A) \equiv \partial \underline{F}. \tag{2.18}$$

We are now prepared to define the *adjoint* of $F = F(X)$ by

$$\bar{F} = \bar{F}(A') \equiv \underline{\partial F} * A' = \partial F * A'. \tag{2.19a}$$

This implies immediately that

$$B * \bar{F}(A) = \underline{F}(B) * A, \tag{2.19b}$$

an expression that is commonly used to define the adjoint of a linear function $\underline{F}$. The adjoint has the properties

$$P(\bar{F}(A)) = \bar{F}(A), \tag{2.20}$$

$$\langle\bar{F}(A)\rangle_r = \underline{\partial_{\bar{r}} F} * A = \partial_{\bar{r}} F * A, \tag{2.21}$$

$$\bar{F}(A + B) = \bar{F}(A) + \bar{F}(B), \tag{2.22}$$

$$\bar{F}(\lambda A) = \lambda\bar{F}(A) \quad \text{if } \lambda = \langle\lambda\rangle. \tag{2.23}$$

The derivative obeys the sum and product rules

$$\partial_{\bar{r}}(F+G) = \partial_{\bar{r}}F + \partial_{\bar{r}}G, \tag{2.24}$$

$$\partial_{\bar{r}}(FG) = \dot{\partial}_{\bar{r}}\dot{F}G + \dot{\partial}_{\bar{r}}F\dot{G}, \tag{2.25a}$$

$$\dot{F}\dot{\partial}_{\bar{r}}\dot{G} = \dot{F}\dot{\partial}_{\bar{r}}G + F\dot{\partial}_{\bar{r}}\dot{G}. \tag{2.25b}$$

If $F(X) = G(X')$ and $X' = f(X)$, then the transformation of the derivative induced by the change of variables $X \to X'$, is given by

$$\partial_{\bar{r}} = \langle\partial_X\rangle_r = \bar{f}(\langle\partial_{X'}\rangle_r) = \bar{f}(\partial'_{\bar{r}}). \tag{2.26a}$$

and the *chain rule* for the derivative can be written

$$\partial_{\bar{r}}F = \dot{\partial}_{\bar{r}}G(\dot{f}) = \bar{f}(\dot{\partial}'_{\bar{r}})\dot{G}. \tag{2.26b}$$

Note that the dot on the right side of (2.26b) indicates the variable to be differentiated more directly than the prime does on the right of (2.26a). More important, note that, because of (2.15), $\partial_{\bar{r}}$ can be replaced by ∂ in (2.24), (2.25) and (2.26).

Throughout the above discussion of derivative, differential and adjoint, we have dealt with a function $F = F(X)$ defined for $X = \Sigma_{r=0}^n X_{\bar{r}}$ in $\mathcal{G}(I)$. But a glance at the results shows that they all obtain if $F = F(X)$ is defined for $X = X_{\bar{r}}$ in $\mathcal{G}^r(I)$. Indeed, we have already studied the special case of a vector variable $X = X_{\bar{1}} = x$ in Section 2-1. The essential assumption in our discussion is that our functions be defined on a linear space, so the above results can be adapted quite readily to functions defined on any linear subspace of $\mathcal{G}(I)$. The most general set on which differentiation can be defined is called a manifold. In later chapters we examine vector manifolds in considerable detail.

Now that we have completed our summary of the general features of multivector differentiation, some commentary is in order. We have shown that every continuously differentiable function $F = F(X)$ determines two linear functions for each value of X, the differential $\underline{F} = \underline{F}(X, A)$ and the adjoint $\bar{F} = \bar{F}(X, A')$. We have adopted the underbar-overbar notations for differential and adjoint to emphasize that these functions are of comparable and complementary significance. This is reflected in their respective definitions (2.2) and (2.19a). Indeed, from the definitions it is clear that they are are the only two linear functions that can be formed from F using the derivative ∂ and the scalar product. The key role of the multivector derivative is obvious. Besides providing us with simple coordinate-free definitions of the differential and adjoint functions, it greatly aids manipulations with these functions, as our formulation of the chain rule already shows.

We should examine the relation of our formalism to the usual notions of scalar differential calculus. Let $X = X(\tau)$ be a multivector valued function of a scalar variable τ, or, as we shall say a curve. For such a function our general multivector derivative reduces exactly to the usual scalar derivative. Since the argument of $X(\tau)$ is a scalar, we have $\langle\partial_\tau\rangle_r X = 0$ for $r \neq 0$, so (2.14) gives us

$$\partial_\tau X = \langle\partial_\tau\rangle X = \frac{\mathrm{d}X}{\mathrm{d}\tau}. \tag{2.27a}$$

The differential of $x = x(\tau)$ is

$$\underline{X}(\lambda) = \lambda * \partial X = \lambda \partial_\tau X = \lambda \frac{\mathrm{d}X}{\mathrm{d}\tau}. \tag{2.27b}$$

Thus, for scalar variables, the differential differs from the derivative only by a multiplicative factor. The adjoint of $X = X(\tau)$ is

$$\overline{X}(A) = \partial_\tau X * A = \left(\frac{\mathrm{d}X}{\mathrm{d}\tau}\right) * A. \tag{2.27c}$$

The chain rule applied to $F = F(X(\tau))$ becomes

$$\frac{\mathrm{d}F}{\mathrm{d}\tau} = \overline{X}(\partial_\tau)F(\tau) = \left(\frac{\mathrm{d}X}{\mathrm{d}\tau}\right) * \partial_X F(X). \tag{2.27d}$$

If $X = x(\tau)$ is scalar valued, then only $\alpha = \langle A \rangle$, the scalar part of A, contributes to (2.27c), and the adjoint reduces to $\overline{X}(\alpha) = \alpha\, \mathrm{d}x/\mathrm{d}\tau$. The chain rule (2.27d) can then be written in the usual form for a change of scalar variables: $\mathrm{d}F/\mathrm{d}\tau = \mathrm{d}x/\mathrm{d}\tau\ \mathrm{d}F/\mathrm{d}x$. Thus, in the scalar differential calculus, differentials and adjoints are identical, and they differ trivially from derivatives. Clearly, the single concept of a derivative in elementary differential calculus is generalized to three distinct but related concepts by Geometric Calculus.

We have assembled a list of basic multivector derivatives needed in applications. In the following list we assume that $X = F(X)$ is the identity function on some linear subspace of $\mathcal{G}(I)$ with dimension d, and we use the notation $\not{A}$ to denote the projection of a multivector A into that subspace. Also, obvious singularities at $X = 0$ are understood to be excluded.

$$A * \partial_X X = \dot{\partial}_X \dot{X} * A = \not{A}, \tag{2.28a}$$

$$A * \partial_X X^\dagger = \dot{\partial}_X \dot{X}^\dagger * A = \not{A}^\dagger, \tag{2.28b}$$

$$\partial_X X = d, \tag{2.29}$$

$$\partial_X |X|^2 = 2X^\dagger, \tag{2.30}$$

$$A * \partial_X X^k = \not{A} X^{k-1} + X \not{A} X^{k-2} + \ldots + X^{k-1} \not{A}, \tag{2.31}$$

$$\partial_X |X|^k = k|X|^{k-2} X^\dagger, \tag{2.32}$$

$$\partial_X \log|X| = \frac{X^\dagger}{|X|^2}, \tag{2.33}$$

$$A * \partial_X \{|X|^k X\} = |X|^k \left\{\not{A} + k \frac{A * X^\dagger X}{|X|^2}\right\}, \tag{2.34}$$

$$\partial_X \{|X|^k X\} = |X|^k \left\{d + k \frac{X^\dagger X}{|X|^2}\right\}. \tag{2.35}$$

We omit proofs of these formulas, because they are essentially the same as proofs of corresponding formulas in the preceding section.

If $X = \Sigma_{r=0}^{n} X_{\bar{r}}$ is defined on the whole of $\mathcal{G}(I)$, and we write $\partial = \partial_X$, then (2.28a) and (2.29) can be written

$$A * \partial X = \dot{\partial}\dot{X} * A = P(A), \tag{2.36a}$$

$$A * \partial_{\bar{r}} X = \dot{\partial}_{\bar{r}} \dot{X} * A = P(A_{\bar{r}}), \tag{2.36b}$$

$$\partial_{\bar{r}} X = \partial X_{\bar{r}} = \partial_{\bar{r}} X_{\bar{r}} = \binom{n}{r}, \tag{2.37a}$$

$$\partial X = \sum_r \binom{n}{r} = 2^n. \tag{2.37b}$$

For $P(A) = A$ we have, in addition, the general result,

$$\partial_{\bar{s}} A_{\bar{r}} X_{\bar{s}} = \sum_J \langle a^J \rangle_s A_{\bar{r}} \langle a_J \rangle_s = \Gamma_s^r A_{\bar{r}}, \tag{2.38a}$$

$$\partial_{\bar{s}} A X = \partial A X_{\bar{s}} = \sum_{r=0}^{n} \Gamma_s^r A_{\bar{r}}, \tag{2.38b}$$

where the scalar coefficients Γ_s^r are given by

$$\Gamma_s^r = \sum_{k=0}^{K} (-1)^{rs-k} \binom{r}{k} \binom{n-r}{s-k}, \tag{2.38c}$$

with $K \equiv \frac{1}{2}(r + s - |r - s|)$ and the convention that $\binom{i}{j} \equiv 0$ if $j > i$. The formulas (2.38a) and (2.38c) can be proved by first using (1-1.20) to get

$$\partial_{\bar{s}} A_{\bar{r}} X_{\bar{s}} = \sum_{k=0}^{K} (-1)^{rs-k} \partial_{\bar{s}} \langle X_{\bar{s}} A_{\bar{r}} \rangle_{r+s-2k},$$

and then establishing the result

$$\partial_{\bar{s}} \langle X A_{\bar{r}} \rangle_m = \langle A_{\bar{r}} \partial_{\bar{s}} \rangle_m X = \binom{r}{k} \binom{n-r}{s-k} A_{\bar{r}}\, \delta_{r+s-2k}^m. \tag{2.39}$$

Here the indicies on the Kroenecker delta are understood to be nonnegative integers and, of course, the result vanishes unless $m = r + s - 2k$. Two important special cases of (2.39) are

$$\partial_{\bar{s}} X \wedge A_{\bar{r}} = A_{\bar{r}} \wedge \partial_{\bar{s}} X = \binom{n-r}{s} A_{\bar{r}}, \tag{2.40}$$

$$\partial_{\bar{s}} X \cdot A_{\bar{r}} = A_{\bar{r}} \cdot \partial_{\bar{s}} X = \left\{ \begin{array}{ll} \binom{r}{s} A_{\bar{r}} & \text{if } 0 < s \leqslant r \\ \binom{n-r}{s-r} A_{\bar{r}} & \text{if } 0 < r \leqslant s \end{array} \right\}. \tag{2.41}$$

It suffices to prove (2.39) for a unit r-blade $A_r = a_1 a_2 \ldots a_r$ factored into a product of r orthogonal unit vectors. Choosing an orthonormal basis including these vectors, we can write the left side of (2.39) in the form

$$\partial_{\bar{s}}\langle XA_r\rangle_m = \partial_{\bar{s}}\langle X_{\bar{s}}A_r\rangle_m = \sum_J \langle a^J\rangle_s \langle\langle a_J\rangle_s A_r\rangle_m$$

$$= \sum_{j_1<\ldots<j_s} a^{j_s}\ldots a^{j_2}a^{j_1}\langle a_{j_1}a_{j_2}\ldots a_{j_s}a_1a_2\ldots a_r\rangle_m .$$

The indicated sum is now easily performed. Each nonzero term is identical to the others, so all we need to do is count them. For $m = r + s - 2k$, a term is zero unless $a_{j_1}a_{j_2}\ldots a_{j_s}$ has exactly k vectors in common with $a_1a_2\ldots a_r$, and there are $\binom{r}{k}\binom{n-r}{s-k}$ distinct ways this can happen, which gives the coefficient on the right side of (2.39). Note that, for A_r simple, we can write

$$\begin{aligned}\partial_{\bar{s}}X_{\bar{s}} &\equiv \partial_{\bar{s}}(X_{\bar{s}}A_r)A_r^{-1},\\ &= \partial_{\bar{s}}X_{\bar{s}}\wedge A_rA_r^{-1} + \partial_{\bar{s}}\langle X_{\bar{s}}A_r\rangle_{r+s-2}A_r^{-1} +\\ &\quad + \ldots + \partial_{\bar{s}}\langle X_{\bar{s}}A_r\rangle_{|r-s|+2}A_r^{-1} + \partial_{\bar{s}}X_{\bar{s}}\cdot A_rA_r^{-1},\end{aligned} \tag{2.42a}$$

which, by (2.37a) and (2.39), is termwise equivalent to the combinatorial identity

$$\binom{n}{s} = \binom{r}{0}\binom{n-r}{s} + \binom{r}{1}\binom{n-r}{s-1} + \ldots + \binom{r}{s-1}\binom{n-r}{1} + \binom{r}{s}\binom{n-r}{0}. \tag{2.42b}$$

2-3. Factorization and Simplicial Derivatives

In this section we explore an important relation between the theory of functions of several vector variables and the theory of functions of a single multivector variable. This leads us to the concept of simplicial derivative, and, as we shall see in Chapter 3, results which are very useful in the theory of linear functions.

The question arises as to when a multivector derivative can be factored into a product of derivatives. To answer this question, we apply the chain rule (2.26) to the function of two variables $G(A\wedge B)$; thus,

$$\begin{aligned}\partial_A G(A\wedge B) &= \dot{\partial}_A(\dot{A}\wedge B) * \partial_U G_U(A\wedge B)\\ &= \dot{\partial}_A \underline{G}(A\wedge B, \dot{A}\wedge B).\end{aligned} \tag{3.1a}$$

By the way, it should be noted that use of the differential notation is a convenient way to indicate unambiguously the point at which a derivative is evaluated, for example

$$\partial_U G_U(A\wedge B) = \partial_X G(X)\big|_{X=A\wedge B}.$$

Now, using the product rule (2.25) and the notation (2.11) for the second differential, we get from (3.1a),

$$\partial_B \partial_A G(A \wedge B) = \dot{\partial}_B \dot{\partial}_A (\dot{A} \wedge \dot{B}) * \partial_U G_U(A \wedge B) + \\ + \dot{\partial}_B (A \wedge \dot{B}) * \partial_V \dot{\partial}_A (\dot{A} \wedge B) * \partial_U G_{UV}(A \wedge B). \tag{3.1b}$$

Equations (3.1a, b) are actually more general than they appear, for no property of the outer product was used to get them, so they remain valid if $A \wedge B$ is replaced by some more general function $f(A, B)$.

Suppose G is defined on $\mathcal{G}^r(I)$, so

$$G(X) = G(\langle X \rangle_r). \tag{3.2a}$$

Then, if $A = \langle A \rangle_s$ in (3.1), we must have $B = \langle B \rangle_{r-s}$, and we can use (1-1.25b), (2.36) and (2.41) to evaluate the derivatives:

$$\partial_A (A \wedge B) * \partial_X = \partial_A A \cdot (B \cdot \partial_X) = B \cdot \partial_X,$$

$$\partial_B \partial_A (A \wedge B) * \partial_X = \partial_B B \cdot \partial_X = \binom{r}{r-s} \partial_X.$$

So (3.1a, b) assume the forms

$$\partial_A G(A \wedge B) = B \cdot \partial_U G_U(A \wedge B), \tag{3.2b}$$

$$\partial_B \partial_A G(A \wedge B) = \binom{r}{r-s} \partial_U G_U(A \wedge B) + \\ + (-1)^{s(r-s)} A \cdot \partial_V B \cdot \partial_U G_{UV}(A \wedge B). \tag{3.2c}$$

Consider a linear multivector function $L = L(X)$. Using linearity, we can evaluate the first and second differentials of L; thus,

$$\underline{L} = L_U(X) = U * \partial_X L(X) = L(U * \partial_X X) = L(U), \tag{3.3a}$$

$$L_{UV}(X) = 0. \tag{3.3b}$$

Clearly, a function is linear if and only if it is identical to its own differential. Moreover, its second differential vanishes. Hence, for a linear function, (3.2c), relieved of the restriction (3.2a), assumes the simple form

$$\partial_B \partial_A L(A \wedge B) = \binom{r}{r-s} \partial_{\bar{r}} L. \tag{3.4}$$

This is the 'factorization theorem' for the derivative of a linear function. Of course, $\partial_{\bar{r}} L = \partial L$ if L satisfies (3.2a).

If we use (3.4) to 'factor out' the derivative of a vector, we get

$$\partial_B\, \partial_{a_1} L(a_1 \wedge B) = r\, \partial_{\bar{r}} L\,.$$

Repeating the process r times we get the complete factorization

$$\begin{aligned}\partial_r \ldots \partial_2\, \partial_1 L(a_1 \wedge \ldots \wedge a_r) &= r!\, \partial_{\bar{r}} L \\ &= \partial_r \wedge \ldots \wedge \partial_2 \wedge \partial_1 L(a_1 \wedge \ldots \wedge a_r),\end{aligned} \tag{3.5}$$

where we have adopted the abbreviation $\partial_k = \partial_{a_k}$. We have also expressed in (3.5) that all multivector parts of $\partial_r \ldots \partial_2\, \partial_1$ except $\langle \partial_r \ldots \partial_2\, \partial_1 \rangle_r = \partial_r \wedge \ldots \wedge \partial_2 \wedge \partial_1$ give zero when they operate on L because of the skew symmetry of the argument of L.

The result (3.5) suggests that we define the *simplicial variable*

$$a_{(r)} \equiv a_1 \wedge \ldots \wedge a_r, \tag{3.6a}$$

and define the *simplicial derivative*

$$\partial_{(r)} \equiv (r!)^{-1}\, \partial_{a_r} \wedge \ldots \wedge \partial_{a_2} \wedge \partial_{a_1}. \tag{3.6b}$$

Then (3.5) can be written

$$\partial_{(r)} L(a_{(r)}) = \partial_{\bar{r}} L. \tag{3.6c}$$

Thus, for linear functions the r-vector derivative is equivalent to the simplicial derivative. Therefore, we can use results from the preceding section to evaluate simplicial derivatives. For example, we find that

$$\partial_{(r)} a_{(r)} = \partial_{\bar{r}} X_{\bar{r}} = \binom{n}{r} \tag{3.7a}$$

and

$$\partial_{(r)} (a_{(r)})^2 = (r+1) a_{(r)}. \tag{3.7b}$$

An important general result is the following *factorization theorem* for simplicial derivatives, which is an easy consequence of (3.4):

$$\begin{aligned}\binom{r+s}{s} \partial_{(r+s)} L(a_{(r+s)}) &= \partial_{(r)} \wedge \partial_{(s)} L(a_{(r)} \wedge a_{(s)}) \\ &= \partial_{(r)}\, \partial_{(s)} L(a_{(r)} \wedge a_{(s)}).\end{aligned} \tag{3.8}$$

Note that the skew symmetry of the argument in (3.8) ensures that $\partial_{(r)}\, \partial_{(s)} = \partial_{(r)} \wedge \partial_{(s)}$ when operating on L, and, because of (3.6c), this implies that $\partial_B\, \partial_A = \partial_B \wedge \partial_A$ in (3.4). Let $F_r = F_r(a_1, a_2, \ldots, a_r)$ be a linear function of r vector variables. According to (3.3a), the linearity of F_r can be expressed by

$$a_k \cdot \partial_k F_r = F_r, \quad k = 1, 2, \ldots, r, \tag{3.9}$$

where ∂_k differentiates the kth argument of F_r. From F_r we can obtain a new function

$$G_r = \frac{1}{r!}(a_1 \wedge \ldots \wedge a_r) \cdot (\partial_r \wedge \ldots \wedge \partial_1)F_r = a_{(r)} \cdot \partial_{(r)} F_r. \tag{3.10}$$

The function $G_r = G_r(a_1, \ldots, a_r)$ is obviously skewsymmetric in its arguments. Moreover, if F_r is already skewsymmetric, then G_r is identical with F_r; that is,

$$F_r = a_{(r)} \cdot \partial_{(r)} F_r. \tag{3.11}$$

This can be proved by using (1-1.25b) and (1-1.38) along with skew-symmetry and (3.9) as follows:

$$\begin{aligned}
&(a_1 \wedge \ldots \wedge a_r) \cdot (\partial_r \wedge \ldots \wedge \partial_1)F_r \\
&\quad = (a_1 \wedge \ldots \wedge a_{r-1}) \cdot [a_r \cdot (\partial_r \wedge \ldots \wedge \partial_1)] F_r \\
&\quad = \sum_{k=1}^{r} (-1)^{r-k} (a_1 \wedge \ldots \wedge a_{r-1}) \cdot (\partial_r \wedge \ldots \check{\partial}_k \ldots \wedge \partial_1) a_r \cdot \partial_k F_r \\
&\quad = r(a_1 \wedge \ldots \wedge a_{r-1}) \cdot (\partial_{r-1} \wedge \ldots \wedge \partial_1) a_r \cdot \partial_r F_r \\
&\quad = \ldots = r!\, a_1 \cdot \partial_1 a_2 \cdot \partial_2 \ldots a_r \cdot \partial_r F_r = r!\, F_r.
\end{aligned}$$

Thus, the operator $a_{(r)} \cdot \partial_{(r)}$ is the *skew-symmetrizer* of an r-linear function.

Let $\alpha_r = \alpha_r(a_1, \ldots, a_r)$ be an alternating linear r-form as defined in Section 1-4. The simplicial derivative of α_r results in an r-vector

$$A_r^{\dagger} \equiv \partial_{(r)} \alpha_r. \tag{3.12}$$

So, according to the results of the preceding paragraph,

$$\alpha_r = (a_1 \wedge a_2 \wedge \ldots \wedge a_r) \cdot A_r^{\dagger}. \tag{3.13}$$

This establishes the 'canonical form' for α_r, which we had assumed in Section 1-4.

Chapter 3

Linear and Multilinear Functions

This chapter shows the advantages of developing the theory of linear and multilinear functions on finite dimensional spaces with Geometric Calculus. The theory is sufficiently well developed here to be readily applied to most problems of linear algebra.

Many recent books develop a major part of linear algebra by general 'operator techniques' which are independent of matrix representations. These operator techniques have become fairly standard, and they fit perfectly into the present formalism, so it will be unnecessary to discuss them here, and we can use well-known results of such techniques without proof.

General operator techniques are often impotent or awkward in specific computations. In such cases, matrix algebra is usually employed. But matrix algebra has the drawback of requiring a specific choice of frames, which frequently is quite irrelevant to the problem at hand. Moreover, the method of matrices is so tailored to the theory of linear functions that it is not readily applied to the representation of multilinear or nonlinear functions. These defects can be removed if matrix calculus is supplanted by Geometric Calculus as the principal computational tool in the theory of linear functions.

In this chapter we show explicitly that many common applications of matrix algebra are better performed by Geometric Algebra. However, we do not mean to suggest that matrix algebra should be eliminated altogether; we propose only that it should be subordinated to Geometric Algebra. In some problems, initial data appears naturally in a matrix format, so matrices can hardly be avoided. But we have already shown in Section 1-3 how matrices can be handled efficiently within the purview of Geometric Algebra.

Comparison of the theory of vector functions with the theory of scalar functions helps explain the utility of Geometric Algebra. In the theory of scalar-valued functions of a scalar variable, the elementary operations of addition and multiplication naturally play a special role, because they are essential to the very definition of the scalars. With these two fundamental operations the most important scalar functions are constructed, notably the polynomials and functions defined by infinite series. In contrast, multiplication of vectors is not ordinarily defined in treatments of vector functions, but scalar multiplication is, so it is necessary to deal with sets of scalars (matrices) to construct even the simplest functions. However, in Geometric Algebra multiplication is as much a part of the notion of vector as is addition. As in the scalar algebra, which is a subalgebra of Geometric Algebra, the most basic vector functions can be constructed by addition and multiplication of vectors, in particular, and of multivectors, in general.

Obviously the most general linear function of a vector variable which can be constructed by multiplication alone is $f(a) = AaB$ where A and B are any multivectors. If we use addition as well, then we can construct

$$f(a) = \sum_k A_k a B_k . \tag{0.1}$$

In fact, every linear transformation can be expressed in this form. This will be obvious after we have examined some important special cases, so we will not bother with a proof until we consider a more general theorem in Section 3-9. It will become evident also that a matrix representation of a linear transformation can be regarded as a special case of (0.1).

Many different selections of A_k and B_k in (0.1) determine the same function $f(a)$. The classical eigenvalue problem can be interpreted as the problem of finding the simplest selection. But we shall see that there are often easier and more transparent ways to make the 'canonical selection' than the usual method of forming the 'secular determinant' and solving the characteristic polynomial at once. Indeed, with Geometric Algebra it is often possible to write down the canonical form for a linear transformation immediately from initial data.

In the first three sections of this chapter, we use Geometric Algebra to develop general methods for characterizing linear functions which are valuable supplements to standard operator techniques. Our approach to linear functions is multilinear from the beginning, of course! The geometric product itself is multilinear; hence it is linear in *each* of its factors. It is impossible to develop the linear theory independently of the multilinear theory. Even matrix algebra is fundamentally multilinear, for the matrix product is a linear function of *each* matrix. The only question is how best to integrate the linear and multilinear theories. The answer promoted here is 'By employing the geometric product from the beginning'.

In Sections 3-4, 5, and 6, we construct various canonical forms for linear transformations and carry out the necessary computations without using matrices. Few of our results are new in the sense that equivalent results cannot be found in the literature. What is new is the fact that here, for the first time, powerful results which have long languished in the anterooms of mathematics have been admitted to the dignified place they deserve in the august chambers of linear algebra. For example, the canonical forms for orthogonal transformations developed in Section 3-5 are so simple and so useful that an efficient theory of linear transformations ought to make them readily available. This is possible only if the geometric product is an integral part of the theory. Representations of orthogonal transformations by Clifford numbers equivalent to ours can be found in references [Ch], [Po], [Hu] and [Ri], but, being only indirectly related to the standard apparatus of linear algebra, their practical value has been severely limited. An elementary but detailed treatment of the canonical forms for orthogonal transformations along with applications to physics will be given in the sequel to this book, *New Foundations for Classical Mechanics*.

The fact that our treatment of linear algebra employs the Euclidean inner product $x \cdot y$ from the beginning requires some justification, because it is at variance with recent mathematical practice. Modern expositions of linear algebra usually try to push the subject as far as possible without an inner product. To do this they exploit the concept of *dual space*. The dual space $\mathscr{A}_n^*$ of a vector space $\mathscr{A}_n$ is the space of 1-forms on $\mathscr{A}_n$, that is, the space of all linear mappings of $\mathscr{A}_n$ into the scalars. The space $\mathscr{A}_n^*$ is isomorphic to $\mathscr{A}_n$, but this isomorphism is not unique unless $\mathscr{A}_n$ is endowed with a specific metric or inner product. It might appear, therefore, that by a premature introduction of the inner product our method sacrifices a significant degree of flexibility and generality. On the contrary, we hold that delay in introducing an inner product merely complicates the development of linear algebra unnecessarily.

In Section 2-3 we proved that in Geometric Algebra every 1-form $\alpha(x)$ can be expressed with the inner product by $\alpha(x) = a \cdot x$. In writing $a \cdot x$ we have tacitly committed ourselves to admitting the existence of a set of orthogonal unit vectors, while the form $\alpha(x)$ makes no such commitment. But this costs us nothing, because we have not specified which vectors are the units, and we are not required to do so any sooner than other approaches. Defining forms without an inner product is like defining the real numbers without distinguishing a unit element. Specification of the unit scalar determines a scale for the real numbers; a scale is essential if the distributive rule is to be used to reduce multiplication to repeated addition. To describe relations among real numbers which are independent of scale, no one is so foolish as to ask for a number system without a unit element; one simply uses ratios or replaces specific numerals by less specific letters. The unit element is essential to the very notion of real number and is indispensible in computations. We regard an inner product as no less essential to the notion of a vector and equally indispensible in computations.

There is no way to carry out computations in linear algebra without utilizing an inner product in one form or another, whatever name it is given. We use the inner product explicitly to construct and compose linear transformations. Matrix algebra implicitly utilizes the inner product when summing over indices in the composition of matrices. The significant issue is not when to introduce the inner product, but how to do it in the most efficient way.

Use of the Euclidean inner product in computations and construction of functions by no means commits us to a Euclidean metric. Indeed, as we show in Section 3-7, the Euclidean inner product helps us achieve a unified theory of bilinear forms and isometries. Every bilinear form can be obtained from the Euclidean inner product by a linear transformation. This helps us to find and represent isometries of a given bilinear form by using the canonical forms for linear transformations already constructed from the (Euclidean) geometric product.

And, of course, it greatly facilitates comparison of one class of isometries with another.

For certain purposes it is desirable to work directly with nonEuclidean metrics without relating them to a Euclidean metric. The Lorentz metric, for example, is used in spacetime physics, because it has a direct physical significance. Section 3-8 uses pseudo-Euclidean Geometric Algebra to characterize the orthogonal group of spaces with any signature. It introduces a new and simpler definition of spinor and develops canonical forms for Lorentz transformations, for which important physical applications can be found in references [H7–10].

In Section 3-9 we begin the study of general linear multivector functions. We consider in detail the classification of linear bivector functions (called biforms) on bivector spaces of any signature, including the so-called Petrov classification of the Weyl conformal tensor. Another generalization of linear algebra is considered in Section 3-10, where multilinear functions are discussed.

3-1. Linear Transformations and Outermorphisms

This section develops the general properties of outermorphisms, exploiting, in particular, the simplicial derivative. The general study of linear transformations is greatly simplified by introducing outermorphisms at the outset. Many properties of a linear transformation are most easily ascertained or expressed by its corresponding outermorphism. The determinant of a transformation is an example. It can, of course, be expressed as a simple function of a matrix representation of the transformation. But other significant properties of a transformation are not so easily expressed by matrices, because they involve minors of the transformation matrix and seem so complicated that they are hardly ever used. The outermorphism provides a means of representing and analyzing such properties. The theory also produces a useful expression for the inverse of a nonsingular linear transformation without using matrices. Many other applications of outermorphisms appear in subsequent sections.

We use the term *linear transformation* exclusively to refer to a vector-valued function of a vector variable, especially to distinguish it from other kinds of linear multivector functions. If f is a linear transformation defined on an n-dimensional vector space $\mathscr{A}_n$, then

$$\underline{f}(a) = a \cdot \partial_x f(x) = f(a \cdot \partial_x x) = f(a) \tag{1.1}$$

for every element a of $\mathscr{A}_n$. Thus a linear transformation $f = f(x)$ is equivalent to its own differential $\underline{f} = \underline{f}(x)$ on $\mathscr{A}_n$. But we saw in Section 2-1 that $\underline{f}$ is defined on *all* vectors, those vectors orthogonal to $\mathscr{A}_n$ being in its null space. So $\underline{f}$ automatically defines an extension of f to the space $\mathscr{G}^1$ of all vectors.

Because of (1.1), the composite gf of linear transformations g and f is given directly by the chain rule (2-1.14); thus,

$$gf(a) \equiv g(f(a)) = f(a) \cdot \partial g. \tag{1.2}$$

If f is a linear transformation of $\mathscr{A}_n$ to $\mathscr{A}_n$, then the transformation f^r obtained by an r-fold composition of f with itself can be written

$$f^r = f \cdot \partial f^{r-1} = f^{r-1} \cdot \partial f, \quad \text{where } f^1 \equiv f. \tag{1.3}$$

In matrix algebra, the composite of linear transformations is computed by summing over indices of associated matrices. Instead, (1.2) shows that the composite of linear transformations can be computed without introducing coordinates by using the directional derivative; of course, this can be reduced to matrix composition should the need arise, but the use of matrices is often quite uncalled for, as we shall see.

The linear transformation $f = f(a)$ induces a linear mapping $\underline{f}(A)$ of every multivector A in $\mathscr{G}(\mathscr{A}_n)$, which we call the *outermorphism* of $\overline{f}$, or the *differential outermorphism* of f when we want to distinguish it from the adjoint outermorphism defined later. The outermorphism $\underline{f}$ is defined by

$$\underline{f}(A) = \underline{f}\left(\sum_r A_{\bar{r}}\right) = \sum_r \underline{f}(A_{\bar{r}}), \tag{1.4a}$$

$$\underline{f}(A_{\bar{0}}) = \underline{f}(\langle A \rangle) \equiv \langle A \rangle, \tag{1.4b}$$

$$\underline{f}(A_{\bar{r}}) \equiv A_{\bar{r}} \cdot \partial_{(r)} f_{(r)} = A * \partial_{(r)} f_{(r)} \quad \text{for } r > 0, \tag{1.4c}$$

where $\partial_{(r)}$ is the simplicial derivative defined by (2-3.6b) and

$$f_{(r)} \equiv f(a_1) \wedge f(a_2) \wedge \ldots \wedge f(a_r). \tag{1.5}$$

Actually, $\underline{f}(A)$ is well defined by (1.4) for any A. The operator $A * \partial_{(r)}$ in (1.4c) projects $\bar{A}$ into $\mathscr{G}^r(\mathscr{A}_n)$. Hence,

$$\underline{f}(P(A)) = \underline{f}(A), \tag{1.6}$$

where $P(A)$ is the projection of A into $\mathscr{G}(\mathscr{A}_n)$.

From (1.4c), it follows immediately, by virtue of (2-3.11), that

$$\underline{f}(a_1 \wedge a_2 \wedge \ldots \wedge a_r) = \underline{f}(a_1) \wedge \underline{f}(a_2) \wedge \ldots \wedge \underline{f}(a_r). \tag{1.7a}$$

Differentiating this by $A * \partial_{(s)} \equiv (s!)^{-1} A * (\partial_s \wedge \ldots \wedge \partial_2 \wedge \partial_1)$ and $B * \partial_{(r-s)} \equiv ((r-s)!)^{-1}\, B * (\partial_r \wedge \ldots \wedge \partial_{s+2} \wedge \partial_{s+1})$ for $s \leqslant r$, we establish the more general relation,

$$\underline{f}(A \wedge B) = \underline{f}(A) \wedge \underline{f}(B), \tag{1.7b}$$

which holds for any pair of multivectors A and B.

It follows from (1.7) that an outermorphism is grade-preserving, that is,

$$\underline{f}(\langle A\rangle_r) = \langle \underline{f}(A)\rangle_r. \tag{1.8}$$

Moreover,

$$\underline{f}(A^\dagger) = [\underline{f}(A)]^\dagger. \tag{1.9}$$

It also follows from (1.7) and (1.2) that the outermorphism of a composite transformation equals the composite of the outermorphisms. Thus the composition of outermorphisms is given directly by the general chain rule (2-2.10) for multivector functions.

The fundamental property (1.7) explains our choice of the term 'outermorphism', for it shows that the outer product relation is 'preserved' by the mapping. However, there is no corresponding 'innermorphism', and we shall see that the inner product is not generally 'preserved' by an outermorphism.

Since the extension of a linear transformation to its differential outermorphism is unique and always well defined, we always denote it by the simple underbar notation we use for differentials. The differential $\underline{f}(a)$ can be distinguished from its extension to an outermorphism $\underline{f}(A)$ by the appearance of a vector in its argument.

If the linear transformation f defined on $\mathscr{A}_n$ has values in an m-dimensional vector space $\mathscr{A}'_m$, then the differential outermorphism of f maps $\mathscr{G}(\mathscr{A}_n)$ into $\mathscr{G}(\mathscr{A}'_m)$. This can be expressed by the equation

$$P'(\underline{f}(A)) = \underline{f}(A), \tag{1.10}$$

where P' is the projection operator for $\mathscr{G}(\mathscr{A}'_m)$. The adjoint $\overline{f}(a)$ of f maps $\mathscr{A}'_m$ into $\mathscr{A}_n$, and its extension to an outermorphism $\overline{f}(A)$ maps $\mathscr{G}(\mathscr{A}'_m)$ into $\mathscr{G}(\mathscr{A}_n)$. In fact, $\overline{f}(A)$ is exactly the adjoint of the outermorphism $\underline{f}(A)$, as defined for multivector functions by (2-2.19). We can define $\overline{f}(A)$ immediately by (2-2.19), or we can define it directly in terms of f by writing, in parallel with (1.4),

$$\overline{f}(A) = \overline{f}\left(\sum_r A_{\bar{r}}\right) = \sum_r \overline{f}(A_{\bar{r}}), \tag{1.11a}$$

$$\overline{f}(\langle A\rangle) = \langle A\rangle, \tag{1.11b}$$

$$\overline{f}(A_{\bar{r}}) \equiv \partial_{(r)} f_{(r)} \cdot A_{\bar{r}} = \partial_{(r)} f_{(r)} * A \quad \text{for } r > 0. \tag{1.11c}$$

We will refer to $\overline{f}(A)$ as the *adjoint outermorphism* of f, because it is the outermorphism of the adjoint of f.

It is easily proved that

$$\overline{f}(P'(A)) = P(\overline{f}(A)) = \overline{f}(A), \tag{1.12}$$

$$\overline{f}(A\wedge B) = \overline{f}(A)\wedge\overline{f}(B), \tag{1.13}$$

and, of course, that $\overline{f}$ is grade-preserving as well.

Although the inner product is not, in general, preserved by the differential and adjoint outermorphisms, it does provide important general relations between them, namely

$$A_{\bar{r}} \cdot \bar{f}(B_{\bar{s}}) = \bar{f}[\underline{f}(A_{\bar{r}}) \cdot B_{\bar{s}}] \quad \text{for } r \leqslant s, \tag{1.14a}$$

$$\underline{f}(A_{\bar{r}}) \cdot B_{\bar{s}} = \underline{f}[A_{\bar{r}} \cdot \bar{f}(B_{\bar{s}})] \quad \text{for } r \geqslant s. \tag{1.14b}$$

The order of factors in (1.14) can be reversed by using (1.9). For the important special case $r = s$, both (1.14a) and (1.14b) reduce to

$$A_{\bar{r}} \cdot \bar{f}(B_{\bar{r}}) = \underline{f}(A_{\bar{r}}) \cdot B_{\bar{r}}, \tag{1.15}$$

which, of course, is just a special case of the general relation (2-2.19b).

We can prove (1.14a) as follows

$$\begin{aligned}
A_r \cdot \bar{f}(B_s) &= A_r \cdot \partial_{(s)}\underline{f}(a_{(s)}) \cdot B_s \\
&= \partial_{(s-r)}\underline{f}(a_{(s-r)} \wedge A_r) \cdot B_s \\
&= \partial_{(s-r)}\{\underline{f}(a_{(s-r)}) \wedge \underline{f}(A_r)\} \cdot B_s \\
&= \partial_{(s-r)}\underline{f}(a_{(s-r)}) \cdot (\underline{f}(A_r) \cdot B_s) \\
&= \bar{f}(\underline{f}(A_r) \cdot B_s),
\end{aligned}$$

where we have used (1.11), (2-3.2b), (1.7) and (1-1.25b). To prove (1.14b), we use (1-1.40) and note that, because of the skew symmetry of $f_{(r)}$, all the $\binom{r}{s}$ terms of the expansion are identical to one another, so

$$\begin{aligned}
\underline{f}(A_r) \cdot B_s &= A_r \cdot \partial_{(r)} f_{(r)} \cdot B_s \\
&= \tbinom{r}{s} A_r \cdot \partial_{(r)} B_s \cdot (f_{r-s+1} \wedge \ldots \wedge f_r) f_1 \wedge \ldots \wedge f_{r-s} \\
&= A_r \cdot (\partial_{(s)} \wedge \partial_{(r-s)}) B_s \cdot f_{(s)} f_{(r-s)} \\
&= A_r \cdot (\bar{f}(B_s) \wedge \partial_{(r-s)}) f_{(r-s)} \\
&= [A_r \cdot \bar{f}(B_s)] \cdot \partial_{(r-s)} f_{(r-s)} = \underline{f}(A_r \cdot \bar{f}(B_s)).
\end{aligned}$$

A linear transformation f of $\mathscr{A}_n$ to $\mathscr{A}'_n$ is said to be *nonsingular* if $\underline{f}(I) \neq 0$, where $I = \langle I \rangle_n$ is the unit pseudoscalar of $\mathscr{A}_n$. Now,

$$\underline{f}(I) = I \cdot \partial_{(n)} f_{(n)} = I\, \partial_{(n)} f_{(n)},$$

and

$$\bar{f}(I') = \partial_{(n)} f_{(n)} \cdot I' = \partial_{(n)} f_{(n)} I',$$

where I' is the unit pseudoscalar for $\mathscr{A}'_n$. Hence, f is nonsingular if and only if the quantity

$$\partial_{(n)}f_{(n)} = I^\dagger \underline{f}(I) = \overline{f}(I')I'^\dagger \tag{1.16}$$

does not vanish.

The *determinant* of f is a scalar quantity denoted by det (f) and defined by the equation

$$\underline{f}(I) = I' \det(f). \tag{1.17a}$$

Using (1.16), we find also that

$$\overline{f}(I') = I \det(f). \tag{1.17b}$$

The sign of the determinant depends on the relative orientations assigned to I and I'; this will be arbitrary unless fixed by some other transformation. For example, we can take $I' = I$ if $\mathscr{A}'_n = \mathscr{A}_n$. In this case, we have from (1.16) and (1.17),

$$\det(f) = \partial_{(n)}f_{(n)} = \partial_{(n)} \cdot f_{(n)} = I^\dagger \underline{f}(I) = I^\dagger \overline{f}(I), \tag{1.18}$$

and, as will be seen in the next section, det (f) is identical to the determinant of the matrix of f. In general, we have

$$(\det f)^2 = |\,\partial_{(n)}f_{(n)}|^2. \tag{1.19}$$

The quantity $\partial_{(n)}f_{(n)}$ is a generalization of det (f). The determinant describes only induced changes of relative scale and orientation of a pseudoscalar, while $\partial_{(n)}f_{(n)}$ describes the relative change of direction as well, as (1.16) shows.

For a nonsingular transformation the differential and adjoint outermorphisms are related by duality. Specifically, if A is a multivector in $\mathscr{G}(\mathscr{A}_n) = \mathscr{G}(I)$ and A' is in $\mathscr{G}(\mathscr{A}'_n) = \mathscr{G}(I')$, then

$$A = \overline{f}(A') \quad \text{iff } \underline{f}(AI) = A'\underline{f}(I) = A'I' \det(f), \tag{1.20a}$$

$$A' = f(A) \quad \text{iff } \overline{f}(A'I') = A\overline{f}(I') = AI \det(f). \tag{1.20b}$$

These results follow from (1.14) and (1.17). For example,

$$\begin{aligned}\underline{f}(IA) &= \underline{f}(I\overline{f}(A')) = \underline{f}(I \cdot \overline{f}(A'))\\ &= \underline{f}(I) \cdot A' = \underline{f}(I)A' = I'A' \det(f).\end{aligned}$$

We can solve (1.20b) to get an expression for the inverse $\underline{f}^{-1}$ of the differential outermorphism in terms of its adjoint, namely,

$$\underline{f}^{-1}(A') = \overline{f}(A'I')\,[\overline{f}(I')]^{-1} = \frac{\overline{f}(A'I')I^{-1}}{\det(f)}. \tag{1.21a}$$

Of course, for vector arguments (1.21) yields the inverse of the linear transformation f. This result is more general than the expression for f^{-1} which we got from the matrix of f in Section 1-3. In a similar way, we get from (1.20a),

$$\overline{f}^{-1}(A) = [\underline{f}(I)]^{-1}\underline{f}(AI) = (I' \det(f))^{-1}\underline{f}(AI). \tag{1.21b}$$

3-2. Characteristic Multivectors and the Cayley–Hamilton Theorem

The trace and the determinant of a matrix play an important role in matrix algebra, because they are properties of a linear transformation which are independent of its matrix representation. They are only the simplest examples of a set of intrinsic properties of a linear transformation which are represented in this section by *characteristic multivectors*.

We expect that important applications for characteristic multivectors can be found above and beyond what is presented in this section. But the value of these quantities is definitely established here by relating them to the 'characteristic polynomial'.

The proof of the celebrated Cayley–Hamilton theorem presented in this section differs from other proofs in the literature in that it does not appeal to any other result of linear algebra. It does not even use the notion of an eigenvector or require any prior knowledge of the characteristic polynomial. Of course, it is the Geometric Calculus that makes the proof so elementary.

Let f be a linear transformation of $\mathscr{A}_n$ to $\mathscr{B}_m$. We call the simplicial derivatives $\partial_{(r)}f_{(r)}$ of the r-blades $f_{(r)} = f(a_1) \wedge f(a_2) \wedge \ldots \wedge f(a_r)$ the *characteristic multivectors* of f. As established in Section 2-3, since $\underline{f}$ is linear, we can replace the simplicial derivative by a multivector derivative in our definition of the characteristic multivectors; thus

$$\partial_{(r)}f_{(r)} = \partial_{(r)}\underline{f}(a_{(r)}) = \partial_{A_{\bar{r}}}\underline{f}(A_{\bar{r}}) = \partial_{\bar{r}}\underline{f}_{\bar{r}} = \partial_{\bar{r}}\underline{f}. \tag{2.1}$$

But simplicial derivatives will be most convenient for our purposes in this section. The $\partial_{(r)}f_{(r)}$ obviously describe intrinsic properties of f, but they can be related to well-known quantities in matrix theory by expressing them in terms of frames. Using (1.1), (2-1.5), and (1-3.27a), we find immediately that, if $\{a_k\}$ is a frame in $\mathscr{A}_n$ with image $\{b_k = f(a_k)\}$ in $\mathscr{B}_m$, then

$$\partial f = a^k f(a_k) = a^k b_k. \tag{2.2}$$

The scalar part of (2.2) is called the *trace* of f (or of the matrix $a^j \cdot b_k$);

$$\partial \cdot f = \operatorname{Tr}(f) = \operatorname{Tr}(a^j \cdot b_k). \tag{2.3}$$

Generalizing the derivation of (2.2), we easily find, from the definitions (1.5) and (2-3.6b), that

$$\partial_{(r)}f_{(r)} = a^{j_r} \wedge \ldots \wedge a^{j_1} b_{j_1} \wedge \ldots \wedge b_{j_r}, \tag{2.4}$$

where the sum over repeated indices is restricted by the condition $0 < j_1 < \ldots < j_r \leqslant n$. We may choose the dimension m of $\mathscr{B}_m$ to be the rank of f; it is clear, then, from (2.4) that $\partial_{(m)} f_{(m)} \neq 0$, while $\partial_{(r)} f_{(r)} = 0$ for $r > m$. For the case $r = n$, (2.4) becomes

$$\partial_{(n)} f_{(n)} = a^n \wedge \ldots \wedge a^1 b_1 \wedge \ldots \wedge b_n = A^n B_n. \tag{2.5}$$

The scalar part of (2.5) is the determinant of the matrix $a^k \cdot b_j$;

$$\partial_{(n)} \cdot f_{(n)} = A^n \cdot B_n = \det (a^k \cdot b_j); \tag{2.6}$$

however, this quantity is equivalent to the determinant of f only if $\mathscr{B}_m$ is a subspace of $\mathscr{A}_n$, in which case $\partial_{(n)} f_{(n)} = \partial_{(n)} \cdot f_{(n)}$. As was noted in the last section, the magnitude of the determinant of f is

$$|\det (f)| = |\partial_{(n)} f_{(n)}|.$$

This quantity can be expressed as the determinant of a matrix by embedding $\mathscr{A}_n$ and $\mathscr{B}_m$ in a larger vector space $\mathscr{E}_p$ and extending f to a transformation of $\mathscr{E}_p$ to $\mathscr{E}_p$. But such is an example of the unnecessary complications introduced when linear transformations are represented by matrices. With Geometric Algebra, we can describe a transformation of $\mathscr{A}_n$ to $\mathscr{B}_m$ completely without any reference to some larger vector space in which $\mathscr{A}_n$ and $\mathscr{B}_m$ are embedded.

If f is the sum of two linear transformations g and h, that is, if

$$f(a) = g(a) + h(a), \tag{2.7a}$$

then the characteristic multivectors of f are given by

$$\partial_{(r)} f_{(r)} = \sum_{s=0}^{r} \partial_{(s)} \wedge \partial_{(r-s)} g_{(r-s)} \wedge h_{(s)}, \tag{2.7b}$$

with the understanding that $\partial_{(0)} g_{(0)} \equiv 1$. This important result can be established by straightforward computation:

$$\begin{aligned}\partial_{(r)} f_{(r)} &= \partial_{(r)} (g_1 + h_1) \wedge (g_2 + h_2) \wedge \ldots \wedge (g_r + h_r) \\ &= \frac{1}{r!} \partial_r \wedge \ldots \wedge \partial_1 \Big\{ g_1 \wedge \ldots \wedge g_r + \binom{r}{1} g_1 \wedge \ldots \wedge g_{r-1} \wedge h_r + \\ &\quad + \binom{r}{2} g_1 \wedge \ldots \wedge g_{r-2} \wedge h_{r-1} \wedge h_r + \cdots + \\ &\quad + \binom{r}{r-1} g_1 \wedge h_2 \wedge \ldots \wedge h_r + h_1 \wedge \ldots \wedge h_r \Big\} \\ &= \partial_{(r)} g_{(r)} + \partial_r \wedge \partial_{(r-1)} g_{(r-1)} \wedge h_r + \cdots + \partial_{(r)} h_{(r)}.\end{aligned}$$

Let us apply (2.7) to the function

$$F(a) = f(a) - \lambda a, \tag{2.8a}$$

where λ is a scalar and $f = f(a)$ has its values in $\mathscr{A}_n$. Since $\partial_{(n)}$ and $F_{(n)}$ are pseudoscalar, with the help of (2-2.40), we find

$$\partial_{(n)}F_{(n)} = \langle\partial_{(n)}F_{(n)}\rangle = \left\langle \sum_{s=0}^{n} \partial_{(s)} \wedge \partial_{(n-s)}(-\lambda a)_{(n-s)} \wedge f_{(s)} \right\rangle$$

$$= \sum_{s=0}^{n} (-\lambda)^{n-s} \langle \partial_{(s)}\, \partial_{(n-s)} a_{(n-s)} \wedge f_{(s)} \rangle$$

$$= \sum_{s=0}^{n} (-\lambda)^{n-s} \langle \partial_{(s)} f_{(s)} \rangle.$$

Thus,

$$C_f(\lambda) \equiv \partial_{(n)}F_{(n)} = \partial_{(n)} * F_{(n)} = \sum_{s=0}^{n} (-\lambda)^{n-s}\, \partial_{(s)} * f_{(s)}, \tag{2.8b}$$

where $\partial_{(0)} * f_{(0)} \equiv 1$. As we shall see, $C_f(\lambda)$ is the so-called *characteristic polynomial* of the linear transformation f.

Matrix algebra makes good use of the trace and the determinant, but without the help of Geometric Algebra, it is hardly able to formulate, let alone exploit, the other intrinsic features of a linear transformation so directly described by characteristic multivectors. The fact that $C_f(\lambda)$ given by (2.8b) is the usual characteristic polynomial for a linear transformation of $\mathscr{A}_n$ into $\mathscr{A}_n$ follows immediately from the fact that, by (2.6), $\partial_{(n)}F_{(n)}$ is the determinant of the matrix of F. The outstanding feature of (2.8b) is that it explicitly identifies the coefficients of the characteristic polynomial as the scalar parts of the characteristic multivectors. Of course, matrix algebra reveals no such simple expression for the coefficients. That the expression is by no means trivial is amply demonstrated by the following.

The Cayley–Hamilton theorem states that a linear transformation f from $\mathscr{A}_n$ to $\mathscr{A}_n$ satisfies its own characteristic equation, that is, according to (2.8b),

$$\sum_{s=0}^{n} (-1)^{n-s} \partial_{(s)} * f_{(s)} f^{n-s} = 0, \tag{2.9}$$

where $f^0(a) = a$ and $f^r = f^r(a)$ is the r-fold transformation defined by (1.3). We prove (2.9) with the help of (2-3.2b), (1.3) and (1-1.40), by decomposing the last term of (2.9) into the negative of the others:

$$\begin{aligned}
\partial_{(n)} * f_{(n)} f^0(a) &= a\, \partial_{(n)} f_{(n)} = a \cdot \partial_{(n)} f_{(n)} \\
&= \partial_{(n-1)} f_{(n-1)} \wedge (a \cdot \partial f) = \partial_{(n-1)} \cdot [f_{(n-1)} \wedge f(a)] \\
&= \partial_{(n-1)} \cdot f_{(n-1)} f(a) - (n-1)\, \partial_{(n-1)} \cdot (f_1 \wedge \ldots \wedge f_{n-2} \wedge f) f_{n-1} \\
&= \partial_{(n-1)} \cdot f_{(n-1)} f - (\partial_{n-1} \wedge \partial_{(n-2)}) \cdot (f_{(n-2)} \wedge f) f_{n-1} \\
&= \partial_{(n-1)} \cdot f_{(n-1)} f - \partial_{(n-2)} \cdot f_{(n-2)} f \cdot \partial_{n-1} f_{n-1} + \\
&\quad + (\partial_{n-2} \wedge \partial_{(n-3)}) \cdot (f_{(n-3)} \wedge f) f_{n-2} \cdot \partial_{n-1} f_{n-1} \\
&= \cdots \\
&= \partial_{(n-1)} \cdot f_{(n-1)} f(a) - \partial_{(n-2)} \cdot f_{(n-2)} f^2(a) + \cdots + (-1)^{n-1} f^n(a).
\end{aligned}$$

This exhibits the Cayley–Hamilton theorem as no more than a differential identity in the 'vector analysis' of linear functions.

Of course, if $f(a) = \lambda a$, we have, from (2.9),

$$C_f(\lambda) \equiv \sum_{s=0}^{n} (-\lambda)^{n-s} \partial_{(s)} * f_{(s)} = 0. \tag{2.10}$$

Thus, the real roots of the characteristic equation are eigenvalues of f. Note that (2.10) was obtained without using determinants.

The coefficients of the characteristic polynomial can be expressed in terms of the traces $\mathrm{Tr}\,(f^r) = \partial \cdot f^r$ of the r-fold transformation f^r by using the recursion formula

$$\partial_{(s)} * f_{(s)} = \frac{1}{s} \sum_{r=1}^{s} (-1)^{r+1} (\partial_{(s-r)} * f_{(s-r)})\, \partial \cdot f^r. \tag{2.11}$$

Our proof of (2.11) is quite similar to our proof of the Cayley–Hamilton theorem; thus, using (1-1.40) and (1-1.42),

$$\begin{aligned}
\partial_{(s)} \cdot f_{(s)} &= \frac{1}{s} (\partial_{(s-1)} \wedge \partial_1) \cdot (f_1 \wedge f_{(s-1)}) \\
&= \frac{1}{s} \{\partial_1 \cdot f_1\, \partial_{(s-1)} \cdot f_{(s-1)} - [(\partial_{(s-1)} \cdot \dot{f}_1) \wedge \dot{\partial}_1] \cdot f_{(s-1)}\} \\
&= \frac{1}{s} \{\partial \cdot f\, \partial_{(s-1)} \cdot f_{(s-1)} - (\partial_{(s-2)} \wedge \partial_1) \cdot (f_1^2 \wedge f_{(s-2)})\} = \cdots \\
&= \frac{1}{s} \{\partial \cdot f\, \partial_{(s-1)} \cdot f_{(s-1)} - \partial \cdot f^2\, \partial_{(s-2)} \cdot f_{(s-2)} + \cdots + (-1)^{s+1} \partial \cdot f^s\}.
\end{aligned}$$

Now let $\lambda_1, \ldots, \lambda_n$ be the roots of the characteristic polynomial (2.10); we will assume here that the λ_k's are real or formally complex numbers, although in Section 3-5 and 3-9 we will see that geometric algebra enables us to identify the complex numbers with multivectors and so supply them with a geometric interpretation. The characteristic polynomial can be written

$$C_f(\lambda) = \prod_{k=1}^{n} (\lambda - \lambda_k) = \lambda^n - (\lambda_1 + \cdots + \lambda_n)\lambda^{n-1} + \cdots + (-1)^n \lambda_1 \ldots \lambda_n. \tag{2.12}$$

Comparing coefficients of λ^k in (2.10) and (2.12) gives us the relationships

$$\partial_{(k)} \cdot f_{(k)} = \sum_{1 \leqslant i_1 < \ldots < i_k \leqslant n} \lambda_{i_1} \ldots \lambda_{i_k} \tag{2.13}$$

between the characteristic scalars $\partial_{(k)} \cdot f_{(k)}$ and the symmetric products of the eigenvalues of f. Note also that (2.13) together with (2.11) implies

$$\partial \cdot f^k = \lambda_1^k + \cdots + \lambda_n^k. \tag{2.14}$$

We can distinguish two general approaches to the study of a linear transformation f. The *many-point* approach studies the effect of f on relations among several points, while the *many-fold* approach studies the effect of repeated applications of f. Clearly $\partial_{(r)} f_{(r)}$ characterizes a single mapping of r points, while f^r describes the r-fold mapping of a single point. To date, the many-fold approach to linear transformations has been systematically developed by many authors, while the many-point approach has hardly been recognized, probably because it has lacked an adequate mathematical formulation. But we have already seen that Geometric Calculus greatly facilitates the characterization of many-point properties. So now the many-point approach can be systematically developed.

Since they deal with one and the same subject, the many-point and the many-fold approaches are certainly interrelated and to some degree equivalent. This is shown expressly by Eqns. (2.9) and (2.11). Nevertheless, for a given problem one approach may be much simpler than the other. Thus, though Eqn. (2.11) shows that the coefficients of the characteristic polynomial can be completely expressed in terms of the many-fold traces Tr (f^r), the simpler many-point expression $\partial_{(r)} * f_{(r)}$ certainly suggests that the characteristic polynomial is best studied by the many-point approach.

The many-fold approach deals with the scalar part of $\partial_{(r)} f_{(r)}$ as a coefficient of the characteristic polynomial, however, it has no straightforward way of comprehending the information about f contained in the nonscalar parts of $\partial_{(r)} f_{(r)}$. Therefore, we expect that further study of $\partial_{(r)} f_{(r)}$ will give us new insight into the theory of linear transformations.

3-3. Eigenblades and Invariant Spaces

This short section is intended to show how useful the differential and adjoint outermorphisms can be in the study of invariant spaces of a linear transformation. We particularly wish to stress the value of the general relations (1.14) in this approach. Although other authors have found equivalent relations ([Bo], [Wh]), they have not exploited them or pointed out their usefulness. We are confident that much more can be achieved along the lines set down here. Of course, this simple approach is possible only because the essentials of Geometric Calculus have been developed before linear transformations are studied in detail.

In accordance with common usage we say that a vector a is an *eigenvector* of the linear transformation f with *eigenvalue* α if

$$f(a) = \alpha a. \tag{3.1}$$

Except in Section 3-5, where a different possibility is considered, we always take eigenvalues to be scalar (real) quantities.

The differential outermorphism $\underline{f}$ defined in Section 3-1 provides us with a straightforward generalization of the notion of eigenvector. Let $A = \langle A \rangle_k$ be a *k-blade*. We say that A is an *eigenblade* of f with *eigenvalue* α if

$$\underline{f}(A) = \alpha A. \tag{3.2}$$

This reduces to (3.1) if A is a vector (or 1-blade).

An eigenblade with nonvanishing eigenvalue determines a subspace of vectors which is *invariant* under f. The proof is easy. Recall from Section 1-2 that $\mathscr{A} = \mathscr{G}^1(A)$ is the space of all vectors b for which $A \wedge b = 0$. By (1.7) and (3.2),

$$\underline{f}(A \wedge b) = \underline{f}(A) \wedge \underline{f}(b) = \alpha A \wedge \underline{f}(b).$$

So if $\alpha \neq 0$, then $A \wedge b = 0$ implies $A \wedge \underline{f}(b) = 0$, which says that $f(b) = \underline{f}(b)$ is in $\mathscr{A}$. Using the same argument, it is easy to prove that the sum and intersection of invariant spaces (related to blades by (1-2.23) and (1-2.35)) are themselves invariant spaces.

The adjoint outermorphism $\overline{f}$ introduced in Section 3-1 provides a generalization of eigenvector complementary to (3.2). We say that an r-blade B is a *right eigenblade* of f with *eigenvalue* β if

$$\overline{f}(B) = \beta B. \tag{3.3}$$

To distinguish (3.2) from (3.3), it is sometimes convenient to call the eigenblade A a *left eigenblade*.

If a left eigenblade A is also a right eigenblade, we say that A is a *proper blade*, and call the corresponding eigenvalue a *proper value*.

Suppose that the linear transformation f is defined on the vector space $\mathscr{A} = \mathscr{G}^1(I)$ with unit pseudoscalar I. With (1.17), it is trivial to show that if I is an eigenblade of f, it is also a proper blade of f with proper value equal to the determinant of the transformation, that is,

$$\underline{f}(I) = I \det(f) = \overline{f}(I). \tag{3.4}$$

It follows that f maps $\mathscr{A}$ into $\mathscr{A}$. For the sake of brevity, we suppose that (3.4) holds throughout this section.

We say that an eigenblade of f is *irreducible* under f if it cannot be factored into an outer product of blades which are also eigenblades of f.

Consider the factorization

$$I = A_1 \wedge A_2 \wedge \ldots \wedge A_m \tag{3.5}$$

of the pseudoscalar I into irreducible eigenblades of f: $A_1, A_2, \ldots, A_m$. By (1-2.25), the factorization (3.5) corresponds exactly to the usual decomposition of a vector space $\mathscr{A}$ into a direct sum of invariant subspaces $\mathscr{A}_1 = \mathscr{G}^1(A_1), \ldots, \mathscr{A}_m = \mathscr{G}^1(A_m)$. However, eigenblades are considerably more convenient to deal with than invariant subspaces because of their richer algebraic properties, which, for instance, make possible the eigenblade Eqn. (3.2). To prove our point, we show below how some important theorems pertaining to invariant subspaces can be simply stated and proved, using eigenblades.

Given (3.5), we can prove that the characteristic polynomial $C_{f_k}(\lambda)$ of f restricted to $\mathscr{A}_k = \mathscr{G}^1(A_k)$ divides the characteristic polynomial of f on $\mathscr{A} = \mathscr{G}^1(I)$. To do this, we introduce the auxiliary function $F(x) = f(x) - \lambda x$. Comparing the expression (2.8b) for the characteristic polynomial of f with the expression (3.4) for the differential of the pseudoscalar, we see that

$$\underline{F}(I) = C_f(\lambda) I. \tag{3.6}$$

It is slightly more difficult to prove

$$\underline{F}(A_k) = C_{f_k}(\lambda) A_k, \tag{3.7}$$

i.e., that A_k is an eigenblade of F with the eigenvalue $C_{f_k}(\lambda)$. The key step in the proof is to show that

$$A_k \cdot \dot{\partial}_x F(x) = A_k \cdot \partial_x F[P_k(x)] = A_k \cdot \partial_x P_k(F[P_k(x)])$$

where $P_k(x) \equiv x \cdot A_k A_k^{-1}$, so $F[P_k(x)]$ is the 'restriction of F to $\mathscr{A}_k = \mathscr{G}^1(A_k)$'; after that, the proof of (3.7) is essentially the same as the proof of (3.6). Having established (3.6) and (3.7), the relation $\underline{F}(I) = \underline{F}(A_1) \wedge \ldots \wedge \underline{F}(A_m)$ obtained from (1.7) immediately gives the desired factorization of the characteristic polynomial:

$$C_f(\lambda) = C_{f_1}(\lambda) C_{f_2}(\lambda) \ldots C_{f_m}(\lambda). \tag{3.8}$$

If $\underline{f}(A) = \alpha A$ and $\overline{f}(B) = \beta B$, then the general Eqns. (1.14) imply immediately that

$$\beta \underline{f}(A \cdot B) = \alpha A \cdot B \quad \text{if grade } A \geqslant \text{grade } B, \tag{3.9a}$$

$$\alpha \overline{f}(A \cdot B) = \beta A \cdot B \quad \text{if grade } B \geqslant \text{grade } A. \tag{3.9b}$$

A number of important results follow almost trivially from these equations. For instance, if A is taken to be the pseudoscalar I, then (3.9a) is seen to imply that for a nonsingular transformation the dual of a right eigenblade B is a left eigenblade $B \cdot I = BI$ with eigenvalue α/β. Applying this to the factorization (3.5), we find the 'dual factorization'

$$I = A^1 \wedge A^2 \wedge \ldots \wedge A^m, \tag{3.10}$$

where the A^k are right eigenblades defined by

$$A^k = (-1)^{\epsilon_k} A_1 \wedge \ldots \wedge \check{A}_k \wedge \ldots \wedge A_m I^\dagger \tag{3.11}$$

with $\epsilon_k = (\text{grade } A_k)(\text{grade } A_1 + \cdots + \text{grade } A_{k-1})$. Obviously,

$$A_j \cdot A^k = \delta_k^j. \tag{3.12}$$

Further, the eigenvalue α_k of each blade A_k is equal to that of the 'dual blade' A^k, that is

$$\alpha_k = (A_k)^{-1}\underline{f}(A_k) = (A^k)^{-1}\overline{f}(A^k). \tag{3.13}$$

Another easy but important consequence of (3.9) is the fact that the projection $b_{\|} \equiv b \cdot AA^{-1}$ of any eigenvector b into the 'eigenspace' $\mathcal{G}^1(A)$ of a *proper* blade A with nonzero proper value is an eigenvector with the same eigenvalue as b. Thus, if $\underline{f}(A) = \alpha A = \overline{f}(A)$ and $\underline{f}(b) = \beta b$, then by (3.9),

$$f(b_{\|}) = f[(b \cdot A) \cdot A^{-1}] = \frac{\beta}{\alpha} f[\overline{f}(b \cdot A) \cdot A^{-1}]$$

$$= \frac{\beta}{\alpha}(b \cdot A) \cdot \underline{f}(A^{-1}) = \beta(b \cdot A) \cdot A^{-1} = \beta b_{\|}. \tag{3.14}$$

This property is useful in the classification of degenerate eigenvalues.

While the present discussion is certainly incomplete, we hope it is sufficient to show the utility of the eigenblade concept in connection with Geometric Algebra.

3-4. Symmetric and Skew-symmetric Transformations

Any linear transformation $f = f(x) = \underline{f}(x)$ can be uniquely expressed as a sum of symmetric and skewsymmetric transformations; more specifically,

$$f(x) = f_+(x) + f_-(x), \tag{4.1a}$$

$$f_+(x) \equiv \tfrac{1}{2}(\underline{f}(x) + \overline{f}(x)) = \partial_x(\tfrac{1}{2}x \cdot f(x)), \tag{4.1b}$$

$$f_-(x) \equiv \tfrac{1}{2}(\underline{f}(x) - \overline{f}(x)) = \tfrac{1}{2}x \cdot (\partial \wedge f). \tag{4.1c}$$

These equations follow easily from the definitions of differential and adjoint:

$$\underline{f}(x) \equiv x \cdot \partial_y f(y) = \partial_y y \cdot f(x) = f(x),$$

$$\overline{f}(x) \equiv \partial_y x \cdot f(y).$$

For example, Eqn. (4.1c) is a consequence of the identity (2-1.42).

A linear transformation $f = f(x)$ is said to be *symmetric* if it satisfies the following set of equivalent conditions

$$\underline{f} = \overline{f} \quad \text{or} \quad f_-(x) = 0 \quad \text{for all } x, \tag{4.2a}$$

$$x \cdot f(y) = y \cdot f(x) \quad \text{for all } x \text{ and } y, \tag{4.2b}$$

$$\partial \wedge f = 0. \tag{4.2c}$$

Obviously, (4.2a) follows from (4.2c) by (4.1c). Conversely, (4.2c) can be obtained from (4.2a) by differentiating (4.1c). The equivalence to (4.2b) can be established by 'dotting' (4.1c) by y, and by differentiating (4.2b).

By well-known methods it can be shown that a symmetric linear transformation $f = f(x)$ can be given the unique 'spectral decomposition'

$$f(x) = \alpha_1 P_1(x) + \alpha_2 P_2(x) + \cdots + \alpha_m P_m(x), \tag{4.3a}$$

where

$$\alpha_1 < \alpha_2 < \ldots < \alpha_m, \tag{4.3b}$$

and $P_k = P_k(x)$ is the projection into the space of proper vectors of f with proper value α_k. As explained in Section 1-2, the geometric product can be used to give the projections the explicit form

$$P_k(x) = x \cdot A_k^\dagger A_k = \tfrac{1}{2}(x - (-1)^{\epsilon_k} A_k^\dagger x A_k), \tag{4.3c}$$

where ϵ_k = grade A_k is equal to the 'degeneracy' of the eigenvalue α_k. The 'completeness' of the spectral form can be expressed by the equation

$$I = A_1 A_2 \ldots A_m = A_1 \wedge A_2 \wedge \ldots \wedge A_m; \tag{4.3d}$$

this is a factorization of the unit pseudoscalar I (which determines the domain of f) into a product of orthogonal unit proper blades $A_1, \ldots, A_m$. Each proper blade A_k can in turn be expressed as a product of ϵ_k orthonormal proper vectors a_{jk}; that is,

$$A_k = a_{1k} a_{2k} \ldots a_{\epsilon_k k}. \tag{4.4a}$$

By (1.7),

$$f(a_{jk}) = \bar{f}(a_{jk}) = \alpha_k a_{jk} \tag{4.4b}$$

implies

$$\underline{f}(A_k) = \bar{f}(A_k) = \alpha_k^{\epsilon_k} A_k. \tag{4.4c}$$

On substituting (4.3c) into (4.3a), f can be written in the alternative canonical form

$$f(x) = \tfrac{1}{2}\alpha_0 x - \frac{1}{2} \sum_{k=1}^{m} \alpha_k (-1)^{\epsilon_k} A_k^\dagger x A_k, \tag{4.5a}$$

where

$$\alpha_0 = \alpha_1 + \alpha_2 + \cdots + \alpha_m. \tag{4.5b}$$

If a linear transformation $C_A = C_A(x)$ is orthogonal as well as symmetric, then all its proper values must have unit magnitude, and (4.5a) reduces to the simple canonical form

$$C_A(x) \equiv (-1)^\epsilon A^\dagger x A, \tag{4.6}$$

where A is a unit blade of grade ϵ, and A-space is the space of all proper vectors with proper value -1. We call (4.6) a *conjugation in A-space*, because it generalizes 'complex conjugation' (a reflection in the 'imaginary axis' of the complex plane). The decomposition (4.3) expresses f as a *superposition of projections*, while (4.5) alternatively expresses f as a *superposition of conjugations*.

A symmetric linear transformation is said to be *positive* if all its proper values are positive. By choosing A in (4.6) to be the product of all A_k in (4.3) associated with negative proper values, one proves immediately that any symmetric linear transformation can be written as a commuting composite of a positive symmetric transformation $S = S(x)$ and a conjugation $C_A = C_A(x)$; that is

$$f(x) = S(C_A(x)) = C_A(S(x)) = (-1)^{\epsilon} A^{\dagger} S(x) A. \tag{4.7}$$

A linear transformation $f = f(x)$ said to be *skew* (or skew-symmetric) if it satisfies either of the equivalent conditions

$$f(x) = \underline{f}(x) = -\overline{f}(x) \quad \text{or} \quad 2f_{+}(x) = \partial x \cdot f(x) = 0, \tag{4.8a}$$

and

$$y \cdot f(x) = -x \cdot f(y) \quad \text{or} \quad x \cdot f(x) = 0 \quad \text{for all } x, y. \tag{4.8b}$$

As proved in Section 2-3, the skew bilinear form (4.8b) can be written

$$y \cdot f(x) = (y \wedge x) \cdot F = y \cdot (x \cdot F),$$

where F is a *unique* bivector. Since this holds for an arbitrary vector y it follows that f has the *canonical form*

$$f(x) = x \cdot F = x \times F, \tag{4.9}$$

where $x \times F$ is the commutator product defined in Section 1-1. According to (2-1.38), the derivative of (4.9) is

$$\partial f = \partial \wedge f = 2F. \tag{4.10}$$

Thus a skew linear transformation is completely determined by its curl.

Instead of characterizing a skew transformation directly, it is easier to characterize its curl. Every bivector F can be expressed as a sum of distinct commuting blades, that is,

$$F = F_1 + F_2 + \cdots + F_m, \tag{4.11a}$$

where

$$F_j F_k = F_k F_j = F_k \wedge F_j \quad \text{for } k \neq j, \tag{4.11b}$$

$$F_k^2 = -|F_k|^2 < 0. \tag{4.11c}$$

All properties of a bivector F are easily ascertained after it has been reduced to the '*orthogonal form*' (4.11a). However, as we shall prove later, this orthogonal decomposition is not unique if distinct blades have the same magnitude.

Frequently a bivector is given as a sum of nonorthogonal blades. For instance, if the skew transformations $f(a_k) = a_k \cdot F$ of basis vectors a_k are given, then

$$F = \tfrac{1}{2}\partial \wedge f = \tfrac{1}{2}a^k \wedge f(a_k) = \tfrac{1}{2}f_{jk}a^k \wedge a^j, \tag{4.12}$$

where $f_{jk} = a_j \cdot f(a_k)$ is the matrix of the transformation. The chief problem is to compute the F_k of (4.11) from the expression for F given by (4.12).

To solve for the F_k it is convenient to introduce multivectors C_k of grade $2k$ by the equation

$$C_k \equiv \frac{1}{k!}\langle F^k\rangle_{2k} = \sum_{i_1 < \ldots < i_k} F_{i_1}F_{i_2}\ldots F_{i_k}, \tag{4.13}$$

where $k = 1, 2, \ldots, m$. The right side of (4.13) is obtained by substituting (4.11a) into the left. The left side of (4.13) can be evaluated from any given expression for F, such as (4.12). Then (4.13) can be regarded as a set of m equations to be solved for each F_k in terms of given C_k's. To begin with, the squares of the F_k can be found as the roots of the mth order polynomial

$$\sum_{k=0}^{m} \langle C_k^2\rangle(-\lambda)^{m-k}, \tag{4.14}$$

as is readily verified by using (4.13) to express the coefficients $\langle C_k^2\rangle$ in terms of the F_k and comparing with the factored form $(F_1^2 - \lambda)(F_2^2 - \lambda)\ldots(F_m^2 - \lambda)$ of the polynomial. The roots $\lambda_k \equiv F_k^2$ of (4.14) having been determined, Eqn. (4.13) can be replaced by a set of m linear bivector equations for the m unknowns F_k; thus, from (4.13)

$$C_{k-1} \cdot C_k = \sum_i F_i \sum_{i_1 < \ldots < i_{k-1}}{}' \lambda_{i_1}\lambda_{i_2}\ldots\lambda_{i_{k-1}}, \tag{4.15}$$

where $k = 1, 2, \ldots, m$, and the prime on Σ' is to denote that the ith term is to be deleted from the sum. Equation (4.15) can be solved by standard procedure if the λ_k are all distinct. If the λ_k are not all distinct, the m equations are not linearly independent and additional properties of F are needed to determine the F_k.

As an example, we give the solution for $m = 2$, the simplest nontrivial case.

$$\lambda_i = F_i^2 = \tfrac{1}{2}(-|F|^2 \pm (|F|^4 - |F \wedge F|^2)^{1/2}). \tag{4.16}$$

Since $C_1 \cdot C_2 = C_1C_2$ for this case, and if $\lambda_2 \neq \lambda_1$,

$$F_1 = \frac{C_1C_2 - \lambda_1C_1}{\lambda_2 - \lambda_1} = \frac{F}{1 + \tfrac{1}{2}\lambda_1^{-1}F \wedge F}. \tag{4.17}$$

The corresponding expression for the blade F_2 is obtained by interchanging subscripts in (4.17). It is instructive to check (4.16) and (4.17) by inserting $F = F_1 + F_2$.

To demonstrate the utility of these results, let us use them to find an orthogonal form for the bivector

$$F = a(b + \mu d) + cd,$$

where μ is a scalar and a, b, c, d are mutually orthogonal vectors with $a^2 = c^2 = d^2 = 1$. First we calculate

$$F \wedge F = \beta I = 2abcd,$$

where $\beta = |F \wedge F|$ and I is a unit 4-vector ($I^2 = 1$). Then we calculate

$$\beta^2 = (F \wedge F)^2 = 4a^2b^2c^2d^2 = 4b^2$$

and

$$\alpha \equiv |F|^2 = \mu^2 + b^2 + 1.$$

On substituting into (4.16) and (4.17), we get the desired blades in terms of the given F and the parameters μ^2 and b^2. To work out the details for a numerically simple example, choose $\alpha^2 - \beta^2 = 4$ and $\alpha = 3$. Then $\beta^2 = 4b^2 = 5$ and $\mu^2 = \frac{3}{4}$. So (4.16) gives us $2\lambda_1 = -1$ and $2\lambda_2 = -2$. Using this in (4.17), we obtain

$$F_1 = \frac{F}{1 - \beta I} = \frac{F}{(1 - \beta I)} \frac{(1 + \beta I)}{(1 + \beta I)} = \frac{F(1 + \beta I)}{1 - \beta^2}.$$

But

$$\beta IF = 2abcd\,(ab + \mu ad + cd) = -2(b^2cd + \mu bc + ab).$$

So

$$F_1 = -\frac{F}{4}(1 + \beta I) = \frac{1}{4}\left(ab + \frac{3}{2}cd + \sqrt{3}\,bc - \frac{\sqrt{3}}{2}ad\right).$$

This factors into the form

$$F_1 = \tfrac{1}{8}(a - \sqrt{3}\,c)(2b - \sqrt{3}\,d),$$

showing explicitly that F_1 is a blade, as required. For the other blade, we find

$$F_2 = \tfrac{1}{4}F(5 + \beta I) = \tfrac{1}{8}(\sqrt{3}\,a + c)(5d + 2\sqrt{3}\,b).$$

Of course, the problem of 'orthogonalizing' a bivector is equivalent to the problem of putting the corresponding skew transformation in canonical form. The polynomial (4.14) is just the characteristic polynomial (2.8b) for $f(x) = x \cdot F$, and comparing (4.13), (4.14) and (2.8b) we conclude that

$$\begin{aligned} \partial_{(2k)} \cdot f_{(2k)} &= (-1)^k \quad \langle C_k^2 \rangle = (-1)^k \quad \left\langle \left[\frac{1}{k!}\langle F^k \rangle_{2k}\right]^2 \right\rangle, \\ \partial_{(2k+1)} \cdot f_{(2k+1)} &= 0, \end{aligned} \tag{4.18}$$

a result which is more difficult to obtain by carrying out the differentiation directly. The $\lambda_k = F_k^2$ are proper values of the symmetric transformation f^2; for if a_k is a vector satisfying $a_k \wedge F_k = 0$, then by (4.11b) $a_k \cdot F_j = 0$ for $j \neq k$, so by using (4.11a) in (4.9) one shows easily that

$$f^2(a_k) = (a_k \cdot F) \cdot F = F_k^2 a_k. \tag{4.19}$$

Beginning with this observation, it is a simple matter to prove that every blade F_k can be written

$$F_k = |F_k| a_k b_k, \tag{4.20}$$

where $\{a_1, \ldots, a_m, b_1, \ldots, b_m\}$ is a set of orthonormal proper vectors of f^2 with nonvanishing proper values.

We now supply the promised proof that *the decomposition of a bivector into orthogonal blades is not unique if distinct blades have the same magnitude* with an argument due to Bernard Jancewicz (personal communication). In accordance with (4.20), for the case $m = 2$, the orthogonal decomposition of a bivector F can be put in the form

$$F = \alpha a_1 b_1 + \beta a_2 b_2,$$

where $\alpha = |F_1|$ and $\beta = |F_2|$. Our proof has its clearest and most elegant form if we use a result proved in the next section, namely, that any rotation of the proper vectors a_k and b_k can be written in the form

$$a_k' = R^\dagger a_k R \text{ and } b_k' = R^\dagger b_k R.$$

Consider a rotation specified by

$$R = R_1 R_2 = e^{a_1 a_2 \theta/2}\, e^{b_1 b_2 \phi/2} = \exp\left[\tfrac{1}{2}(a_1 a_2 \theta + b_1 b_2 \phi)\right].$$

Then,

$$\alpha a_1' b_1' + \beta a_2' b_2' = R^\dagger F R = F R^2 = F R_1^2 R_2^2.$$

This provides us with a different decomposition of F into commuting blades $\alpha a_1' b_1'$ and $\beta a_2' b_2'$ if

$$R^\dagger F R = F R^2 = F,$$

in other words, if F is an *eigenbivector* of the rotation determined by R. More explicitly, this condition can be written

$$\begin{aligned}\alpha a_1 b_1 + \beta a_2 b_2 &= (\alpha a_1 b_1 + \beta a_2 b_2)\, e^{a_1 a_2 \theta}\, e^{b_1 b_2 \phi} \\ &= (\alpha a_1 a_2 + \beta a_2 b_2)(\cos\theta + a_1 a_2 \sin\theta)(\cos\phi + b_1 b_2 \sin\phi).\end{aligned}$$

Expanding this last line we find that the condition can be satisfied for *arbitrary* $\theta = \phi$ if and only if $\alpha = \beta$. Thus, we have proved that the decomposition of a bivector into orthogonal blades is not unique if any two of the blades are *degenerate* in the sense that they have the same magnitude. This nonuniqueness for degenerate blades is similar to the degeneracy of distinct eigenvectors with the same eigenvalues.

The differential transformation of a bivector induced by $f(a) = a \cdot F$ is easily found from

$$\underline{f}(a \wedge b) = f(a) \wedge f(b) = (a \cdot F) \wedge (b \cdot F)$$

and

$$(a \wedge b) \cdot (F \wedge F) = a \cdot [2(b \cdot F) \wedge F] = 2(a \wedge b) \cdot FF + 2(a \cdot F) \wedge (b \cdot F).$$

Thus,

$$\underline{f}(a \wedge b) = \tfrac{1}{2}(a \wedge b) \cdot (F \wedge F) - (a \wedge b) \cdot FF. \tag{4.21}$$

Similarly the specific form of the differential $\underline{f}(a \wedge b \wedge c)$ can be ascertained from

$$\begin{aligned}(a \wedge b \wedge c) \cdot (F \wedge F \wedge F) &= (a \wedge b) \cdot [3F \wedge F \wedge (c \cdot F)] \\ &= 3a \cdot [2F \wedge (b \cdot F) \wedge (c \cdot F) + F \wedge F (b \wedge c) \cdot F] \\ &= 3!\, f(a) \wedge f(b) \wedge f(c) + 3F \wedge [(a \wedge b \wedge c) \cdot (F \wedge F)],\end{aligned}$$

the last term being obtained by noting that

$$\begin{aligned}(a \wedge b \wedge c) \cdot (F \wedge F) &= (a \wedge b) \cdot [2F \wedge (c \cdot F)] \\ &= 2a \cdot [(b \cdot F) \wedge (c \cdot F) + F(b \wedge c) \cdot F] \\ &= 2[(a \wedge b) \cdot Fc \cdot F + (c \wedge a) \cdot Fb \cdot F + (b \wedge c) \cdot Fa \cdot F].\end{aligned}$$

Continuing in this manner, the differential and adjoint transformations of an arbitrary r-vector A_r are found to be

$$\begin{aligned}\underline{f}(A_r) = (-1)^r \overline{f}(A_r) &= \sum_{k=p}^{r} (-1)^{r-k} (r-k)!\, C_{r-k} \wedge (A_r \cdot C_k) \\ &= \frac{1}{r!} A_r \cdot \langle F^r \rangle_{2r} + \frac{1}{(r-1)!} F \wedge (A_r \cdot \langle F^{r-1} \rangle_{2r-2}) + \\ &\quad + \ldots + \frac{(-1)^{r-p}}{p!} \langle F^{r-p} \rangle_{2(r-p)} \wedge (A_r \cdot \langle F^p \rangle_{2p}),\end{aligned} \tag{4.22}$$

where the C_k are defined by (4.13) for $k > 0$, $C_0 \equiv 1$, and $p = r/2$ if r is even, while $p = \frac{1}{2}(r+1)$ if r is odd.

The differential transformation (4.21) is a symmetric bivector function of a bivector variable, and the problem of solving the eigenbivector equation

$$\underline{f}(F_k) = |F_k|^2 F_k \tag{4.23}$$

is just the problem of orthogonalizing a bivector. Thus skew transformations of vectors correspond to symmetric transformations of bivectors.

To evaluate characteristic multivectors and carry out other computations, it is convenient to have formulas for derivatives of skew transformations. We list some without proof for reference purposes.

$$\partial f \wedge A_r = F \times A_r + 2F \wedge A_r, \tag{4.24}$$

$$(A_r \wedge \partial)f = A_r \times F + 2F \wedge A_r. \tag{4.25}$$

If $r \geqslant 2$,

$$\partial f \cdot A_r = F \times A_r + 2F \cdot A_r, \tag{4.26}$$

$$A_r \cdot \partial f = A_r \times F + 2F \cdot A_r, \tag{4.27}$$

$$\partial A_r f = 2(-1)^r (A_r \wedge F - A_r \cdot F), \tag{4.28}$$

$$f^{k+1}(x) = f^k(x) \cdot F = -F \cdot (x \wedge \partial_y) f^k(y), \tag{4.29}$$

$$\partial \cdot f^{k+1} = F \cdot (\partial \wedge f^k), \tag{4.30}$$

$$\partial \wedge f^{k+1} = \partial \wedge (f^k \cdot F) = (F \cdot \partial) \wedge f^k, \tag{4.31}$$

$$\partial f^{k+1} = F \cdot \partial f^k, \tag{4.32}$$

$$\partial f^{2k} = \partial \cdot f^{2k}, \tag{4.33}$$

$$\partial f^{2k+1} = \partial \wedge f^{2k+1}. \tag{4.34}$$

The above formulas can be used to evaluate derivatives of powers of f by iteration. The same end can be achieved more directly by eliminating inner products in favor of geometric products to get f^k in the form

$$f^k(x) = \frac{1}{2^k} \sum_{r=0}^{k} (-1)^r \binom{k}{r} F^r x F^{k-r}. \tag{4.35}$$

From (2-1.40),

$$\partial_x F^r x = nF^r - 4 \sum_{s=0}^{s_m} s \langle F^r \rangle_{2s}, \tag{4.36}$$

where $s_m = r$ if r even, or $s_m = r - 1$ if r odd. Applying (4.36) to (4.35) we get, for $k > 0$,

$$\partial f^k = \frac{-4}{2^k} \sum_{r=0}^{k} (-1)^r \binom{k}{r} \sum_{s=1}^{s_m} s \langle F^r \rangle_{2s} F^{k-r}, \tag{4.37}$$

the terms nF^r from (4.36) having been cancelled by virtue of the identity

$$\sum_{r=0}^{k} \binom{k}{r} (-1)^r = 0 \quad \text{for } k > 0. \tag{4.38}$$

Many other terms in (4.37) cancel, because ∂f^k is a bivector if k is odd and a scalar if k is even. Thus, though (4.37) is explicit, it is unwieldy.

3.5. Normal and Orthogonal Transformations

The skew symmetric part f_- of a linear transformation f is quite similar to the imaginary part of a complex number, since it always produces a vector (possibly zero) orthogonal to any it operates on. If the symmetric part f_+ of f commutes with the skew part, then the decomposition (4.1) of f into f_+ and f_- corresponds to the decomposition of a complex number into real and imaginary parts. We wish to show how geometric algebra enhances this correspondence and so provides insight into the role of complex numbers in linear algebra.

A linear transformation $f = f_+ + f_-$ is said to be *normal* if it satisfies the following equivalent conditions.

$$f_+(f_-(x)) = f_-(f_+(x)), \tag{5.1a}$$

$$\overline{f}\underline{f}(A) = \underline{f}\overline{f}(A), \tag{5.1b}$$

$$\langle \underline{f}(A)\underline{f}(B)\rangle = \langle \overline{f}(A)\overline{f}(B)\rangle \tag{5.1c}$$

for any vector x and multivectors A, B. Actually it suffices to assume that (5.1) holds for vectors; the general statement of (5.1) for arbitrary multivectors can then be proved with the help of (1.7). Equivalence of (5.1b) and (5.1c) can be established by using (2-2.19b) to get the sequence of equations

$$\langle \underline{f}(A)\underline{f}(B)\rangle = \langle A\overline{f}\underline{f}(B)\rangle = \langle A\underline{f}\overline{f}(B)\rangle = \langle \overline{f}(A)\overline{f}(B)\rangle.$$

Proof that (5.1a) is equivalent to (5.1b) requires only the definitions of f_+ and f_-.

The properties of normal transformations are easily found from the known properties of symmetric and skew transformations. As shown below, the commutivity (5.1a) of f_+ and f_- implies that every eigenblade of f_- is an eigenblade of f_+. But we know that the irreducible eigenblades of f_+ are proper vectors, while, outside its nullspace, the irreducible blades of f_- are proper bivectors. Hence the irreducible eigenblades of any normal transformation are proper blades of grade 1 and 2.

According to the results of the preceding section, for *any* linear transformation f, we can write

$$\partial \wedge f = \partial \wedge f_- = \sum_{k=1}^{n} \beta_k i_k \equiv 2F, \tag{5.2a}$$

where the i_k are bivectors with the properties

$$i_k^2 = -1, \quad i_j i_k = i_k i_j, \tag{5.2b}$$

and

$$\underline{f}_-(i_k) = \beta_k^2 i_k. \tag{5.2c}$$

In the sum (5.2a) at least one of the β_k must vanish if n is odd, and the nonvanishing terms are identical in pairs but otherwise distinct.

Operating on (5.2c) with the differential of f_+ and using (5.1a) we find immediately that $\underline{f}_+(i_k)$ is an eigenblade of $\underline{f}_-$ with eigenvalue β_k^2, but, if the β_k^2 are distinct, only bivectors proportional to i_k have this property, so we can write

$$\underline{f}_+(i_k) = \alpha_k^2 i_k, \tag{5.3}$$

where α_k^2 is some scalar. Thus, the i_k-plane is invariant under f_-, f_+ and $f = f_+ + f_-$. Since every symmetric transformation has a complete set of eigenvectors, f_+ must have an eigenvector a_k in the i_k plane. Again, using (5.1a), we find that $f_-(a_k)$ is also an eigenvector of f_+ with the same eigenvalue as a_k. It follows that every vector in the i_k-plane is an eigenvector of f_+ with, to be consistent with (5.3) and (4.5a), eigenvalue α_k.

Thus, for any normal transformation f, there exists an orthonormal frame of vectors a_k which are eigenvectors of f_+,

$$f_+(a_k) = \alpha_k a_k, \tag{5.4a}$$

and satisfying, in accordance with (5.2),

$$f_-(a_k) = a_k i_k \beta_k = -i_k a_k \beta_k. \tag{5.4b}$$

Combining (5.4a) and (5.4b), one obtains

$$f(a_k) = a_k \lambda_k = \lambda_k^\dagger a_k, \tag{5.5a}$$

$$\bar{f}(a_k) = \lambda_k a_k = a_k \lambda_k^\dagger, \tag{5.5b}$$

where

$$\lambda_k = \alpha_k + i_k \beta_k = \rho_k\, \mathrm{e}^{i_k |\theta_k|} = \rho_k\, \mathrm{e}^{\theta_k}, \tag{5.6a}$$

$$\rho_k = |\lambda_k| > 0, \quad 0 \leqslant |\theta_k| \leqslant \pi, \tag{5.6b}$$

$$\theta_k = |\theta_k| i_k, \tag{5.6c}$$

and

$$\lambda_j \lambda_k = \lambda_k \lambda_j. \tag{5.7}$$

The λ_k with $\beta_k \neq 0$ are identical in pairs but otherwise distinct. The exponential in (5.6a) can be defined by a power series.

$$\mathrm{e}^{\theta_k} = \sum_{r=0}^{\infty} \frac{\theta_k^r}{r!} = \cosh \theta_k + \sinh \theta_k = \cos |\theta_k| + i_k \sin |\theta_k|. \tag{5.8}$$

And note that, for $|\theta_k| = 0$ or π (i.e. for $\alpha_k = \rho_k$ or $\alpha_k = -\rho_k$) the corresponding i_k is not (and need not be) defined. The above equations lead to the following 'spectral decomposition' of a normal transformation:

$$f(x) = \sum_{k=1}^{n} P_k(x)\lambda_k = \sum_k P_k(x)\rho_k \, e^{\theta_k}, \tag{5.9a}$$

where

$$P_k(x) = x \cdot a_k a_k. \tag{5.9b}$$

Equations (5.5) and (5.6) suggest that λ_k be regarded as a 'complex eigenvalue' associated with eigenvector a_k. Indeed, the λ_k are roots of the characteristic equation for f, differing from the usual complex eigenvalues of a normal transformation only in using the bivectors i_k as roots of the equation $i^2 = -1$ in place of the usual $\sqrt{-1}$ which is supposed to be a scalar. But this small difference alters the notion of eigenvalue considerably. Thus, for $\beta_k \neq 0$, Eqn. (5.5a) implies that under f the i_k-plane undergoes a homothetic transformation, every vector in the plane being dilated by the factor ρ_k and rotated 'through' an oriented angle $\theta_k = |\theta_k| i_k$. Equation (5.5b) shows that the 'conjugate eigenvalue' $\lambda_k^\dagger$ is associated with the 'conjugate transformation' $\overline{f}$, and multiplication rotates vectors of the i_k-plane in the 'opposite direction'.

It should be noted that, without further conditions, the equation

$$f(a) = a\lambda \tag{5.10}$$

cannot be regarded as an 'eigenvalue equation'; for every vector a satisfies such an equation, as can be seen by taking $\lambda = a^{-1}f(a)$. The additional condition

$$\overline{f}(a) = \lambda a = a^{-1}f(a)a \tag{5.11}$$

only implies that a is an eigenvector of f_+. If λ is to be regarded as an eigenvalue, some restriction must be made on its bivector part. It seems natural to require that

$$\underline{f}(\lambda) = \overline{f}(\lambda). \tag{5.12}$$

However, the preceding discussion shows that any transformation with n orthogonal vectors satisfying (5.10-12) is a normal transformation. So it is not immediately obvious how best to generalize the notion of eigenvalues along the lines above to apply to arbitrary linear transformations.

The discussion leading to the spectral form (5.9) for a normal transformation f provides a systematic procedure for finding the eigenvectors and 'complex eigenvalues' if f is given, for instance, in the form of a matrix. First, one takes the curl of f and finds its m orthogonal blades, for instance, by the procedure outlined in Section 3-4. One has then

$$\partial \wedge f = \sum_{k=1}^{2m} \beta_k i_k = 2 \sum_{k=1}^{m} \beta_{2k} i_{2k}, \tag{5.13a}$$

where

$$\beta_{2k} = \beta_{2k-1} \quad \text{and} \quad i_{2k} = i_{2k-1}. \tag{5.13b}$$

Factoring i_{2k} into a product of orthonormal vectors,

$$i_{2k} = a_{2k-1}a_{2k}, \tag{5.14}$$

one has the $2m$ eigenvectors $a_1, a_2, \ldots, a_{2m}$ with complex eigenvalues $\lambda_k = \alpha_k + i_k\beta_k$. The $i_k\beta_k$ are known from (5.13a), and the $\alpha_{2k} = \alpha_{2k-1}$ are easily found from

$$f_+(a_{2k}) = \alpha_{2k}a_{2k}. \tag{5.15}$$

The remaining eigenvectors $a_{2m+1}, \ldots, a_{n-1}, a_n$ of the desired orthonormal set satisfy

$$f(a_k) = f_+(a_k) = \alpha_k a_k \quad (2m < k \leqslant n), \tag{5.16}$$

so they can be found by well-known methods used for symmetric transformations.

The eigenvectors of a normal transformation can also be found by relying primarily on the methods used for symmetric transformations. To show this, note that

$$a \wedge f(a) - \bar{f}(a) \wedge a = 2a \wedge f_+(a); \tag{5.17}$$

also note that

$$\underline{f}(\bar{f}(a) \wedge a) - \bar{f}(a \wedge \underline{f}(a)) = 2\underline{\bar{f}}(a) \wedge f_+(a), \tag{5.18a}$$

where

$$\underline{\bar{f}}(a) \equiv \bar{f}\underline{f}(a) = \underline{f}\bar{f}(a). \tag{5.18b}$$

Comparing, one sees that the conditions (5.10-12) for a to be an eigenvector obtain if and only if (5.17) and (5.18a) vanish simultaneously, that is, if and only if a is an eigenvector of the two symmetric transformations f_+ and $\underline{\bar{f}}$ simultaneously. Thus, to find the eigenvectors of f, it will suffice to find the simultaneous eigenvectors of f_+ and $\underline{\bar{f}}$, except in the (unusual) case of 'degenerate complex roots' (i.e. when $\alpha_k = \alpha_j$ and $\beta_k = \beta_j \neq 0$ for distinct i_k and i_j), when, as already mentioned in the discussion of skew transformations, special methods are required. Once the eigenvectors a_k have been found, the bivector parts of the corresponding eigenvalues are easily found from $\beta_k i_k = a_k \wedge f(a_k)$.

Geometric Algebra admits especially simple canonical forms for orthogonal transformations. A linear transformation R defined on a vector space $\mathscr{A}_n$ is said to be *orthogonal* if

$$\underline{R}^{-1}(A) = \bar{R}(A), \tag{5.19a}$$

or equivalently, if

$$\langle \underline{R}(A)\underline{R}(B)\rangle = \langle AB\rangle \tag{5.19b}$$

for all multivectors A, B in $\mathcal{G}(\mathcal{A}_n)$. Actually, it is easy to show that (5.19) holds in general if it is assumed only to hold for vectors. For the sake of brevity, we suppose here that the range of R is the same as its domain; this restriction affects only the interpretation and not the form of our results.

Since an orthogonal transformation is normal, the spectral decomposition (5.9) obtains. However, condition (5.19a) applied to (5.5a) implies that $\rho_k = 1$ in (5.6a) for all k, so the n 'eigenvectors' a_k of an orthogonal transformation f satisfy

$$R(a_k) = a_k\, e^{\theta_k} = e^{-\theta_k} a_k. \tag{5.20}$$

To derive simple canonical forms for an orthogonal transformation, we take note of the algebraic properties of the eigenvectors and eigenvalues. Suppose for the moment that the number of distinct eigenvectors with 'negative eigenvalues' ($e^{\theta_k} = e^{i_k\pi} = -1$) is even. Then there is an even number $2m$ of eigenvectors with eigenvalues $e^{\theta_k} \neq 1$, and they can be ordered so that, as in (5.14),

$$\begin{aligned} a_{2k-1}a_{2k} &= i_{2k-1} = i_{2k} = |\theta_{2k}|^{-1}\theta_{2k}, \\ \theta_{2k} &= \theta_{2k-1} \qquad \text{for } k = 1, \ldots, m. \end{aligned} \tag{5.21}$$

It should be noted that there is a trivial ambiguity in the pairing of eigenvectors with negative eigenvalues if there are more than two of them. Now, for $k = 1, \ldots, m$ and $j = 1, \ldots, n$,

$$\begin{aligned} a_j i_{2k} &= -i_{2k} a_j \qquad \text{if } j = 2k-1, 2k, \\ a_j i_{2k} &= i_{2k} a_j \qquad \text{if } j \neq 2k-1, 2k. \end{aligned} \tag{5.22}$$

Hence, by (5.8),

$$\begin{aligned} e^{-\theta_{2k}/2} a_j\, e^{\theta_{2k}/2} &= a_j\, e^{\theta_{2k}} \qquad \text{if } j = 2k-1, 2k, \\ e^{-\theta_{2k}/2} a_j\, e^{\theta_{2k}/2} &= a_j \qquad \text{if } j \neq 2k-1, 2k. \end{aligned} \tag{5.23}$$

Recalling that, by (5.7),

$$e^{\theta_j} e^{\theta_k} = e^{\theta_k} e^{\theta_j}, \tag{5.24}$$

and using (5.23), one finds easily from (5.9) that

$$R(x) = \sum_{k=0}^{m} P_{2k}(x)\, e^{\theta_{2k}} = e^{-\theta/2} x\, e^{\theta/2}, \tag{5.25a}$$

where

$$P_{2k}(x) \equiv x \cdot i_{2k} i_{2k}^{\dagger} = x \cdot \theta_{2k} \theta_{2k}^{-1}, \tag{5.25b}$$

$$P_0(x) = x - \sum_{k=1}^{m} P_{2k}(x), \tag{5.25c}$$

$$\theta = \sum_{k=1}^{m} \theta_{2k}. \tag{5.25d}$$

The exponential $e^{\theta/2}$ of the m-bladed bivector $\frac{1}{2}\theta$ can be defined, as usual, by a power series

$$e^{\theta/2} \equiv \sum_{r=0}^{\infty} \frac{(\frac{1}{2}\theta)^r}{r!} = e^{\theta_2/2} e^{\theta_4/2} \ldots e^{\theta_{2m}/2}. \tag{5.26}$$

Equivalence of the left and right sides of (5.26) follows from the commutivity (5.24) of the factors.

Using (5.26) in (5.25a), one can write R as the composite

$$R(x) = R_m \ldots R_2 R_1(x), \tag{5.27a}$$

where

$$R_k(x) = e^{-\theta_{2k}/2} x\, e^{\theta_{2k}/2}. \tag{5.27b}$$

is a rotation through an angle $|\theta_{2k}|$ in the i_{2k}-plane, that is, a rotation 'through' a *directed angle* θ_{2k}. The interpretation of R_k as a rotation in a plane is fairly obvious from its spectral decomposition; in any case, it is discussed in more detail in reference [H4]. Equation (5.27a) expresses a general rotation as a composite of rotations in *orthogonal* planes, the 'orthogonality' of the rotations being expressed by their commutivity,

$$R_j R_k = R_k R_j, \tag{5.28}$$

as follows at once from (5.24). Thus the right side of (5.25a) is the *canonical form for a rotation* as a function of the m-bladed *directed angle* θ.

Our derivation of (5.25a) proves that every rotation can be put in that canonical form. Conversely, every bivector θ determines a rotation by (5.25a). The right side of Eqn. (5.25a) generalizes Hamilton's quaternion formulation of rotations. It has been derived before; we first learned about it from Marcel Riesz [R1]. However, in spite of its considerable advantages over the corresponding matrix representation of rotations, it has been little used, because it has not been integrated into a systematic general theory of linear transformations. The present treatment aims to rectify that unfortunate state of affairs.

Another representation of a rotation can be obtained by using the fact that the exponential of a simple bivector can be factored into a product of unit vectors; thus,

$$e^{\theta_{2k}/2} = u_{2k-1}u_{2k}. \tag{5.29}$$

This factorization is not unique; any pair of unit vectors in the θ_{2k}-plane separated by an angle $\frac{1}{2}\theta_{2k}$ will do. Substitution of (5.26) and (5.29) in (5.25a) yields

$$R(x) = u_{2m} \ldots u_2u_1xu_1u_2 \ldots u_{2m} = C_{2m} \ldots C_2C_1(x), \tag{5.30}$$

where

$$C_k(x) = -u_kxu_k. \tag{5.31}$$

We call the linear transformation (5.31) a *simple reflection*. It is discussed further in Section 8 of reference [B0]. Comparing (5.29) and (5.31) with (5.27b), it is clear that

$$C_{2k}C_{2k-1} = R_k. \tag{5.32}$$

So by (5.28), (5.30) gives the decomposition of a rotation into $2m$ simple reflections which commute in pairs.

We now consider the case of an orthogonal transformation $R' = R'(x)$ with an odd number $p = 2m + 1$ of eigenvalues different from one. At least one of these, say the pth, must be negative; so all the eigenvectors a_k of R' are also eigenvectors of the simple reflection

$$C_p(x) = -a_pxa_p. \tag{5.33}$$

For, by the anticommutivity of orthogonal vectors,

$$C_r(a_k) = a_k \quad \text{for any } k \neq r$$

and of course,

$$C_p(a_p) = -a_p.$$

If the first $2m$ eigenvectors are arranged as before, it is evident that R' can be expressed as a commuting composite of (5.30) and (5.33). Thus

$$R'(x) = RC_p(x) = C_pR(x) = C_pC_{p-1} \ldots C_2C_1(x) \tag{5.34}$$

or, with $u_p = a_p$,

$$R'(x) = (-1)^p u_p \ldots u_2u_1xu_1u_2 \ldots u_p. \tag{5.35}$$

Obviously, (5.30) can be regarded as a special case of (5.35) if p is allowed to be even as well as odd. Thus, *any orthogonal transformation can be expressed in the form* (5.35), and we have proved, by the way, a theorem of Cartan's that every orthogonal transformation can be expressed as a product of at most n elementary reflections. Also note that if the u_k are orthogonal, we can take $u_k = a_k$ and R' reduces to a conjugation as defined by (4.6).

To summarize, we have shown that every orthogonal transformation $R = R(x)$ can be written in the canonical form

$$R(x) = (-1)^p U^\dagger x U, \tag{5.36a}$$

where

$$UU^\dagger = 1, \tag{5.36b}$$

and

$$\langle U \rangle_p \neq 0 \quad \text{but} \quad \langle U \rangle_k = 0 \quad \text{for } k > p. \tag{5.36c}$$

If p is even, R is said to be a *rotation*, and if p is odd, R is said to be a *reflection*. The multivector U can always be expressed as a product of p vectors as in (5.35), and if p is even, U can be expressed as the exponential of a bivector as in (5.25a). (See Section 8 for further discussion from a more general viewpoint.) It should be mentioned that there are many other decompositions of U which are useful for various purposes. Given (5.36), it is possible to express any parametrization of an orthogonal transformation as some algebraic decomposition of U.

The differential and adjoint of an orthogonal transformation can be obtained from (5.36) simply by multiplication. Thus, if A_r is an r-vector

$$\underline{f}(A_r) = (-1)^{rp} U^\dagger A_r U, \tag{5.37a}$$

$$\overline{f}(A_r) = (-1)^{rp} U A_r U^\dagger. \tag{5.37b}$$

Further, if B_s is an s-vector, then

$$\underline{f}(A_r B_s) = \underline{f}(A_r)\underline{f}(B_s), \tag{5.38}$$

while a similar relation obviously holds for the adjoint.

If an orthogonal transformation R is given, say, in the form of a matrix, its canonical form (5.36a) can be found by methods already mentioned. But the systematic procedure is important enough to merit review. First, one computes the curl of R. By (5.13) and (5.21),

$$\partial \wedge R = 2 \sum_{k=1}^{m} i_{2k} \sin |\theta_{2k}|. \tag{5.39}$$

The m orthogonal blades of $\frac{1}{2}\partial \wedge R$ are precisely the non-vanishing 'imaginary parts' of the eigenvalues of R. Solving (5.39) for the i_{2k} and $|\theta_{2k}|$, we get

$$e^{\theta/2} = \prod_{k=1}^{m} e^{|\theta_{2k}| i_{2k}/2}. \tag{5.40}$$

If R has no negative eigenvalues, then $U = e^{\theta/2}$ and R is a rotation given by (5.36) with $p = 2m$. In any case, the number ϵ of orthogonal eigenvectors with negative eigenvalues can be found immediately from the divergence of R. For,

$$\partial \cdot R = \sum_{k=1}^{n} \cos|\theta_{2k}| = 2 \sum_{k=1}^{m} \cos|\theta_{2k}| + n - 2m - 2\epsilon. \tag{5.41}$$

If $\epsilon \neq 0$, the problem of determining U has at least been reduced to the problem of finding the eigenvectors of a symmetric transformation in the space of vectors orthogonal to the $i_2 \wedge i_4 \wedge \ldots \wedge i_{2m}$ space. If $a_{2m+1}, a_{2m+2}, \ldots, a_{2m+\epsilon}$ are any set of orthonormal eigenvectors of this symmetric transformation with negative eigenvalues, then, of course,

$$U = e^{\theta/2} A_\epsilon = A_\epsilon e^{\theta/2} \tag{5.42}$$

where $A_\epsilon = a_{2m+1}a_{2m+2} \ldots a_{2m+\epsilon}$. And R is given by (5.36a) with $p = 2m + \epsilon$.

3-6. Canonical Forms for General Linear Transformations

The appropriate choice of a canonical form for an arbitrary linear transformation depends on its intended use. We do not discuss the classical Jordan form here, because we have not learned how to use Geometric Algebra to any decisive advantage in its formulation, though the developments in Section 3-3 are promising. A rather different approach was taken in our construction of the spectral form fot a normal transformation. It was based on the separation of a transformation into symmetric and skew-symmetric parts. Similarly, canonical forms for any linear transformation f can be based on the relation of the eigenvectors of f_+ to the blades of $\partial \wedge f$. We will not pursue this approach here, though it is clear that Geometric Algebra would be helpful since it provides such an economical formulation of skew transformations.

For a non-normal transformation f the analogy of f_+ and f_- to the real and imaginary parts of a complex number fails because f_+ and f_- do not commute. However, the analogy with complex numbers is preserved by the so-called *polar decomposition*, which expresses f as a composite of a symmetric transformation S and a rotation R, that is,

$$f(x) = RS(x) = U^\dagger S(x)U, \tag{6.1}$$

where U is a unitary spinor. Standard proofs of the polar decomposition are simple enough without appeal to Geometric Algebra. But computation of the polar decomposition is facilitated by the 'spinor representation' of the rotation in (6.1). For example, let $f = f(x)$ be a linear transformation in a two-dimensional space. Then (6.1) assumes the simpler form $f(x) = S(x)U^2$. Since S is a symmetric transformation, its curl vanishes, so

$$\partial f = (\partial \cdot S)U^2, \tag{6.2a}$$

from which we find

$$\partial \cdot S = |\partial f|, \tag{6.2b}$$

$$U^2 = \frac{\partial f}{|\partial f|} = e^{I\theta}, \tag{6.2c}$$

$$\tan\theta = \frac{I^\dagger \partial \wedge f}{\partial \cdot f}, \tag{6.2d}$$

where I is the unit pseudoscalar and θ is the angle of rotation. It should be mentioned that we can choose $\partial \cdot S \geqslant 0$ as in (6.2b), because a negative sign can always be absorbed in U^2 as a rotation through an angle π. Finally, we get the symmetric part of f from

$$S(x) = U^2 f(x). \tag{6.2e}$$

Unfortunately, this method is not so automatic when applied to transformations in higher dimensions.

As a significant application of (6.2), consider the function

$$f(x) = x + x \cdot ab. \tag{6.3a}$$

We find

$$\partial f = 2 + ab, \tag{6.3b}$$

$$\tan\theta = \frac{a \wedge b}{(2 + a \cdot b)I}, \tag{6.3c}$$

$$U^2 = \frac{2 + ab}{(4 + 4a \cdot b + a^2 b^2)^{1/2}}, \tag{6.3d}$$

$$S(x) = \frac{(2 + a \cdot b)x + x \cdot ab + x \cdot ba + a \cdot xb^2 a}{(4 + 4a \cdot b + a^2 b^2)^{1/2}}. \tag{6.3e}$$

In this case, the restriction to two dimensions is easily removed to get a more general result.

Every linear transformation, hence every canonical form, can be expressed as a composite of elementary transformations of the types R_a and S_{ab} defined by

$$R_a(x) = -axa^{-1}, \tag{6.4}$$

$$S_{ab}(x) = x + x \cdot ab, \tag{6.5}$$

where a and b are non-zero vectors parametrizing the transformations. The elementary reflection R_a has already been discussed. The transformation S_{ab} is a *shear* in the $a \wedge b$-plane if $a \cdot b = 0$, and if $a \wedge b = 0$, S_{ab} is a *strain* along a.

To prove that any linear transformation can be expressed as a product of types R_a and S_{ab}, it is sufficient to prove that every symmetric transformation can be composed of strains in orthogonal directions and appeal to the polar form (6.1), for we have already proved that every rotation can be composed of elementary reflections.

Elementary operations on rows and columns of matrices are, of course, simply related to R_a and S_{ab}. This is easily established by choosing an orthonormal frame of vectors a_k and defining the functions

$$P_{ij}(x) \equiv -R_{a_{ij}}(x) = a_{ij} x a_{ij} \quad \text{with } a_{ij} \equiv (2)^{-1/2}(a_i + a_j), \tag{6.6}$$

and,

$$S_{ij}^{\alpha}(x) \equiv S_{\alpha a_i a_j}(x) = x + \alpha x \cdot a_i a_j. \tag{6.7}$$

Operating on the set of vectors $\{a_k\}$, the reflection P_{ij} simply permutes the ith and jth vector, while S_{ii}^{α} multiplies the ith vector by $(1 + \alpha)$. For fear of overdoing the obvious, we say no more on the subject.

3-7. Metric Tensors and Isometries

Every linear transformation $g(y)$ of $\mathscr{A}_n = \mathscr{G}^1(I)$ into itself determines a unique bilinear form

$$g(x, y) = x \cdot g(y). \tag{7.1}$$

Conversely, from the bilinear form $g(x, y)$ we get the linear transformation

$$g(y) = \partial_x g(x, y). \tag{7.2}$$

Thus, a bilinear form is equivalent to a linear transformation.

A bilinear form can be regarded as an 'inner product' $x \circ y$ of vectors, where

$$x \circ y \equiv g(x, y) = x \cdot g(y). \tag{7.3}$$

Because of its conciseness, the notation $x \circ y$ is convenient for a bilinear form which is used repeatedly, but $g(x, y)$ and $x \cdot g(y)$ are better notations when a variety of forms is being considered.

A bilinear form $g(x, y)$ or its equivalent linear transformation $g(y)$ is said to be a *metric tensor* if

$$x \circ x = g(x, x) = x \cdot g(x) \tag{7.4}$$

is regarded as determining a *length* $|x \circ x|^{1/2}$ for each vector x in $\mathscr{A}_n$. The study of a bilinear form $g(x, y)$ is equivalent to the study of a 'metric' or 'metric structure' g on $\mathscr{A}_n$.

A bilinear form or metric on $\mathscr{A}_n$ is said to be *nonsingular* if for every x in $\mathscr{A}_n$ there exists a vector y such that $g(x, y) \neq 0$, that is, if the corresponding metric tensor $g(y)$ is nonsingular. It is not difficult to prove that any singular linear transformation on $\mathscr{A}_n$ can be expressed as the composite of a nonsingular transformation with a projection into a subspace of $\mathscr{A}_n$. So we can restrict our attention to nongsingular metrics without loss of generality.

A linear transformation $f = f(x)$ is said to be an *isometry* of a metric tensor g if it leaves the 'inner product' (7.3) invariant, that is, if

$$f(x) \circ f(y) = f(x) \cdot gf(y) = x \cdot g(y) = x \circ y, \tag{7.5a}$$

or equivalently, if

$$\bar{f}gf(y) = g(y). \tag{7.5b}$$

We wish to explain how Geometric Algebra can best be used to characterize the isometries of any given metric tensor. Since a bilinear form can be uniquely expressed as a sum of symmetric and skewsymmetric parts, we can consider each part separately. We discuss isometries of skew bilinear forms first.

We already know that any skew bilinear form can be written $x \cdot g(y) = x \cdot (y \cdot G) = (x \wedge y) \cdot G$ where G is a bivector. Therefore any isometry f of such a form must satisfy

$$f(x) \wedge f(y) \cdot G = \underline{f}(x \wedge y) \cdot G = (x \wedge y) \cdot \bar{f}(G) = (x \wedge y) \cdot G, \tag{7.6a}$$

or,

$$\bar{f}(G) = G. \tag{7.6b}$$

Isometries of a given skew bilinear form are commonly called *symplectic transformations*. The group of all such isometries is called the *symplectic group*. According to (7.6), the symplectic group can equally well be regarded as the group of outermorphisms which leave a given bivector invariant. With Geometric Algebra at our disposal, this latter view is the simplest, because we can determine the structure of the group directly from its action on G in (7.6) without examining its effect on arbitrary vectors. To begin with, we know that G can be uniquely expressed as the sum of commuting blades:

$$G = G_1 + G_2 + \cdots + G_m, \tag{7.7a}$$

$$G_j G_k = G_k G_j. \tag{7.7b}$$

We can assume that

$$G_k^2 = -1 \tag{7.7c}$$

without disturbing the structure of the symplectic group. It follows that

$$g^2(x) = (x \cdot G) \cdot G = \sum_{k=1}^{m} (x \cdot G_k)G_k = -x,$$

from which it is clear that g is nonsingular only if $n = 2m$, and

$$g^{-1}(x) = -g(x) = -x \cdot G. \tag{7.8}$$

Using (7.8) in connection with (7.5b), it is easy to prove that

$$fg\bar{f}(y) = g(y), \tag{7.9}$$

which is to say that if f is symplectic, then so is the adjoint of f. It follows that we can replace (7.6) by

$$\underline{f}(G) = G. \tag{7.10}$$

The determinant of f can be computed directly from (7.10); consider

$$\langle[\underline{f}(G)]^m\rangle_{2m} = \langle G^m\rangle_{2m}. \tag{7.11}$$

Each nonvanishing term on the right side of (7.11) can be written

$$G_1 \wedge G_2 \wedge \ldots \wedge G_m = G_1 G_2 \ldots G_m = I, \tag{7.12}$$

where I is the unit pseudoscalar of $\mathscr{A}_n = \mathscr{A}_{2m}$. Since $\underline{f}$ is an outermorphism, the nonvanishing terms on the left of (7.11) can be written

$$\underline{f}(G_1) \wedge \underline{f}(G_2) \wedge \ldots \wedge \underline{f}(G_m) = \underline{f}(G_1 \wedge \ldots \wedge G_m) = \underline{f}(I). \tag{7.13}$$

Hence,

$$\underline{f}(I) = I. \tag{7.14}$$

Thus the determinant of a symplectic transformation must be unity.

The results of Section 3-5 enable us to find those symplectic transformations which are orthogonal by inspection. If f is orthogonal, then, according to (5.36) and (5.37), (7.10) can be written as an explicit algebraic equation

$$U^\dagger G U = G, \tag{7.15}$$

where $U^\dagger U = 1$. Reflections are excluded by (7.14); this can also be proved directly from (7.15) by showing that (7.15) cannot be satisfied if U is a vector. The rotation (7.15) can be factored into at most m rotations in orthogonal planes, as is best expressed by the factorization

$$U = U_1 U_2 \ldots U_m \tag{7.16a}$$

$$U_j U_k = U_k U_j. \tag{7.16b}$$

The irreducible symplectic rotations are of two types, those that leave the blades of G unchanged and those that map one blade of G into another. The spinors for the first type satisfy the equation

$$U_k^\dagger G_j U_k = G_j \quad \text{for all } j, \tag{7.17a}$$

which has the solution

$$U_k = e^{\theta_k G_k/2}. \tag{7.17b}$$

Thus, the symplectic transformation $f(x) = U_k^\dagger x U_k$ is merely a rotation in the G_k-plane through an angle θ_k. An irreducible rotation which interchanges the jth and kth blades of G satisfies the equations

$$U_{jk}^\dagger G_j U_{jk} = G_j (U_{jk})^2 = G_k, \tag{7.18a}$$

$$U_{jk}^\dagger G_i U_{jk} = G_i \quad \text{if } i \neq j, k. \tag{7.18b}$$

The transformation (7.18a, b) is induced by a symplectic rotation

$$U_{jk}^\dagger e_j U_{jk} = e_k, \tag{7.18c}$$

which rotates a unit vector e_j in the G_j-plane into a unit vector e_k in the G_k-plane. We can factor G_k into a product of orthogonal vectors

$$G_k = e_k e_k', \tag{7.19a}$$

where

$$e_k' \equiv e_k \cdot G = e_k \cdot G_k = e_k G_k. \tag{7.19b}$$

Using the corresponding factorization for G_j, we can write the solution to (7.18) in the form

$$U_{jk} = \exp\left[\tfrac{1}{4}\pi(e_j e_k + e_j' e_k')\right] = \exp\left(\tfrac{1}{4}\pi e_j e_k\right) \exp\left(\tfrac{1}{4}\pi e_j' e_k'\right). \tag{7.20}$$

Clearly (7.20) is the spinor which rotates e_j' into e_k' and e_j into e_k. Now the most general 'symplectic spinor' (7.16a) can be composed of single factors of the type (7.17b) and pairs of factors of the type (7.20).

A factorization of type (7.19) of each blade G_k of G provides us with a basis $e_1, \ldots, e_k, e_1', \ldots, e_k'$ for $\mathscr{A}_{2m}$, which we call a *symplectic basis*. The basis satisfies the relations

$$e_j \circ e_k = 0 = e_j' \circ e_k', \tag{7.21a}$$

$$e_j' \circ e_k = \delta_{jk} = -e_j \circ e_k', \tag{7.21b}$$

where, of course, $x \circ y = x \cdot g(y) = (x \wedge y) \cdot G$ is the 'symplectic inner product'.

Our analysis of symplectic rotations shows that any symplectic basis can be rotated into any other symplectic basis. Hence, any symplectic transformation can be expressed as the composite of a rotation and a transformation f which leaves a vector e_k in each G_k-plane invariant. The most general such transformation satisfies the equations

$$f(e_k) = e_k, \tag{7.22}$$

$$f(e_j') = \sum_{i=1}^{m} (\alpha_{ji}e_i + \beta_{ji}e_i'), \tag{7.23}$$

where the scalars α_{ji} and β_{ji} are restricted by the condition that the relations (7.21) be invariant. Hence,

$$f(e_j') \cdot f(e_k) = \beta_{jk} = \delta_{jk},$$

$$f(e_j') \cdot f(e_k') = \sum_i (\alpha_{ji}e_i \cdot e_k' + e_j' \cdot e_i\alpha_{ki}) = -\alpha_{jk} + \alpha_{kj} = 0,$$

and (7.23) reduces to

$$f(e_j') = e_j' + \sum_k \alpha_{jk}e_k, \quad \text{where } \alpha_{jk} = \alpha_{kj}. \tag{7.24}$$

Of course, $x = \Sigma_k x_k e_k + x_k' e_k'$, where $x_k' = x \cdot e_k' = x \cdot g(e_k) = x \cdot e_k$, so, from (7.22) and (7.24), we have

$$\begin{aligned} f(x) &= x + \sum_{j,k} \alpha_{jk} x \cdot e_j e_k \\ &= x + \sum_k \alpha_k x \cdot e_k e_k + \sum_{j<k} \alpha_{jk}(x \cdot e_j e_k + x \cdot e_k e_j), \end{aligned} \tag{7.25}$$

where $\alpha_k \equiv \alpha_{kk}$. It is easily verified that (7.25) can be factored into a composite of commuting symplectic transformations

$$f_k(x) \equiv x + \alpha_k x \cdot e_k e_k, \tag{7.26a}$$

$$f_{jk}(x) = x + \alpha_{jk}(x \cdot e_j e_k + x \cdot e_k e_j). \tag{7.26b}$$

Thus,

$$f(x) = \prod_{j<k} f_j f_{jk}(x) \tag{7.26c}$$

where Π indicates product. To sum up, every symplectic transformation can be expressed as the composite of a rotation and a transformation of the type (7.26).

The transformation (7.26a) should be recognized as a shear in the G_k-plane. Any rotation in the G_k-plane can be expressed as a composite of shears, as can be proved by the method used to compute the polar decomposition in Section 6. Indeed, it can be proved ([Ar], p. 137) that any symplectic transformation can be obtained as a composite of symplectic transformations of the type

$$f_a(x) = x + \alpha aa \cdot x, \tag{7.27}$$

where a is a vector and α is a scalar. The transformation (7.27) is called a *symplectic transvection* by Artin [Ar]. A symplectic basis can always be found which gives (7.27) the form (7.25).

Our remaining considerations in this section concern symmetric bilinear forms. A form $g(x, y)$ is said to be *positive definite* or *Euclidean* if $g(x, x) > 0$ for all $x \neq 0$. The form $g(x, y) = x \cdot g(y)$ is positive definite if and only if there exists a nonsingular symmetric linear transformation $h(x)$, such that

$$g(x, y) = h(x) \cdot h(y) = x \cdot h^2(y). \tag{7.28}$$

This follows easily from the fact that since $g = h^2$ is a symmetric function, its 'square root' h can be found by well-known methods.

The isometries of (7.28) are essentially the same as the isometries of $x \cdot y$, namely the orthogonal transformations, which we have already completely characterized in Section 5. For to each isometry $R(x)$ of $x \cdot y$ there obviously corresponds exactly one isometry $h^{-1}Rh$ of (7.28). We can regard $h(x)$ as a mere change of scale or choice of units in the underlying vector space $\mathscr{A}_n$, and this evidently does not alter the structure of the isometry group. Disregarding the positive symmetric transformations which we have just analyzed, it is clear from (4.7) that the most general nonsingular metric tensor which remains to be considered has the form

$$g(y) = (-1)^k Q^\dagger y Q, \tag{7.29}$$

where Q is a blade of grade k and $Q^\dagger Q = 1$. The expression (7.29) is the general canonical form for a symmetric orthogonal linear transformation. The function (7.29) determines the inner product

$$x \circ y \equiv (-1)^k x \cdot (Q^\dagger y Q) = (-1)^k \langle x Q^\dagger y Q \rangle. \tag{7.30}$$

The k-blade Q determines a k-dimensional subspace $\mathscr{G}^1(Q)$ of $\mathscr{A}_n$. Recalling (1-2.8), we can write

$$y = y_\perp + y_\parallel, \tag{7.31a}$$

where

$$y_\parallel \equiv y \cdot QQ^\dagger \tag{7.31b}$$

is the component of y in $\mathscr{G}^1(Q)$, and

$$y_\perp \equiv y \wedge QQ^\dagger \tag{7.31c}$$

is the component of y orthogonal to all vectors in $\mathscr{G}^1(Q)$. From our study of orthogonal transformations in Section 3-5, we know that the transformation (7.29) reverses the direction of vectors in $\mathscr{G}^1(Q)$ and leaves vectors orthogonal to $\mathscr{G}^1(Q)$ unchanged. Hence,

$$g(y) = g(y_\perp) + g(y_\parallel) = y_\perp - y_\parallel. \tag{7.32}$$

So the inner product (7.30) can be written

$$x \circ y = x \cdot (y_\perp - y_\parallel) = x_\perp \cdot y_\perp - x_\parallel \cdot y_\parallel. \tag{7.33}$$

Clearly we can find at most k linearly independent vectors with the property $x \circ x < 0$ and at most $n - k$ linearly independent vectors with the property $x \circ x > 0$. We say that (7.30) is a *metric with signature* $(n - k, k)$. The metric is referred to as *Euclidean* if $k = 0$ or $k = n$, and *pseudoEuclidean* otherwise.

If h is any nonsingular transformation of $\mathscr{A}_n$ which is not an isometry of g, then instead of (7.5b), we have the operator equation,

$$\bar{h}gh = g', \tag{7.34}$$

which can be regarded as a change of metric from g to g'. We have considered such a change of the Euclidean metric in connection with (7.28). The important feature of (7.34) is that g' has the same signature as g, so nonEuclidean metrics cannot be obtained from Euclidean metrics by transformations of $\mathscr{A}_n$. This is known as *Sylvester's Law of Inertia*. We will not bother to prove it.

Isometries of the pseudoEuclidean metric (7.30) can be characterized by the method we used to characterize symplectic transformations. For example, a rotation or reflection $R(x)$ is an isometry of (7.30) if and only if $\underline{R}(Q) = \pm Q$. But we have adequately illustrated this method already, so we consider a different approach in the next section.

3-8. Isometries and Spinors of PseudoEuclidean Spaces

In the preceding section we saw how to characterize any metric tensor and its isometries with the Euclidean Geometric Algebra. In this section we take a more direct approach, expressing the isometries of a symmetric metric tensor with signature (p, q) in terms of its associated pseudoEuclidean Geometric Algebra. In Section 3-7 we regarded metric tensors with different signatures as different kinds of functions on a single vector space. Here the metric tensor is identified with the inner product which is regarded as part of the very notion of vector, so the study of metric tensors with different signatures becomes the study of different kinds of vector spaces.

This section employs the nomenclature and properties of pseudoEuclidean Geometries Algebras developed in Section 1-5. Let $\mathscr{A}_{p,q} = \mathscr{G}^1(I)$ be a $(p+q)$-dimensional vector space with a nonsingular unit pseudoscalar I of signature (p, q).

The inner product $x \cdot y$ of vectors in $\mathscr{A}_{p,q}$ defines a metric tensor on $\mathscr{A}_{p,q}$ isomorphic to the metric tensor $x \cdot y$ defined by (7.30). A linear transformation $f = f(x)$ of $\mathscr{A}_{p,q}$ into itself, is said to be an *orthogonal transformation* of $\mathscr{A}_{p,q}$ if

$$f(x) \cdot f(y) = x \cdot y, \tag{8.1}$$

for each x and y in $\mathscr{A}_{p,q}$. The orthogonal transformations of $\mathscr{A}_{p,q}$ are, of course, the isometries of the inner product. The group of all orthogonal transformations of $\mathscr{A}_{p,q}$ is called the *orthogonal group* of $\mathscr{A}_{p,q}$ and denoted by $\mathrm{O}(\mathscr{A}_{p,q})$ or, more briefly, by $\mathrm{O}(p, q)$.

The simplest kind of isometry is a *simple reflection* along some vector u:

$$R_u(x) = -uxu^{-1}. \tag{8.2}$$

This transformation reverses the direction of all vectors collinear with u and leaves the hyperplane of vectors orthogonal to u invariant. It is thus an isometry of a one-dimensional subspace of $\mathscr{A}_{p,q}$, and a simpler isometry is obviously not to be found. Equation (8.2) holds for $u^2 < 0$ as well as $u^2 > 0$, but a simple reflection along a null vector is impossible.

Every isometry $f = f(x)$ of $\mathscr{A}_{p,q}$ can be expressed as a composite of at most n simple reflections, that is, there exist vectors $u_1, u_2, \ldots, u_k$, where $k \leqq n$, such that

$$f(x) = (-1)^k u_k \ldots u_2 u_1 x u_1^{-1} u_2^{-1} \ldots u_k^{-1}. \tag{8.3}$$

A proof of this result for a nonsingular metric of any signature (and, by the way, allowing the scalars to be a finite number field) is given by Artin [Ar, pp. 129–130]. The proof is by induction on the dimension of $\mathscr{A}_{p,q}$, and makes no use of the representation (8.2) of a simple reflection in terms of the geometric product. We will not reproduce it.

The representation of reflections by the geometric product in (8.3) tells us much more than the fact that every isometry can be generated by reflections. Writing $U = u_k \ldots u_2 u_1$, we immediately obtain from (8.3) the general result that every isometry $f = f(x)$ of $\mathscr{A}_{p,q}$ can be expressed in the form

$$f(x) = (-1)^k U x U^{-1}. \tag{8.4}$$

Multivectors which can be factored into a product of vectors are so important that they deserve a class name. Reviving and generalizing somewhat a term from Hamilton's quaternion calculus which has fallen into disuse, we call them *versors*. A versor which can be factored into a product of k vectors will be called a *k-versor*. It is convenient to define a *o-versor* to be a scalar. Naturally, a versor will be called an *even*(odd) *versor* if it is an even(odd) multivector. We have already seen the importance of k-blades, which are a special kind of versor. The importance of the general concept of versor is brought to light by Eqn. (8.4) in which, by construction, the multivector U must be a nonsingular (hence invertible) versor.

If U is taken to be the pseudoscalar I in (8.4), then

$$IxI^{-1} = x \tag{8.5a}$$

if $n = p + q$ is odd, and

$$IxI^{-1} = -x \tag{8.5b}$$

if n is even. This is a consequence of the equation $I \wedge x = Ix + (-1)^n xI = 0$ satisfied by every vector x in $\mathcal{G}^1(I)$. Since I can be factored into a product of n orthogonal vectors, Eqn. (8.5) describes the isometry generated by simple reflections along n mutually orthogonal directions. Writing $V = UI$ when k is odd and using (8.5b), we see that for $n = p + q$ even we can replace (8.4) by

$$f(x) = VxV^{-1}, \tag{8.6}$$

where V is, of course, a nonsingular versor.

Obviously U can be replaced by a nonzero scalar multiple of U without altering Eqn. (8.4), but, except for this ambiguity in scale and sign, U is uniquely determined by the isometry f. To prove this, suppose there are versors U and V such that $(-1)^k f(x) = UxU^{-1} = VxV^{-1}$. Then $V^{-1}UxU^{-1}V = x$ or $V^{-1}Ux = xV^{-1}U$ where $V^{-1}U$ is even. Comparison with (8.5) shows that $V^{-1}U \neq I$; since only scalars or pseudoscalars (if $p + q$ is odd) can commute with every vector in $\mathcal{G}^1(I)$, we have $V^{-1}U = \alpha$ or $U = \alpha V$, where α is a nonzero scalar. If $|U| = |V| = 1$, we have $U = \pm V$. Hence, to every orthogonal transformation there corresponds two unique unit versors differing only by sign. Equation (8.3) is a special case of (8.4), because the factorization of a k-versor into vectors is not unique unless $k \leqslant 1$.

The multiplicative group of all invertible versors in $\mathcal{G}(\mathcal{A}_{p,q})$ has been dubbed the *Clifford group* of the orthogonal group $O(p, q)$ by Chevalley [Ch]. We call the multiplicative group of unit versors in $\mathcal{G}(\mathcal{A}_{p,q})$ the *versor group Vers* (p, q). We have established that Vers (p, q) is 2 : 1 homomorphic to $O(p, q)$. The problem of describing the structure of $O(p, q)$ and its subgroups is now reduced to an algebraic problem of classifying different kinds of versors.

The subgroup of all orthogonal transformations with determinant one is called the *special orthogonal group* $SO(p, q)$. To express the condition $\det f = 1$ as a property of versors, we factor the pseudoscalar into a product of n vectors, $I = a_1 a_2 \ldots a_n$. Using (8.1) and (8.4), we evaluate the outermorphism $\underline{f}(I)$ as follows,

$$\begin{aligned}\underline{f}(I) &= \underline{f}(a_1 a_2 \ldots a_n) = \underline{f}(a_1 \wedge a_2 \wedge \ldots \wedge a_n)\\ &= f(a_1) \wedge f(a_2) \wedge \ldots \wedge f(a_n) = f(a_1)f(a_2)\ldots f(a_n)\\ &= (-1)^{nk} UIU^{-1}.\end{aligned}$$

Since U can be expressed as a product of vectors, we ascertain from (8.5) that $UI = IU$ if n is odd, and $UI = (-1)^k IU$ if n is even. Hence

$$\underline{f}(I) = (-1)^k I, \tag{8.7}$$

and $\det f = (-1)^k$. So the determinant of an orthogonal transformation is one if and only if the corresponding versor is even. The multiplicative group of all even unit versors in $\mathcal{G}(\mathcal{A}_{p,q})$ is called the *spin group* of $\mathcal{A}_{p,q}$ and denoted by Spin $(\mathcal{A}_{p,q})$ or Spin (p, q). We frequently refer to even versors as *spinors*, though, as we shall see, some spinors are not versors.

A unit k-versor U can obviously be expressed as a product of k unit vectors. If p of these vector factors have positive square and the remaining $(k-p)$ vectors have negative square, then $U^{-1} = (-1)^{k-p}U^\dagger$ and

$$U^\dagger U = (-1)^{k-p}. \tag{8.8a}$$

Hence (8.4) can equally well be written

$$f(x) = (-1)^p UxU^\dagger. \tag{8.8b}$$

Reviving and adapting another old term which has fallen into disuse, we call an even versor S a *rotor* if

$$S^\dagger S = 1. \tag{8.9a}$$

A rotor is a special kind of spinor. From (8.8a) it follows that a rotor has an even number of distinct vector factors with positive (or negative) square. The orthogonal transformation $f = f(x)$ corresponding to a rotor S is called a *rotation* and, according to (8.8b), has the form

$$f(x) = SxS^\dagger. \tag{8.9b}$$

The group of all rotors in $\mathcal{G}(\mathcal{A}_{p,q})$ is called the *rotor group* of $\mathcal{A}_{p,q}$ and denoted by $\text{Spin}^+(p, q)$. By construction, $\text{Spin}^+(p, q)$ is the group of all rotations of $\mathcal{A}_{p,q}$. Clearly, $\text{Spin}^+(p, q)$ is a subgroup of Spin (p, q) and $\text{SO}^+(p, q)$ is the corresponding subgroup of $\text{SO}(p, q)$. For spaces with *Euclidean signature* $(0, n)$ or $(n, 0)$, the spin group is obviously identical to the rotor group.

Our terminology is somewhat unconventional, so a word of explanation is in order. Ordinarily, $\text{SO}(p, q)$ is called the rotation group. The name 'rotation' has been generalized from Euclidean geometry to apply to spaces with arbitrary signature. In the Euclidean case a rotation has determinant one and is continuously connected to the identity. These two properties are not equivalent in the pseudo-Euclidean case. We regard the 'continuity property' as most characteristic of rotations, and, as we shall see, this is precisely what distinguishes $\text{SO}^+(p, q)$ from $\text{SO}(p, q)$.

The identification and classification of the so-called 'classical groups' was greatly influenced by the representation of linear transformations by matrices. But when linear transformations are described instead by Geometric Algebra a different characterization of groups is more natural. For example, the designation of elements of $\text{SO}(p, q)$ as rotations in matrix group theory was probably adopted because determinants are natural in matrix algebra, while the condition for rotations expressed by (8.9) is not so easily expressed with matrices. For another example, the use of complex instead of real numbers simplifies matrix computations considerably, so much attention has been given to the study of the 'unitary groups' associated with a complex bilinear form. On the other hand, in Geometric Algebra a complex scalar field is uncalled for, so the unitary groups assume less significance. Important applications of unitary groups to physics are better based on the spin groups when Geometric Calculus is used. Some relations of the spin and rotor groups are described in [Po]. To understand the results of this section for applications like those in [H9], let us introduce some terminology of physics. *Spacetime* will be regarded as a pseudoEuclidean vector space with signature $(1, 3)$. The orthogonal group $O(1, 3)$ of spacetime is called the *Lorentz group*. The elements of this group are called *Lorentz transformations*. The group $SO^+(1, 3)$ is called the *proper Lorentz group* or the *Lorentz Rotation group*. A Lorentz transformation is said to be *proper* if it belongs to $SO^+(1, 3)$ and *improper* otherwise. $\text{Spin}^+(1, 3)$ is called the *Spin-$\frac{1}{2}$ representation* of the proper Lorentz group.

Our use of the term 'spinor' in reference to elements of $\text{Spin}(p, q)$ is unusual. It is justified by our unusual general definition of 'spinor', which is as follows: We say that an *even* multivector ψ in $\mathcal{G}(\mathscr{A}_{p,q})$ is a *spinor* of $\mathscr{A}_{p,q}$ if for each vector x in $\mathscr{A}_{p,q}$, $\psi x \psi^\dagger$ is also a vector. Versors satisfy this definition, so the question is, how much more general can ψ be? If we write $\psi x \psi^\dagger = \rho y$ where ρ is a scalar and $x^2 = y^2$, it is evident that y can be obtained from x by a transformation of type (8.8), so we can write

$$\psi x \psi^\dagger = \rho U x U^\dagger, \tag{8.10}$$

where U is an even versor. Solving for ρx, we get

$$\lambda x \lambda^\dagger = \rho x,$$

where $\lambda = U^\dagger \psi$. From previous considerations, we know that such an equation can obtain for all x only if λ has a scalar and possibly a pseudoscalar part only. If $p + q$ is odd, the pseudoscalar part must vanish, because λ is even. If $n = p + q$ is even, we write $\lambda = \alpha + \beta I$, where α and β are scalars, and use (8.5b) to get

$$(\alpha + \beta I)x(\alpha + \beta I^\dagger) = (\alpha^2 - II^\dagger \beta^2 + \alpha\beta(I - I^\dagger))x = \rho x.$$

Therefore, (unless $n = 2$) we have $\alpha = 0$, $\beta = 0$ or $I^\dagger = (-1)^{[n(n+1)]/2} I = I$. Also note that $\rho = \alpha^2 - II^\dagger \beta^2$ is always positive when $I^{-1} = (-1)^q I^\dagger = -I^\dagger$, that is, when q and p are odd. We conclude, finally, that every spinor ψ of $\mathscr{A}_{p,q}$ can be written in the form

$$\psi = (\alpha + \beta I)U, \tag{8.11}$$

where U is an even versor, and β necessarily vanishes unless $n = p + q = 4m$ where m is an integer. Thus, a spinor is always an even versor unless $\frac{1}{4}n = \frac{1}{4}(p + q)$ is an integer, in which case a spinor can always be expressed as the sum of two even versors.

At first sight, our definition of spinor appears to be quite different from the conventional definition, but the two have been proved equivalent in the cases of physical interest (see [H2], [H6] and [H9]). Our definition has the advantages of simplicity in its algebraic formulation and its geometrical interpretation. Thus, (8.10) shows that a spinor determines an orthogonal transformation and a dilatation (by a factor ρ).

Now let us return to the problem of classifying orthogonal transformations algebraically. Choose any pair of orthogonal vectors a and b such that $a^2 = 1$ and $b^2 = -1$. Every odd unit versor U can be expressed as the product of a rotor S with either a or b. One simply defines $S = Ua^{-1}$ if $U^\dagger U = 1$ or $S = Ub^{-1}$ if $U^\dagger U = -1$. If instead, U is an even unit versor satisfying $UU^\dagger = -1$, we can write $U = Sab$ where $S = Ub^{-1}a^{-1}$ is a rotor. We note that the set $\{\pm 1, \pm a, \pm b, \pm ab\}$ is a discrete subgroup of the versor group, and we have shown that every element of the versor group is the product of a rotor and an element of this discrete subgroup. This reduces the problem of describing the orthogonal group to the problem of characterizing rotors and rotations.

A rotor which can be expressed as the product of two unit vectors will be called a *simple rotor*, and the corresponding rotation will be called a *simple rotation*. Consider a simple rotor

$$S = ab = a \cdot b + a \wedge b. \tag{8.12a}$$

It determines a rotation

$$R(x) = SxS^\dagger = abxba. \tag{8.12b}$$

Now $S^\dagger S = a^2 b^2 = 1$, hence ab is a rotor if and only if $a^2 = b^2 = \pm 1$. The bivector $a \wedge b$, if it is not zero, determines a plane called the *plane of R or S*. The simple rotors, and, of course, their corresponding rotations and planes are of three distinct types distinguished by $(a \wedge b)^2$ negative, positive or zero. These types are referred to as *elliptic, hyperbolic* and *parabolic* respectively.

A vector in the plane of S satisfies the equation $2a \wedge b \wedge u = (a \wedge b)u + u(a \wedge b) = 0$, or equivalently,

$$v = SuS^\dagger = S^2 u. \tag{8.13}$$

The transformation (8.13) is said to be a *simple rotation of u*. We may assume that $v^2 = u^2 = \pm 1$ in (8.13), so

$$\pm S^2 = vu = v \cdot u + v \wedge u, \tag{8.14}$$

whence

$$S^2 S^{-2} = (v \cdot u)^2 - (v \wedge u)^2 = 1. \tag{8.15}$$

We can write S^2 in the parametric form

$$S^2 = e^{i\theta} = \sum_{k=0}^{\infty} \frac{1}{k!} (i\theta)^k = \pm vu, \tag{8.16}$$

where θ is a scalar and i is a bivector of the rotation plane. Then, for elliptic, hyperbolic and parabolic rotations respectively, we have the special forms

$$S^2 = \cos\theta + i\sin\theta \quad \text{if } i^2 = -1, \tag{8.17a}$$

$$S^2 = \cosh\theta + i\sinh\theta \quad \text{if } i^2 = 1, \tag{8.17b}$$

$$S^2 = 1 + i\theta \quad \text{if } i^2 = 0. \tag{8.17c}$$

Comparison of (8.17) with (8.14) shows that θ is the *rotation angle* in the elliptic and hyperbolic cases. In the parabolic case we have $v \wedge u = i\theta$ where the singular bivector $v \wedge u$ is unique, but its factors i and θ are not unless some additional condition is given. Note that in elliptic case a minus sign can be absorbed in the exponential by redefining the angle, thus, $-e^{i\theta} = e^{i(\pi+\theta)} = e^{i\theta'}$, but this is impossible in the hyperbolic and parabolic cases.

The square root of an exponential is easily computed, so, we get from (8.16) and (8.12a),

$$S = \pm e^{i\theta/2} = ab = \pm(vu^{-1})^{1/2}. \tag{8.18}$$

The exponential representation of a simple rotor shows that as θ tends to zero $S \to \pm 1$ and, according to (8.12b), $R(x) \to x$. Hence every simple rotation is continuously connected to the identity transformation.

Now that the procedure for finding the square root of a simple rotor S is clear, it is easy to prove that a vector a factors S if and only if it is a nonsingular vector in the plane of S. We only note that, if $aS = S^\dagger a = S^{-1}a$, then

$$S = aa^{-1}S = aS^{-1/2}a^{-1}S^{1/2} = ab,$$

where the factor of S 'conjugate' to a is $b = a^{-1}S = S^{-1/2}a^{-1}S^{1/2}$.

The square root of (8.14) can be evaluated in terms of vectors u and v without introducing exponentials. It is easily verified that the result is

$$\pm S = \frac{1 + vu^{-1}}{[2(1 + v\cdot u^{-1})]^{1/2}} = \frac{1 + S^2}{[2(1 + \langle S^2\rangle)]^{1/2}}. \tag{8.19}$$

This form is especially useful in physical applications of Lorentz transformations. For a parabolic rotation $S^2 = 1 \pm v \wedge u$ and (8.19) reduces to

$$\pm S = 1 + \tfrac{1}{2} v \wedge u^{-1}. \tag{8.20}$$

In most applications the negative root can be ignored.

We have found expressions for rotors and rotations as a function of angle and of vectors in the rotation plane. Other parametrizations of rotations are useful for different purposes. We mention only the *Cayley form* for a simple rotor S:

$$S = \frac{1 + B}{1 - B}, \tag{8.21}$$

where B is a bivector. To relate this to (8.18), we solve for B, obtaining

$$B = \frac{S-1}{S+1} = \frac{\pm e^{i\theta/2} - 1}{\pm e^{i\theta/2} + 1} = \frac{a \wedge b}{1 + a \cdot b}. \tag{8.22}$$

Having analyzed simple rotors in detail, we turn to the problem of finding a canonical form for a general rotor. In connection with (8.9) we determined that a rotor has an even number of distinct vector factors with positive (or negative) square. From any given factorization $S = u_{2m} \ldots u_2 u_1$ we can obtain a factorization for the rotor S which has all vector factors with positive square to the right of vector factors with negative square. This can be accomplished by noting that

$$ab = aba^{-1}a = b'a,$$

where $b'^2 = b^2$. By successive transformations of this kind we can achieve the desired order of factors. It follows that any rotor S can be factored into the form

$$S = a_m b_m \ldots a_2 b_2 a_1 b_1, \tag{8.23}$$

where $a_k^2 = b_k^2 = \pm 1$. We have seen that

$$S_k = a_k b_k = e^{i_k \theta_k /2} \tag{8.24a}$$

is a simple rotor. Hence, any rotor S in $\mathcal{G}(\mathcal{A}_{p,q})$ can be expressed as a product of simple rotors

$$S = S_m \ldots S_2 S_1, \tag{8.24b}$$

where $m \leqslant \frac{1}{2}(p+q)$.

We have determined that every simple rotor is continuously connected to 1 or -1 and is connected to both only if it is elliptic. Hence the rotor group $\mathrm{Spin}^+(p, q)$ has at most two connected pieces, and it is connected if p or $q > 1$. It follows that the rotation group $\mathrm{SO}^+(p, q)$ is connected.

The factorization (8.24b) of a rotor is by no means unique. This arbitrariness can be used to advantage as in the following theorem: Every rotation can be uniquely expressed as the composite of a rotation which leaves a given nonsingular vector invariant followed by a simple rotation in a plane containing that vector. The theorem may be re-expressed as an algebraic property of rotors as follows: Given a rotation

$$f(x) = RxR^\dagger \tag{8.25a}$$

and a nonsingular vector u, the rotor R can be expressed as a product

$$R = SU, \tag{8.25b}$$

where S is a simple rotor satisfying

$$Su = uS^\dagger \tag{8.25c}$$

and

$$Uu = uU. \tag{8.25d}$$

Hence, $f = f(x)$ can be expressed as the composite of rotations,

$$f(x) = S(UxU^\dagger)S^\dagger. \tag{8.26}$$

To prove (8.25), we determine S by writing

$$v = RuR^\dagger = SuS^\dagger = S^2u,$$

from which we obtain

$$S = (vu^{-1})^{1/2} = (RuR^\dagger u^{-1})^{1/2}. \tag{8.27}$$

Then we can get U from $U = S^\dagger R$.

We are still faced with the problem of finding a *canonical factorization* of type (8.24b) which best characterizes a rotor and its corresponding rotation. In Section 3-5 we proved that for spaces with Euclidean signature, every rotor S can be expressed as a product of commuting simple rotors, from which it follows that S can be expressed as the exponential of a bivector:

$$S = e^{\theta/2}. \tag{8.28a}$$

Adopting the notation of (8.24), we have

$$\theta = i_1\theta_1 + i_2\theta_2 + \cdots + i_m\theta_m, \tag{8.28b}$$

which expresses the decomposition of bivector θ into orthogonal blades. Since θ is a bivector, and the bivector space $\mathcal{G}^2(\mathcal{A}_n)$ has $\binom{n}{2} = \frac{1}{2}n(n-1)$ dimensions, (8.28a) shows at once that the rotation group is an $\frac{1}{2}n(n-1)$ parameter group.

The *canonical form* (8.28) for a rotor obtains also for spaces with the so-called *Lorentz signatures* $(1, q)$ or $(q, 1)$, with the minor modification

$$S = \pm e^{\theta/2}, \tag{8.29}$$

which is to say that a minus sign cannot always be absorbed in the exponential, a circumstance we have already noted in our study of simple rotor. Our derivation of the canonical form (8.28) in Section 3-5 works without significant change for the case of Lorentz signature, so we will not repeat it. A key step in the derivation is the decomposition of a bivector into orthogonal blades. This is always possible for spaces with Euclidean or Lorentz signature, but not otherwise. The problem of finding canonical forms for rotors of spaces with arbitrary signature is not completely solved.

Our 'rotor representation' of rotations is easily related to the matrix representation by introducing a basis $\{a_k\}$ for $\mathcal{A}_{p,q}$. From (8.9) we have

$$f(a_k) = Sa_kS^\dagger = f_{kj}a^j. \tag{8.30}$$

(Sum over repeated indices.) Hence, the 'matrix elements' of the rotation are

$$f_{kj} = a_j \cdot f(a_k) = \langle a_j S a_k S^\dagger \rangle. \tag{8.31}$$

By evaluating the quantity on the right side of (8.31), we can obtain the rotation matrix from a given rotor.

The inverse problem of finding the rotor from the matrix of a rotation is solved by taking the derivative of f and using (8.30);

$$\partial f = a^k S a_k S^\dagger = a^k f(a_k) = f_{kj} a^k a^j. \tag{8.32}$$

In Section 4 we discussed how to solve this equation for S by decomposing $\partial \wedge f$ into orthogonal blades (which we can do only for Euclidean or Lorentz signatures). Alternatively, for spaces of small dimension it is practical to solve (8.32) directly for S. For example, for $n = p + q = 4$, we can write S as a sum of its homogeneous parts,

$$S = S_{\bar{0}} + S_{\bar{2}} + S_{\bar{4}},$$

and use (2-1.40) to get

$$\partial_x S x = a^k S a_k = 4(S_{\bar{0}} - S_{\bar{4}}).$$

Hence (8.9) gives us

$$\partial f = 4(S_{\bar{0}} - S_{\bar{4}}) S^\dagger.$$

Since $SS^\dagger = 1$,

$$\partial f (\partial f)^\dagger = 16(S_{\bar{0}} - S_{\bar{4}})^2.$$

Hence,

$$S^\dagger = \pm \frac{\partial f}{[\partial f (\partial f)^\dagger]^{1/2}}. \tag{8.33}$$

Since ∂f is expressed in terms of f_{kj} by (8.32), Eqn. (8.33) gives the rotor $S^\dagger$ explicitly as a function of the matrix f_{kj}. Equation (8.33) fails when $S_{\bar{0}} = S_{\bar{4}} = 0$, or equivalently when $\partial f = 0$. In this case $S = S_{\bar{2}}$ in (8.9), which occurs for a simple elliptic rotation through an angle π.

3-9. Linear Multivector Functions

The general theory of linear multivector functions is obviously much more complex than the theory of linear vector functions which occupied us in the first eight sections of this chapter, and much work must be done before the subject can be expounded with any degree of completeness. So we aim here only to suggest a general approach to the subject by considering a few ideas and theorems with nontrivial applications. We concentrate our attention on the algebraic properties of multiforms. We have coined the term 'multiform' to refer to a type of linear function with wide applicability in mathematics. Multiforms are central to our theory of integration in Chapter 7. Chapter 6 shows that the Riemann curvature tensor is a type of multiform which we call a biform in this section. Next to the linear vector functions, the biforms compose the simplest significant class of multiforms. For these reasons, we pay special attention to the algebraic representation of biforms. Among other things, we show how to find canonical forms for tractionless biforms. This provides a complete algebraic classification of curvature tensors for vector manifolds of four dimensions, including the so-called *Petrov classification* of spacetime manifolds, which is of interest in Einstein's geometrical theory of gravitation. The advantages of this simple approach to the Petrov classification will be apparent to anyone familiar with the complexities in the literature.

Let $\mathcal{G} = \mathcal{G}(\mathcal{A}_n)$ be the Geometric Algebra of an n-dimensional vector space $\mathcal{A}_n$, and let $F = F(X)$ be a linear function on $\mathcal{G}$ with values in $\mathcal{G}$. Recalling from Section 2-2 the general definitions of the differential $\underline{F}$ and the adjoint $\overline{F}$ of a multivector function F, we note that the linearity of F entails

$$F = \underline{F}, \tag{9.1}$$

which with

$$\overline{F}(Y) = \partial_X Y * F(X) \tag{9.2}$$

implies

$$Y * F(X) = X * \overline{F}(Y). \tag{9.3}$$

We say that F is *symmetric* if $F = \overline{F}$ and *skewsymmetric* if $F = -\overline{F}$. We can always decompose F into the sum of a symmetric function F_+ and a skewsymmetric function F_- where

$$F_\pm = \tfrac{1}{2}(F \pm \overline{F}). \tag{9.4}$$

These features of linear multivector functions correspond exactly to features of linear vector functions which we noted earlier. To be sure, the geometric algebra $\mathcal{G}$ is a linear space of dimension 2^n, so the algebra of linear functions on $\mathcal{G}$ is isomorphic to the algebra of linear transformations on a vector space with dimension 2^n, and it can be analyzed by the methods we have already considered. But that approach does not exploit the distinctive algebraic structure of $\mathcal{G}$, and it is the effect of linear functions on that structure that interests us most. The effect of the function F on the structure of $\mathcal{G}$ is best determined by expressing F entirely in terms of algebraic operations on multivectors. We call this an *algebraic representation* of F. The problem of finding a *canonical form* for F is the problem of finding its simplest algebraic representation.

The study of how a linear function affects the grade of a multivector, reduces to the study of multiforms. A linear multivector valued function $F = F(X)$ is called a *multiform of degree r* if

$$F(X) = F(\langle X \rangle_r). \tag{9.5}$$

More often than not, the condition (9.5) will be met simply by specifying the space of r-vectors $\mathcal{G}^r = \mathcal{G}^r(\mathcal{A}_n)$ as the domain of F. Note that if F is scalar valued, then it is an 'alternating form of degree r', an 'r-form'. Alternating forms were defined and reduced to canonical form in Section 1-4. We have adopted the term 'multiform' as an abbreviation for the phrase 'alternating form with multivector values' to emphasize that we are dealing with a straightforward generalization of the conventional concept of alternating form.

Any linear multivector function $F = F(X)$ on $\mathcal{G}$ can be expressed as a sum of multiforms by decomposing the argument into its graded parts; thus

$$F(X) = \sum_{r=0}^{n} F_r(X), \tag{9.6a}$$

where

$$F_r(X) \equiv F(\langle X \rangle_r). \tag{9.6b}$$

Therefore, the study of linear multivector functions can be reduced to the study of multiforms. Because of this fact and the important applications of multiforms in later chapters, we concentrate on the analysis of multiforms in the balance of this section. But before moving on, it should be said that for many purposes the decomposition of a linear function into multiform parts is inappropriate. For example, such a decomposition will introduce unnecessary complications when the algebraic representation of the function is simpler than that of its multiform parts, and there are many functions with this property. Indeed, the best way to deal with a given multiform may well be to replace it by a more general function of which it is a part.

Decomposition of a linear function into multiform parts is frequently inappropriate when duality considerations are significant. This brings up a matter of great importance, namely, the role of 'complex numbers' in linear algebra and geometry. In other sections of this book we explain the advantages of interpreting 'imaginary numbers' as bivectors in Geometric Algebra. A little terminology will help us get a more general perspective on complex numbers. Let us refer to an element of the space $\mathcal{G}^r + \mathcal{G}^{n-r}$ as a *complex r-vector*. It is easy to show that any complex r-vector Z can be written in the form

$$Z = X + YI, \tag{9.6c}$$

where $X = \langle X \rangle_r$, $Y = \langle Y \rangle_r$ and I is the unit pseudoscalar. Obviously, the *scalar* imaginary unit i of a 'conventional' m-dimensional 'complex vector space' $\mathcal{V}$ can always be represented as a pseudoscalar I in Geometric Algebra so that multiplication of a 'real vector' in $\mathcal{V}$ by i corresponds to the 'duality rotation' of an r-vector Y into its dual $YI = \langle YI \rangle_{n-r}$.* This requires only that n and r be selected so that the dimensions of $\mathcal{V}$ and $\mathcal{G}^r$ are the same. As a rule, this can be done in a number of different ways, each one of which assigns a different geometrical interpretation to $\mathcal{V}$. We suggest that some such an interpretation is implicit in every significant application of complex vector spaces to geometry and physics, and that much is to gained by making the geometrical role of the unit 'imaginary' explicit. (A detailed analysis of the geometrical role of complex numbers in physics is given in [H9].) An important example is discussed at the end of this section. Finally, to connect these remarks to our discussion of multiforms, let us define a *complex multiform* of degree r to be a linear function of complex r-vectors. This should suffice to indicate how the conventional theory of linear functions on complex vector spaces can be translated into the language of Geometric Calculus with i corresponding to the pseudoscalar and so assuming the geometrical role of 'duality operator'.

Now let us get on with our study of multiforms. Some additional terminology will be helpful. A multiform $F = F(X)$ is said to have *grade s* if

$$F(X) = \langle F(X) \rangle_s. \tag{9.7}$$

A multiform of degree r and grade s is said to be an *s, r-form*. If $r = s$, the multiform is said to be *grade-perserving*. If it maps r-blades into r-blades, a multiform is said to be *blade-preserving*. Obviously, every blade-preserving multiform is grade-preserving.

According to the basic Eqn. (3-1.7), an outermorphism is a blade-preserving linear function. This raises the question: Is a blade-preserving multiform necessarily an outermorphism? The answer is yes, except possibly when $n = 2r$, that is, when the grade of the pseudoscalar is twice the grade of the multiform. (A proof of this fact is given in ref. [We]. We are sure that the treatment in [We] can be simplified and clarified by using Geometric Algebra, but we have not had the occasion to work out the details.) In the exceptional case, a multiform can be expressed as the composite of an outermorphism and a *duality*. A duality is a linear mapping of a multivector into its dual; this is blade-preserving iff $n = 2r$, but it is never an outermorphism.

For the systematic manipulation and classification of multiforms, it is convenient to define operations we call *tractions*. From a multiform $F(X) = F(\langle X \rangle_r)$ of degree r, the operation of *contraction* determines a multiform of grade $r - 1$ defined by

$$\partial_x \cdot F(x \wedge Y), \tag{9.8a}$$

where x is a vector variable and $Y = \langle Y \rangle_{r-1}$. Our definition 'contraction' is in accordance with the conventional use of that term in tensor analysis. However, we introduce a new term for the multiform

$$\partial_x \wedge F(x \wedge Y). \tag{9.8b}$$

* The fact that $i^2 = -1$ where both $I^2 = 1$ and $I^2 = -1$ may occur is quite unnecessary to this correspondence of i with I.

We call it the *protraction* of F. The contraction of an s, r-form is an $(s-1), (r-1)$-form, while its protraction is an $(s+1), (r-1)$-form. Thus, contraction lowers both the degree and grade of a multiform by one unit, while protraction lowers degree and raises grade by one unit. We call the multiform

$$\partial_x F(x \wedge Y) = \partial_x \cdot F(x \wedge Y) + \partial_x \wedge F(x \wedge Y) \tag{9.8c}$$

the *traction* of F. A multiform is said to be *tractionless* if its traction vanishes. The terms *contractionless* and *protractionless* are similarly defined.

We have actually already made good use of tractions in preceding sections. According to our definitions, a linear transformation is a 1, 1-form. In Section 3-4 we proved that a linear transformation is symmetric if and only if it is protractionless (see Eqns. (3-4.2a, c)). In general, however, symmetic multiforms are not necessarily protractionless. This can be seen by examining the 2, 2-form

$$F(x \wedge y) = \tfrac{1}{2}(x \wedge y) \cdot (B \wedge B) - (x \wedge y) \cdot BB, \tag{9.9}$$

where x and y are vectors and B is a bivector. According to (3-4.21), this multiform is the outermorphism of the skewsymmetric linear transformation $f(x) = x \cdot B$. From (9.9) it is easy to see that $F(x \wedge y)$ is symmetric, but with the help of (2-1.38) one gets

$$\partial_x F(x \wedge y) = \tfrac{3}{2} y \cdot (B \wedge B) - y \cdot BB,$$

whence the protraction is

$$\partial_x \wedge F(x \wedge y) = y \cdot (B \wedge B),$$

which does not vanish in general.

Although a symmetric multiform is not necessarily protractionless, a protractionless multiform is necessarily symmetric if it is grade-perserving, that is

$$\partial_x \wedge F(x \wedge Y) = 0 \quad \text{implies } F(X) = \bar{F}(X) \tag{9.10}$$

if $F(X) = F(\langle X \rangle_r) = \langle F(X) \rangle_r$ is a linear function of X. To prove this important theorem and for related computations it is convenient to employ the notations and conventions for the simplicial variable,

$$x_{(r)} \equiv x_1 \wedge x_2 \wedge \ldots \wedge x_r,$$

and the simplicial derivative,

$$\partial_{(r)} = \frac{1}{r!} \partial_r \wedge \ldots \wedge \partial_2 \wedge \partial_1,$$

which were introduced in Section 2-3. This enables us to give a compact form to the following identity which holds for an r, r-form F and r-vector A:

$$A \cdot [\partial_{(r)} \wedge F(x_{(r)})] = (A \cdot \partial_{(1)}) \cdot [\partial_{(r-1)} \wedge F(x_{(r-1)} \wedge x_{(1)})] - $$
$$- (A \cdot \partial_{(2)}) \cdot [\partial_{(r-2)} \wedge F(x_{(r-2)} \wedge x_{(2)})] + \cdots +$$
$$+ (-1)^r (A \cdot \partial_{(r-1)}) \cdot [\partial_1 \wedge F(x_1 \wedge x_{(r-1)})] +$$
$$+ (-1)^{r+1} [F(A) - \bar{F}(A)]. \tag{9.11}$$

This identity can be proved by iteration. The method of proof is amply indicated by the case $r = 2$. With the help of identities (1-1.43) and (1-1.42) we have

$$\begin{aligned} A \cdot [\partial_{(2)} \wedge F(x_{(2)})] &= \tfrac{1}{2} A \cdot [\partial_2 \wedge \partial_1 \wedge F(x_1 \wedge x_2)] \\ &= \tfrac{1}{2} (A \cdot \partial_2) \cdot (\partial_1 \wedge F) + \tfrac{1}{2} \partial_2 \wedge (A \cdot (\partial_1 \wedge F)) \\ &= \tfrac{1}{2} (A \cdot \partial_2) \cdot (\partial_1 \wedge F) + \tfrac{1}{2} \partial_2 \wedge [(A \cdot \partial_1) \cdot F] + \tfrac{1}{2} \partial_2 \wedge \partial_1 (A \cdot F) \\ &= (A \cdot \partial_2) \cdot (\partial_1 \wedge F) - \tfrac{1}{2} A \cdot (\partial_2 \wedge \partial_1) F + \tfrac{1}{2} \partial_2 \wedge \partial_1 A \cdot F, \end{aligned}$$

which agrees with (9.11) when the linearity of F is used. With (9.11) established, the proof of our theorem (9.10) is trivial, for if F is protractionless, all the terms in (9.11) except the last one obviously vanish.

Protractionless multiforms have other properties besides the relation to symmetry which we have just established. If a multiform F is protractionless, then its k-fold contraction $\partial_{(k)} \cdot F(x_{(k)} \wedge Y)$ is also protractionless, that is,

$$\partial_1 \wedge F = 0 \quad \text{implies } \partial_{k+1} \wedge (\partial_{(k)} \cdot F) = 0. \tag{9.12a}$$

The case $k = 1$ is easily proved by using the identity (1-1.42) which gives

$$\partial_2 \cdot (\partial_1 \wedge F) = -\partial_1 \wedge (\partial_2 \cdot F) = \partial_2 \wedge (\partial_1 \cdot F),$$

since $\partial_2 \cdot \partial_1 F = 0$. The same argument gives a similar relation for any k, namely

$$\partial_{k+1} \cdot [\partial_k \wedge (\partial_{(k-1)} \cdot F)] = \partial_{k+1} \wedge [\partial_k \cdot (\partial_{(k-1)} \cdot F)] = k\, \partial_{k+1} \wedge (\partial_{(k)} \cdot F).$$

Whence (9.12a) follows by induction. And from (9.12a) it follows easily that

$$\partial_{(k)} F = \frac{1}{k!} \partial_k \ldots \partial_2 \partial_1 F = \partial_{(k)} \cdot F \tag{9.12b}$$

if $\partial_1 \wedge F = 0$. In a similar way, one proves that $\partial_1 \cdot F = 0$ implies

$$\partial_{k+1} \cdot (\partial_{(k)} \wedge F) = 0 \tag{9.13a}$$

and

$$\partial_{(k)} F = \frac{1}{k!} \partial_k \ldots \partial_2 \partial_1 F = \partial_{(k)} \wedge F. \tag{9.13b}$$

These results can be used in the systematic classification of multiforms.

Any multiform can be decomposed into a part that is tractionless and a part that is not. For a linear transformation this is just the isolation of its symmetric traceless part. Let us see how to carry out the decomposition explicitly for the next case in order of complexity and importance, namely, for a 2, 2-form. We shall be working with 2, 2-forms so frequently that it will be convenient to adopt the compact name *biform* to refer to one. As we shall see in Chapter 5, the Riemann curvature tensor $R(a \wedge b)$ is a biform. In fact it is a protractionless biform, that is,

$$\partial_a \wedge R(a \wedge b) = 0. \tag{9.14}$$

We can analyze the algebraic properties of such biforms without reference to curvature or manifolds, yet the results have implications for differential geometry.

To analyze the tractions of a biform we will need to compute vector and bivector derivative, so let us record the results we need from Chapter 2 for reference. Let a and b be vector variables and let $K = \langle K \rangle_k$ be an arbitrary k-vector. Then

$$\partial_a a \cdot K = kK = K \cdot \partial_a a \quad \text{for } k \geqslant 1, \tag{9.15a}$$

$$\partial_a a \wedge K = (n-k)K = K \wedge \partial_a a \quad \text{for } k \geqslant 0. \tag{9.15b}$$

Now let $F = F(X)$ be an arbitrary biform. According to (1-1.63) the (bivector) derivative $\partial F = \partial_X F(X)$ can be decomposed into 0-vector, 2-vector and 4-vector parts by

$$\partial F = \partial \cdot F + \partial \times F + \partial \wedge F, \tag{9.16}$$

where $\partial \times F$ is the commutator product of ∂ and F. Since $F = F(X)$ is a linear function, the bivector derivative is related to vector derivatives by

$$\partial F = \tfrac{1}{2}\, \partial_b \wedge \partial_a F(a \wedge b) = \tfrac{1}{2}\, \partial_b\, \partial_a F(a \wedge b). \tag{9.17}$$

Derivatives of linear bivector functions of homogeneous grade are given by

$$\partial_X X \cdot K = \frac{k(k-1)}{2} K = K \cdot \partial_X X \quad \text{for } k \geqslant 2, \tag{9.18a}$$

$$\partial_X X \times K = k(n-k)K = K \times \partial_X X \quad \text{for } k \geqslant 1, \tag{9.18b}$$

$$\partial_X X \wedge K = \frac{(n-k)\,(n-k-1)}{2} K = K \wedge \partial_X X \quad \text{for } k \geqslant 0. \tag{9.18c}$$

These equations are easily derived from (9.15a) and (9.15b) by using (9.17). Note the following special cases of (9.18b) and (9.18c) respectively:

$$\partial_X X \times a = \partial_X X \cdot a = (n-1)a = a \cdot \partial_X X, \tag{9.19}$$

$$\partial_X X = \frac{n(n-1)}{2}. \tag{9.20}$$

As a check on the consistency of the above equations it may verified that

$$\partial_X(X \cdot K + X \times K + X \wedge K) = (\partial_X X)K.$$

The decomposition of any biform into independent parts determined by its tractions is completely described by the following theorem: *Any biform* $F = F(X)$ *can be uniquely decomposed into a sum of biforms*

$$F = T + F_0 + F_1 + F_2 + F_3 + F_4, \tag{9.21}$$

where

$$\partial_a T(a \wedge b) = 0, \tag{9.22a}$$

$$F_0(X) = \frac{2X\, \partial \cdot F}{n(n-1)}, \tag{9.22b}$$

$$F_1(X) = \frac{X \cdot \partial f_1}{n-2} \quad \text{with } \partial f_1 = 0, \tag{9.22c}$$

$$F_2(X) = \frac{X \times (\partial \times F)}{2(n-2)}, \tag{9.22d}$$

$$F_3(X) = \frac{(X \cdot \partial) \cdot f_3}{2} \quad \text{with } \partial f_3 = 0, \tag{9.22e}$$

$$F_4(X) = \frac{X \cdot (\partial \wedge F)}{6}. \tag{9.22f}$$

Thus, T is the tractionless part of F and the F_k determine the various tractions of F. In particular, the twofold tractions of F are the 'invariants'

$$\langle \partial F \rangle_0 = \partial \cdot F = \partial F_0, \tag{9.23a}$$

$$\langle \partial F \rangle_2 = \partial \times F = \partial F_2, \tag{9.23b}$$

$$\langle \partial F \rangle_4 = \partial \wedge F = \partial F_4, \tag{9.23c}$$

while $\partial F_1 = \partial F_3 = 0$. A single contraction of F gives a linear vector function

$$f(b) \equiv \partial_a \cdot F(a \wedge b) = \frac{2b}{n}\, \partial \cdot F + f_1(b) + \frac{b \cdot (\partial \times F)}{2}. \tag{9.24}$$

Since $\partial f_1 = \partial_b f_1(b) = 0$, the linear vector function $f_1 = f_1(b)$ is symmetric and traceless. Thus, the right side of (9.24) is the unique decomposition of a linear vector function $f = f(b)$ into symmetric and skewsymmetric parts determined by its trace $\partial \cdot f = 2\partial \cdot F$ and its curl $\partial \wedge f = \partial \times F$. Similarly, a single protraction of F produces a trivector valued function of a vector variable.

$$\partial_a \wedge F(a \wedge b) = \frac{b \wedge (\partial \times F)}{n-2} + f_3(b) + \frac{b \cdot (\partial \wedge F)}{2}, \tag{9.25}$$

where again the tractionless part f_3 has been separated from the rest of the function.

Verification of (9.23), (9.24) and (9.25) is a simple matter of differentiation with the help of (9.15) and (9.18) and a little algebra. For example, to differentiate F_4 we write it in the form

$$F_4(a \wedge b) = (a \times b) \times B = (a \times B) \times b + a \times (b \times B)$$
$$= (a \cdot B) \wedge b + a \wedge (b \cdot B).$$

Then

$$\partial_a \cdot F(a \wedge b) = [\partial_a \cdot (a \wedge B)] \wedge b - b \cdot \partial_a(a \cdot B) + (\partial_a \cdot a)b \cdot B + (b \cdot B) \cdot \partial_a a$$
$$= (n-2)b \cdot B,$$

and

$$\partial_a \wedge F(a \wedge b) = [\partial_a \wedge (a \cdot B)] \wedge b = 2B \wedge b.$$

So

$$\partial_a F_4(a \wedge b) = (n-2)b \cdot B + 2B \wedge b.$$

Differentiating again, we have

$$\partial F_4 = \tfrac{1}{2}\, \partial_b\, \partial_a F_4(a \wedge B) = \frac{(n-2)}{2}\, \partial_b(b \cdot B) + \partial_b(B \wedge b) = 2(n-2)B.$$

To differentiate F_3, we write

$$2F_3(a \wedge b) = [(a \wedge b) \cdot \partial] \cdot f_3 = [ab \cdot \partial - ba \cdot \partial] \cdot f_3 = a \cdot f_3(b) - b \cdot f_3(a).$$

Then

$$2\partial_a \cdot F_3(a \wedge b) = (\partial_a \wedge a) \cdot f_3(b) - b \cdot [\partial \cdot f_3] = 0$$

and

$$2\partial_a \wedge F_3(a \wedge b) = 3f(b) - \partial_a \wedge (b \cdot f_3(a)) = \partial f_3(b),$$

since

$$b \cdot (\partial \wedge f_3) = b \cdot \partial f_3 - \partial \wedge (b \cdot f_3) = 0.$$

Differentiating again, we have

$$\partial F_3 = \tfrac{1}{2}\, \partial_b\, \partial_a F_3 = \tfrac{1}{2}\, \partial f_3 = 0.$$

Differentiation of the other functions procedes in a similar way, but it is easier than the cases we have just considered.

Equation (9.21) can be regarded as a canonical form for a biform, although it leaves us with the problem of finding canonical forms for the tractionless multiforms T, f_1 and f_3. Of course, f_1 is a symmetric linear transformation, and we have already discussed the problem of putting it in canonical form. We shall solve the problem of putting T and f_3 in canonical forms for an important special case later on. It should be realized that there are other canonical forms besides (9.21) which are more appropriate for some purposes. For example, according to (3-5.37), the differential outermorphism of an orthogonal transformation produces a biform

$$F(X) = U^{\dagger} X U, \tag{9.26}$$

where U is a spinor satisfying $U^{\dagger} U = 1$. Clearly (9.26) is already the simplest (canonical) form for such a biform, and a decomposition into parts in accordance with (9.21) would only introduce unnecessary complications.

Equation (9.21) enables us to decompose any biform F into its symmetric part F_+ and its skewsymmetric part F_-; specifically,

$$F_+ = T + F_0 + F_1 + F_4 \tag{9.27a}$$

and

$$F_- = F_2 + F_3. \tag{9.27b}$$

According to (9.10), the symmetry of T, F_0 and F_1 follows from the fact that they are protractionless, although the symmetry of F_0 and F_1 can be verified directly; for example, for any bivectors A and B

$$A \cdot (B \cdot \partial f_1) = [A \cdot (B \cdot \partial)] \cdot f_1 = (A \times B) \cdot (\partial \wedge f_1) + [B \cdot (A \cdot \partial)] f_1$$

implies

$$A \cdot F_1(B) = B \cdot F_1(A) \qquad \text{so } F_1 = \overline{F}_1 .$$

The symmetry of F_4 follows from

$$A \cdot [B \cdot (\partial \wedge F)] = (A \wedge B) \cdot (\partial \wedge F) = B \cdot [A \cdot (\partial \wedge F)] .$$

The skewsymmetry of F_2 follows from the identity

$$A \cdot (B \times C) = -(A \times C) \cdot B .$$

Finally, the skewsymmetry of F_3 can be established by using the identity

$$\begin{aligned}(A \wedge B) \cdot a = (A \wedge B) \times a &= A \wedge (B \times a) + (A \times a) \wedge B \\ &= A \wedge (B \cdot a) + (A \cdot a) \wedge B,\end{aligned}$$

which can be derived from (1-1.57).

Thus,

$$A \cdot F_3(B) = A \cdot [(B \cdot \partial) \cdot f_3] = [A \wedge (B \cdot \partial)] \cdot f_3$$
$$= -[B \wedge (A \cdot \partial)] \cdot f_3 + (A \wedge B) \cdot (\partial \wedge f_3) = -\partial B \cdot F_3(A).$$

This completes the proof of (9.27).

Now let us turn for the moment to the general problem of finding an appropriate algebraic representation for a given multiform. We aim to establish the basic theorem that any multiform F from $\mathcal{G}^r$ to $\mathcal{G}$ can be written in the 'algebraic form'

$$F(X) = \sum_k A_k X B_k, \tag{9.28}$$

where $X = \langle X \rangle_r$ the A_k and B_k are multivectors in $\mathcal{G}$. This generalizes Eqn. (3-0.1) for linear transformations. We can prove (9.28) by expanding F in terms of a basis of blades $\{a^J\}$ for $\mathcal{G}$ and using (9.3); thus,

$$F(X) = \sum_J a_J a^J * F(X) = a_J b^J * X,$$

where $b^J = \bar{F}(a^J) = \Sigma_K \beta^{JK} a_K$ and the β^{JK} are scalars. Introducing multivectors $c^K = \Sigma_J a_J \beta^{JK}$, we have

$$F(X) = \sum_K c^K a_K * X = \sum_K c^K \langle a_K \rangle_r * X. \tag{9.29}$$

We have used the fact that $X = \langle X \rangle_r$ to determine that only the basis vectors $\langle a_K \rangle_r$ for $\mathcal{G}^r$ contribute to the sum, as is expressed on the right side of (9.29). The proof can be completed simply by showing that $\langle a_K \rangle_r * X$ can be expressed in terms of the geometric product. To do this, we factor the blade $\langle a_K \rangle_r$ into vectors, $\langle a_K \rangle_r = a_r \wedge \ldots \wedge a_2 \wedge a_1$, and apply (1-1.25b) r times; thus

$$\langle a_K \rangle_r * X = (a_r \wedge \ldots \wedge a_2 \wedge a_1) \cdot X = a_r \cdot (\ldots a_2 \cdot (a_1 \cdot X) \ldots).$$

Then we apply (1-1.30) r times to 'remove' the dots. For example, if r is even, removal of the first two dots gives

$$a_2 \cdot (a_1 \cdot X) = \tfrac{1}{2} a_2 \cdot (a_1 X - X a_1) = \tfrac{1}{4}(a_2 a_1 X - a_2 X a_1 + a_1 X a_2 - X a_1 a_2).$$

It should be emphasized that the assumption $X = \langle X \rangle_r$ is essential to this last step, so our proof cannot be generalized from multiforms to establish the algebraic form (9.28) for arbitrary multilinear functions.

The algebraic form (9.28) for a multiform is, of course, not unique. The problem of finding a canonical form for F is equivalent to finding the set of A_k and B_k which is simplest in some sense; for example, one could reduce the number of terms in the sum to a minimum, or one could require that the A_k be orthogonal blades and reduce the number to a minimum. If the A_k and B_k are expanded in terms of a basis, then (9.28) can be written in the form

$$F(X) = \sum_{J,K} a_J X a_K \alpha^{JK}, \tag{9.30}$$

where the α^{JK} are scalars. Substitution of (9.30) into (9.2) and use of (1-1.47a) shows that $\alpha^{JK} = \alpha^{KJ}$ if F is symmetric and $\alpha^{JK} = -\alpha^{KJ}$ if F is skew symmetric. The α^{JK} should not be confused with the 'matrix elements' of F, which are given by

$$\begin{aligned} a^M * F(\langle a^N \rangle_r) &= \sum_{J,K} \langle a^M a_J \langle a^N \rangle_r a_K \rangle \alpha^{JK} \\ &= \sum_{J,K} \langle a_K a^M a_J \rangle_r * \langle a^N \rangle_r \alpha^{JK}. \end{aligned} \tag{9.31}$$

Simplifications of the algebraic form (9.30) can be made by a change of basis to simplify the coefficient matrix α^{JK}. But it should be realized that the symmetric part of α^{JK} cannot always be diagonalized in this way because the metric of $\mathcal{G}$ is not necessarily positive definite, since some blades may have negative square while others have positive square.

The sum over all blades in (9.30) can be reduced by half to a sum over all blades with grade $r \leqslant \frac{1}{2}n$ by expressing every blade with grade $r > \frac{1}{2}n$ as the dual of a $(n-r)$-blade. When this has been done, the coefficients α^{JK} are *complex scalars*, so each one can be expressed in the form (9.6) with the unit pseudoscalar I playing the role of 'unit imaginary'. It must be remembered that some multivectors may not commute with complex scalars, because they do not commute with I. As a reminder of this, the α^{JK} have been placed on the right in (9.30).

As a specific application of what we have just learned let us analyze biforms on $\mathcal{G}^2(\mathcal{A}_4)$. As a basis for the algebra $\mathcal{G}(\mathcal{A}_4)$ we choose the scalar 1, four linearly independent vectors a_μ ($\mu = 0, 1, 2, 3$) and three linearly independent bivector blades f_k ($k = 1, 2, 3$). Any element of $\mathcal{G}(\mathcal{A}_4)$ can be written as a linear combination of these eight blades over the 'complex scalars'. In particular, it will be noted that the space of bivectors $\mathcal{G}^2(\mathcal{A}_4)$ is six-dimensional, but it is 'self-dual', which is to say that the dual of a bivector is also a bivector. It follows that cne can choose a basis of blades in $\mathcal{G}^2(\mathcal{A}_4)$ such that three elements are duals of the other three. Hence the six-dimensional space over the 'real' scalars is a three-dimensional linear space over the 'complex scalars' spanned by the blades f_k. Having established this, we can introduce these basis blades into (9.28) and, using the fact that I commutes with X, write any multiform F on $\mathcal{G}^2(\mathcal{A}_4)$ in the form

$$F(X) = A_0 X + X B_0 + \sum_{\mu=1}^{4} a_\mu X A_\mu + \sum_{k=1}^{3} f_k X B_k, \tag{9.32}$$

where $\langle A_\mu \rangle_0 + \langle A_\mu \rangle_4 = 0 = \langle B_k \rangle_0 + \langle B_k \rangle_4$ is required, so the first two terms are distinct from the other terms. Further restrictions must be placed on the multivectors A_0, B_0, A_μ and B_k if $F(X)$ is to be a biform.

We shall prove that any biform $F(X)$ can be put in the form

$$F(X) = X \times B + \sum_{\mu,\nu=1}^{4} a_\mu X a_\nu \beta_{\nu\mu} + \sum_{j,k=1}^{3} f_j X f_k \alpha_{jk}, \tag{9.33a}$$

where $B = \langle B\rangle_2$ and the $\beta_{\mu\nu}$ and the α_{jk} are symmetric matrices of 'complex scalars', that is,

$$\beta_{\mu\nu} = \beta_{\nu\mu}, \qquad \alpha_{jk} = \alpha_{kj}, \tag{9.33b}$$

and $\beta_{\mu\nu} = \langle\beta_{\mu\nu}\rangle_0 + \langle\beta_{\mu\nu}\rangle_4$, $\alpha_{jk} = \langle\alpha_{jk}\rangle_0 + \langle\alpha_{jk}\rangle_4$. To establish (9.33a, b) we merely need to ascertain the restrictions on the various terms in (9.32) required to insure that $F(X) = \langle F(X)\rangle_2$. We see immediately that if the values of $F(X)$ are to be even multivectors, then A_0, B_0 and the B_k must be even while the A_μ must be odd. Further, we see that we must require $\langle B_0\rangle_2 = -\langle A_0\rangle_2 \equiv \frac{1}{2}B$ if the combination of the first two terms in (9.32) is to be a bivector. Then we can write

$$A_0 X + X B_0 = X\alpha + X \times B, \tag{9.34}$$

where $\alpha = \langle\alpha\rangle_0 + \langle\alpha\rangle_4$ is a 'complex scalar'.

Turning to the next group of terms in (9.32), we observe that since A_μ must be an odd multivector and the trivectors are dual to the vectors, we can expand A_μ in terms of a vector basis; thus

$$A_\mu = \sum_{\nu=1}^{4} a_\nu \beta_{\mu\nu},$$

where the $\beta_{\mu\nu}$ are 'complex coefficients'. Restrictions must be placed on the coefficients $\beta_{\mu\nu}$ so that

$$\sum_{\mu=1}^{4} a_\mu X A_\mu = \sum_{\mu,\nu} \alpha_\mu X a_\nu \beta_{\mu\nu} \tag{9.35}$$

is bivectorvalued. To see what these restrictions must be, consider

$$\begin{aligned} a_1 X a_2 &= a_1(X \cdot a_2 + X \wedge a_2) \\ &= a_1 \wedge (X \cdot a_2) + a_1 \cdot (X \wedge a_2) + a_1 \cdot X \cdot a_2 + a_1 \wedge X \wedge a_2 \\ &= a_1 \wedge (X \cdot a_2) + a_2 \wedge (X \cdot a_1) - a_1 \cdot a_2 X + X \cdot (a_2 \wedge a_1) + X \wedge a_1 \wedge a_2. \end{aligned}$$

From this we can conclude that only the symmetric part of the coefficient matrix $\beta_{\mu\nu}$ will contribute to the bivector part of (9.35), so the skewsymmetric part of $\beta_{\mu\nu}$ can be assumed to vanish, and we can write

$$\sum_{\mu,\nu=1}^{4} a_\mu X a_\nu \beta_{\mu\nu} = 2 \sum_{\mu,\nu=1}^{4} a_\mu \wedge (X \cdot a_\nu)\beta_{\mu\nu} - X \sum_{\mu,\nu=1} a_\mu \cdot a_\nu \beta_{\mu\nu}. \tag{9.36}$$

The last term in (9.36) duplicates the term $X\alpha$ in (9.34), so the later has been dropped in combining (9.36) with (9.34) to get (9.33a).

Turning now to the last group of terms in (9.32), we observe that since the B_k must be bivectors they can be expanded in a bivector basis, and we have

$$\sum_{k=1}^{3} f_k X B_k = \sum_{j,k=1}^{3} f_j X f_k \alpha_{jk}. \tag{9.37}$$

To determine possible restrictions on the complex coefficients α_{jk}, consider the bivector identity

$$\begin{aligned} f_1 X f_2 &= f_1 f_2 X + 2 f_1 (X \times f_2) = X f_1 f_2 + 2(f_1 \times X) f_2 \\ &= (f_1 \cdot f_2 + f_1 \wedge f_2) X + (f_1 \times f_2) \cdot X + (f_1 \times f_2) \wedge X + \\ &\quad + f_1 \cdot (X \times f_2) - f_2 \cdot (X \times f_1) + f_1 \wedge (X \times f_2) - f_2 \wedge (X \times f_2) + \\ &\quad + f_1 \times (X \times f_2) + f_2 \times (X \times f_1). \end{aligned}$$

Only the first and the last terms in this expression have bivector values and they are symmetric in f_1 and f_2; the other six terms are skewsymmetric in f_1 and f_2 and none of them have bivector values. Hence, we can assume that $\alpha_{jk} = \alpha_{kj}$ and write (9.37) in the form

$$\sum_{j,k=1}^{3} f_j X f_k \alpha_{jk} = 2 \sum_{j,k} f_j \times (X \times f_k) \alpha_{jk} + X \sum_{j,k} f_j \cdot f_k \alpha_{jk}, \tag{9.38}$$

where we have used the fact that $f_j \wedge f_k = 0$ for the kind of bivector basis we have chosen. This completes our proof that any biform can be written in the form (9.33a) subject to (9.33b).

Equation (9.33a) can be used in the classification of biforms in a number of different ways. For example, a biform $G(X)$ is said to be *selfdual* if

$$G(IX) = IG(X). \tag{9.39a}$$

A biform $H(X)$ is *antiselfdual* if

$$H(IX) = -IH(X). \tag{9.39b}$$

The algebra $\mathcal{G}(\mathcal{A}_4)$ has the unique property that any biform $F(X)$ defined in it can be decomposed into selfdual and antiselfdual parts;

$$F(X) = G(X) + H(X). \tag{9.39c}$$

Since I commutes with bivectors and anticommutes with vectors, from (9.33a) we can see at once that any selfdual biform can be written in the form

$$G(X) = \sum_{j,k=1}^{3} f_j X f_k \alpha_{jk} + X \times B, \tag{9.40a}$$

while any antiselfdual biform can be written

$$H(X) = \sum_{\mu,\nu=1}^{4} a_\mu X a_\nu \beta_{\mu\nu}. \tag{9.40b}$$

As we shall see, these equations yield canonical forms for the biforms when the f_k and the a_μ are chosen to diagonalize the coefficient matrices α_{jk} and $\beta_{\mu\nu}$ in so far as possible.

To relate our algebraic reduction of a biform (9.33a) to the earlier traction-based reduction (9.21), we compute the tractions of (9.33a). For this purpose we need from (2-1.40) the derivatives

$$\partial_a ba = -2b$$

for an vector b and

$$\partial_a Ba = 0$$

for any bivector B. In (9.40a) the selfdual part of $F(X)$ is already separated into symmetric and skewsymmetric parts

$$G(X) = G_+(X) + G_-(X), \tag{9.41a}$$

and the skew part

$$G_-(X) = X \times B \tag{9.41b}$$

can be identified immediately with the part F_2 in (9.22). Now,

$$\partial_a f_j a \wedge b = \partial_a f_j (ab - a \cdot b) = -bf_j,$$

so, from (9.4a),

$$\partial_a G_+(a \wedge b) = -b \sum_{j,k} f_j f_k \alpha_{jk}. \tag{9.42}$$

From this we can conclude that G_+ is tractionless if

$$\partial G_+ = \tfrac{1}{2}\, \partial_b\, \partial_a G_+(a \wedge b) = -2 \sum_{j,k} f_j f_k \alpha_{jk} = -2 \sum_{j,k} f_j \cdot f_k \alpha_{jk} = 0. \tag{9.43}$$

Then we can identify G_+ with the tractionless biform T in (9.22). Thus, the class tractionless biforms in $\mathcal{G}(\mathcal{A}_4)$ is equivalent to the class of selfdual, symmetric 'traceless' biforms.

To compute the traction of an antiselfdual biform $H(X)$, we use

$$\partial_a a_\mu a \wedge b = \partial_a a_\mu (ab - a \cdot b) = -2a_\mu b - ba_\mu = ba_\mu - 2a_\mu \cdot b.$$

Then, from (9.40b) we obtain

$$\partial_a H(a \wedge b) = b \sum_{\mu,\nu} a_\mu a_\nu \beta_{\mu\nu} - 2 \sum_{\mu,\nu} a_\mu \cdot b a_\nu \beta_{\mu\nu}, \tag{9.44}$$

and

$$\partial H = 2 \sum_{\mu,\nu} a_\mu a_\nu \beta_{\mu\nu} = 2 \sum_{\mu\nu} a_\mu \cdot a_\nu \beta_{\mu\nu}. \tag{9.45}$$

Comparing this with (9.23) and (9.24), we see that if $\partial H = 0$ we can identify H with F_1 and F_3 so that in (9.24a)

$$f_1(b) = \partial_a \cdot H(a \wedge b) = -2 \sum_{\mu,\nu} a_\mu \cdot b a_\nu \langle \beta_{\mu\nu} \rangle_0, \tag{9.46a}$$

and, in (9.24b)

$$f_3(b) = \partial_a H(a \wedge b) = -2 \sum_{\mu,\nu} a_\mu \cdot b a_\nu \langle \beta_{\mu\nu} \rangle_4. \tag{9.46b}$$

Finally, we can isolate the term F_0 in (9.23) from nonvanishing values of ∂G_+ and/or ∂H.

We have one task remaining to complete our classification of biforms on $\mathcal{G}^2(\mathcal{A}_4)$, namely, to reduce the coefficient matrices α_{jk} and $\beta_{\mu\nu}$ in (9.40a, b) to their simplest forms. For an antiselfdual biform $H(X)$, the traction of (9.40b) to (9.46a, b) reduces the problem of classifying $H(X)$ to the problem of finding canonical forms for symmetric linear transformations $f_1(b)$ and $f_3(b)I$, a problem which we have discussed at length already. Consequently, we can restrict our attention to selfdual biforms, indeed, to tractionless biforms alone.

We have shown that any tractionless biform $T(X)$ on $\mathcal{G}^2(\mathcal{A}_4)$ can be written in the algebraic form

$$T(X) = \sum_{j,k=1}^{3} f_j X f_k \alpha_{jk}, \tag{9.47a}$$

where $\alpha_{jk} = \langle \alpha_{jk} \rangle_0 + \langle \alpha_{jk} \rangle_4 = \alpha_{kj}$ and

$$\partial T = -2 \sum_{j,k} f_j \cdot f_k \alpha_{jk} = 0. \tag{9.47b}$$

The rest of this section will be devoted to classifying tractionless biforms by reducing (9.47a). The reduction of (9.47a) depends on the metric on $\mathcal{A}_4$. It is trivial for a Euclidean metric on $\mathcal{A}_4$, because the coefficient matrix α_{jk} can always be diagonalized in that case, so let us consider the more interesting case of a spacetime metric (3, 1) on $\mathcal{A}_4$. We refer to the biforms for this case as *spacetime biforms*.

The spacetime metric allows us to choose a basis of orthogonal vectors a_0, a_1, a_2, a_3 such that $a_0^2 = 1$ and $a_k^2 = -1$ ($k = 1, 2, 3$). From these vectors we can construct a basis of orthonormal bivector blades:

$$f_1 = a_1 a_0 = i a_2 a_3, \qquad f_2 = a_2 a_0 = i a_3 a_1, \qquad f_3 = a_3 a_0 = i a_1 a_2, \tag{9.48}$$

where we have used the symbol i instead of I for the unit pseudoscalar, because $i^2 = -1$ and it is the 'imaginary unit' for the 'complex' coefficients. From (9.48) it follows that

$$f_k^2 = 1 \quad \text{for } k = 1, 2, 3, \tag{9.49a}$$

and

$$f_j f_k = -f_k f_j \quad \text{if } j \neq k. \tag{9.49b}$$

Equations (9.49) describes special properties of the f_k we need for our analysis. The relations (9.48) have been mentioned only to show how the properties (9.49) are related to the spacetime metric on $\mathscr{A}_4$.

We are now prepared to classify the tractionless spacetime biforms into the so-called *Petrov types* I, II and III. We say that a biform is of type I if and only if there is a choice of the f_k which diagonalizes the matrix α_{jk}. Thus, for a type I biform, (9.47a) can be reduced to

$$T(X) = \sum_{k=1}^{3} f_k X f_k \alpha_k, \tag{9.50a}$$

where

$$\partial T = \alpha_1 + \alpha_2 + \alpha_3 = 0. \tag{9.50b}$$

For $X = \langle X \rangle_2$ and basis blades f_k satisfying (9.49a, b), one can show that

$$\sum_{k=1}^{3} f_k X f_k = -X. \tag{9.51}$$

This can be used to eliminate f_3 from (9.50a), and (9.50b) can be used to eliminate α_3. Thus, introducing new variables $\mu_1 = 2\alpha_1 + \alpha_2$ and $\mu_2 = 2\alpha_2 + \alpha_1$, we reduce (9.50a) to

$$T(X) = \sum_{k=1}^{2} (f_k X f_k + \tfrac{1}{3} X)\mu_k, \tag{9.52}$$

where $f_j \cdot f_k = \delta_{jk}$. This is the desired *canonical form* for type I biforms.

The canonical form (9.52) shows that type I biforms can be classified into distinct subtypes in a variety of ways corresponding to distinct conditions imposed on the four free parameters in μ_1 and μ_2. For example, there are subtypes for which $\langle\alpha_1\rangle_0 = \langle\alpha_2\rangle_0$ or $\langle\alpha_1\rangle_4 = \langle\alpha_1\rangle_4$ or $\alpha_1 = \alpha_2^\dagger$, etc. Evidently, the simplest subtype is obtained by taking $\mu_2 = 0$, whereupon (9.52) reduces to

$$T_1(X) = (f_1 X f_1 + \tfrac{1}{3}X)\mu_1 . \tag{9.53}$$

By using (9.51), one can easily show that the conditon $\mu_1 = \mu_2$ gives the same subtype as the condition $\mu_2 = 0$.

In types other than I, a null bivector plays a special role. Consider the null bivector

$$f_0 = f_2 + if_3 = a_2(a_1 + a_0). \tag{9.54}$$

Using (9.49) it is readily verified that

$$f_0^2 = 0. \tag{9.55a}$$

Now let f_1' be any bivector with the properties

$$f_1'^2 = 1, \tag{9.55b}$$

$$f_0 \cdot f_1' = 0. \tag{9.55c}$$

It follows that

$$f_0 f_1' = -f_1' f_0 , \tag{9.55d}$$

and using (9.49), it is not difficult to show that

$$f_1' = f_1 + \alpha f_0 \tag{9.56}$$

where α is a complex parameter. It is easy to show that two bivectors with the parametric form (9.56) cannot be mutually orthogonal. Therefore, any set of orthogonal bivectors, one of which is null, cannot contain more than two elements.

A biform is of *Petrov Type II* if and only if the coefficient matrix α_{jk} can be diagonalized by taking a null bivector f_0 as one of the basis elements. For this case, we must be able to construct a canonical form similar to (9.52) from f_0 and f_1' instead of f_1 and f_2. Then, for the *canonical form of a Type II* biform we find

$$\begin{aligned} T(X) &= f_0 X f_0 \mu_0 + (f_1' X f_1' + \tfrac{1}{3}X)\mu_1 \\ &= f_0 X f_0 \mu_0 + ((f_1 + \alpha f_0)X(f_1 + \alpha f_0) + \tfrac{1}{3}X)\mu_1 . \end{aligned} \tag{9.57}$$

The term $\tfrac{1}{3}X\alpha_1$ appears in (9.57) so that

$$\partial T = -2(f_0^2\mu_0 + f_1^2\mu_1) + \tfrac{6}{3}\mu_1 = 0,$$

as follows from the bivector derivatives

$$\partial_X X = 6 \quad \text{and} \quad \partial_X BX = -2,$$

where B is any bivector in $\mathcal{G}^2(\mathcal{A}_4)$.

The canonical form (9.57) shows that the type II biforms have two major subtypes, one with $\mu_1 \neq 0$ and one with $\mu_1 = 0$. Obviously (9.57) reverts to type I if $\mu_0 = 0$. A further subclassification is obtained for $\alpha = 0$ and $\alpha \neq 0$.

The canonical form (9.57) can be expressed in terms of the orthonormal bivector basis (9.48) by substituting

$$f_0 X f_0 = f_2 X f_2 - f_3 X f_3 + i(f_2 X f_3 + f_3 X f_2).$$

This shows explicitly the off-diagonal terms which cannot be eliminated with a basis of type (9.45).

Obviously, the single canonical form

$$T(X) = \sum_{k=1}^{2} (f_k X f_k + \tfrac{1}{3} X f_k^2)\mu_k, \tag{9.58}$$

with $f_1 \cdot f_2 = 0$, suffices for both types I and II, the two types being distinguished by different conditions on f_k^2. Incidentally, with $f_k^2 = -1$, Eqn. (9.58) is also the canonical form for the Euclidean case.

Using f_0 and f_1' we can construct one other tractionless symmetric biform, namely

$$T(X) = (f_0 X f_1' + f_1' X f_0)\mu = (f_0 X f_1 + f_1 X f_0 + 2\alpha f_0 X f_0)\mu. \tag{9.59}$$

This is the canonical form for a biform of *Petrov type III*. The condition $f_0^2 = 0$ prevents this from being diagonalized. This completes the classification of biforms. It can be shown that a linear combination of types II and III does not result in a new type of biform.

A conventional approach to the classification of tractionless spacetime biforms is based on the eigenvalue problem

$$T(f) = \lambda f, \tag{9.60}$$

where f is a bivector and the eigenvalue λ is a complex scalar. The biform T is said to be of *Petrov Type I* if the eigenbivectors span the entire three-dimensional (complex) space $\mathcal{G}^2(\mathcal{A}_4)$; it is of *Type II* if the eigenbivectors span a two-dimensional space and of *Type III* if they span only a one-dimensional space. This classification is equivalent to the one introduced above.

The canonical forms (9.52), (9.57) and (9.59) which we found for the three Petrov types provide us with the solution to the eigenvalue problem. We can easily read off the eigenvectors and eigenvalues simply by inserting the f_k into the canonical forms and using the properties (9.49) and (9.55) to evaluate the results. The results are summarized in the table below.

Petrov Type	*Canonical Form*	*Eigenbivectors*	*Eigenvalues*
I	$(f_1Xf_1 + \frac{1}{3}X)\mu_1 +$	f_1	$\lambda_1 = \frac{2}{3}(2\mu_1 - \mu_2)$
	$+ (f_2Xf_2 + \frac{1}{3})\mu_2$	f_2	$\lambda_2 = \frac{2}{3}(2\mu_2 - \mu_1)$
		f_3	$\lambda_3 = -\frac{2}{3}(\mu_1 + \mu_2)$
II	$(f_1'Xf_1' + \frac{1}{3}X)\mu_1 +$	$f_0 = f_1 + if_2$	$\lambda_0 = -\frac{2}{3}\mu_1$
	$+ f_0Xf_0\mu_0$		
		$f_1' = f_1 + \lambda f_0$	$\lambda_1 = \frac{4}{3}\mu_1$
III	$(f_0Xf_1' + f_1'Xf_0)\mu$	$f_0 = f_1 + if_2$	$\lambda = 0$

The method of algebraic forms exployed here considerably extends and enhances the algebraic machinery available for the study of linear transformations beyond the classical eigenvalue-eigenvector problem. This is particularly evident in the explicit canonical form given to each Petrov type, as exhibited in the above table. In contrast, the solution of the eigenvalue problem for a biform of type II or III does not completely reveal the algebraic structure of the biform. Sobczyk [S2-5] has further extended and developed the method of algebraic forms to show the existence of a principal correlation between a symmetric selfdual biform (9.47), and a symmetric antiselfdual biform (9.40b), a result undetected and unappreciated in the conventional approaches.

Our approach to the Petrov classification with Geometric Calculus should be compared with that of Synge [Sy], who uses conventional mathematical devices. Synge introduces complex numbers formally, because he sees that it simplifies the problem. He does not discover that the geometrical basis for this simplication is the self-duality of the space of bivectors, or that his unit imaginary has assumed the geometrical role of the unit pseudoscalar. Moreover, to achieve his simplification of this particular problem, he is obliged to drastically revise the formalism he uses for other geometrical applications. In contrast, we have seen that the appropriate complex numbers arise automatically when Geometric Calculus is used, and with a clear geometrical interpretation. No special methods are called for in this case, and, as will be seen in subsequent chapters, the same formalism applies as well to any geometrical problem.

3-10. Tensors

This section explains how the conventional theory of tensors can be efficiently integrated into Geometric Calculus. A nomenclature is recommended to maintain contact with conventional terminology and yet suggest how special features of Geometric Calculus can be exploited to simplify and extend the tensor theory. However, no aspects of the theory will be developed here in much detail. The main concern is to establish tensor algebra as the generalization to functions of several variables of conceptions and techniques of linear algebra developed in preceding sections of this chapter.

To anyone with much sophistication in linear algebra, most of this section will appear quite trivial. But these trivial things must be mentioned, because they are useful, indeed, they are essential in many applications, and we are afraid that even some of our sophisticated friends will overlook them.

A geometric function of r variables $T = T(A_1, A_2, \ldots, A_r)$ is said to be *r-linear* if it is a linear function of each argument. We say that T is a *tensor of degree r* (on $\mathcal{A}_n$) if each variable is restricted to some vector space $\mathcal{A}_n$. In the more general case that each variaole is defined on a Geometric Algebra $\mathcal{G}(\mathcal{A}_n)$, we say that T is an *extensor of degree r* (on $\mathcal{G}(\mathcal{A}_n)$). The multiforms defined and discussed in Section 3-9 are extensors of degree one.

If the values of a tensor $T = T(a_1, a_2, \ldots, a_r)$ on $\mathcal{A}_n$ are s-vectors in $\mathcal{G}^s(\mathcal{A}_n)$, we say that T has *grade s* and *rank s + r*. A tensor of grade zero is called a *multilinear form*. The tensor T determines a multilinear form τ of degree (or rank) $r + s$ defined by

$$\tau(a_1, a_2, \ldots, a_{r+s}) = (a_{r+1} \wedge \ldots \wedge a_{r+s}) \cdot T(a_1, \ldots, a_r). \tag{10.1a}$$

Conversely, τ determines T by the equation

$$\begin{aligned} T(a_1, \ldots, a_r) &= (s!)^{-1} \partial_{b_s} \wedge \ldots \wedge \partial_{b_1} \tau(a_1, \ldots, a_r, b_1, \ldots, b_s) \\ &= \partial_B B * T(a_1, \ldots, a_r). \end{aligned} \tag{10.1b}$$

In this sense, T and τ are equivalent representations of a tensor of rank $r + s$. Note that for $r, s = 1$, T is a linear transformation and τ is its associated bilinear form. This important relation in linear algebra has thus been generalized to the relation (10.1b) in multilinear algebra by using the *simplicial derivative* defined in Chapter 2.

We must confess that we are not completely satisfied with our terminology for multilinear functions. We have endeavored to use conventional terminology as much as possible, to help readers identify familiar concepts in our unconventional formalism. Besides, we do not wish to contribute to the unnecessary proliferation of mathematical terminology, which we think is already a serious problem. However, the theory of multilinear functions has developed along so many independent lines, each with its own terminology, that it is impossible to developed a unified theory without tampering with some terms in wide use. Furthermore, the structure of Geometric Algebra suggests new ways for developing and organizing tensor theory, which must be reflected in any appropriate terminology. These things considered, it seems best at the present time not to commit ourselves too heavily to any particular nomenclature.

We have actually abstained from using some terminology we prefer. For example, we would prefer to use the term *r-form* for any multilinear form of degree r. But the term *r-form* is so widely used in modern literature to refer to an *alternating r-form*,* that we have stuck to that sense throughout this book. We have bowed to convention also in our choice of the words 'multiform' and 'extensor' to indicate direct generalizations of the standard concepts of alternating forms and tensors.

Our concept of tensor is somewhat more general than the usual one, because we allow the value of a tensor to be a multivector of mixed grade. This enables us to handle concepts beyond the competence of conventional tensor analysis. Since our tensors can have spinor values, our formalism can reproduce the results of conventional spinor analysis. Spinors play a crucial role in modern theoretical physics, and to deal with them several 'spinor analysis' formalisms have been developed. These formalisms are similar to tensor analysis but not integrated with it, and they rely on 'spinor coordinates' which are difficult to interpret geometrically. In contrast, the present formalism fully integrates tensors with spinors and provides coordinate-free methods for dealing with spinor tensors. However, as this subject is of rather esoteric interest, we will not develop details here. We are more interested just now in seeing how to handle conventional tensor theory.

The Geometric Algebra permits us to construct new tensors from given ones in three basic ways, namely, by *addition, multiplication* and *contraction*. The *sum* of tensors S and T,

$$S(a_1, \ldots, a_q) + T(a_1, \ldots, a_r), \tag{10.2}$$

is obviously a tensor if and only if S and T have the same degree and the same arguments. The geometric *product*

$$ST \equiv S(a_1, \ldots, a_q)T(a_1, \ldots, a_r), \tag{10.3}$$

without restrictions on the arguments, is a tensor of degree $q + r$. Owing to the noncommutivity of the geometric product $S \cdot T$, $S \wedge T$, and $S \times T$ generally yield different tensors of degree $q + r$.

The *contraction* of a tensor T of degree r is a tensor of degree $r - 2$ defined by

$$T(a_1, \ldots, a_{j-1}, \partial_{a_k}, a_{j+1}, \ldots, a_r) = \partial_{a_j} \cdot \partial_{a_k} T(a_1, \ldots, a_r). \tag{10.4}$$

To be more specific, we may say that the tensor (10.4) is the *contraction of T in the jth and kth places*. It is often convenient to form the *traction* of a tensor T (*in the kth place*) defined by

$$\partial_{a_k} T(a_1, \ldots, a_r). \tag{10.5}$$

But note that this can be obtained from the tensor of degree $r + 1$ defined by the product $a_{r+1} T(a_1, \ldots, a_r)$ by contracting the variables a_k and a_{r+1}. The

* Defined in Section 1-4.

definition (10.5) generalizes the definition of traction we gave for multivectors in Section 3-9. Following our practice in Section 3-9, we should refer to the tensor

$$\partial_{a_k} \cdot T(a_1, \ldots, a_k), \tag{10.6}$$

as the *contraction of T in the kth place*. We see little danger of confusion in referring to both (10.4) and (10.6) as contractions, so we will not bother to coin a new word to distinguish them.

As a simple illustration of tensor manipulations, consider the composition of bilinear forms $\sigma(a, b)$ and $\tau(a, b)$ to get another bilinear form

$$\rho(a, b) \equiv \partial_u \cdot \partial_v \tau(a, u)\sigma(v, b) = \tau(a, \partial_v)\sigma(v, b). \tag{10.7}$$

The bilinear forms determine the tensors of degree 1,

$$R(a) = \partial_b \rho(a, b), \qquad S(a) = \partial_b \sigma(a, b), \qquad T(a) = \partial_b \tau(a, b). \tag{10.8}$$

So by differentiating (10.7), we get

$$R(a) = \tau(a, \partial_b)S(b) = T(a) \cdot \partial_b S(b) = S(T(a)). \tag{10.9}$$

This is just the expression for composing linear transformations noted in Section 3-1.

To relate our theory to the conventional covariant formulation of tensor analysis, we restrict our considerations to tensors with scalar values, that is, to multilinear forms. Subject to this restriction, the product (10.3) is just the conventional *tensor product*. The *covariant components* of a rank 3 tensor $\tau(a, b, c)$ with respect to a frame $\{a_k\}$ for $\mathscr{A}_n$ are the quantities

$$\tau_{ijk} \equiv \tau(a_i, a_j, a_k). \tag{10.10a}$$

The *contravariant components* of τ are

$$\tau^{ijk} \equiv \tau(a^i, a^j, a^k), \tag{10.10b}$$

where $\{a^k\}$ is the reciprocal frame. The quantities

$$\tau^{ij}_k \equiv \tau(a^i, a^j, a_k) \tag{10.10c}$$

are called *mixed components*. This suffices to show how the components of any tensor can be determined. Arbitrary values of τ are expressed in terms of components by

$$\tau(a, b, c) = \alpha^i \beta^j \gamma^k \tau_{ijk} = \alpha_i \beta_j \gamma_k \tau^{ijk} = \alpha_i \beta_j \gamma^k \tau^{ij}_k, \tag{10.11}$$

where $\alpha_i = a_i \cdot a$, $\beta_j = a_j \cdot b$, $\gamma_k = a_k \cdot c$ are, respectively, the 'covariant components' of vectors a, b, c, while $\alpha^i, \beta^j, \gamma^k$ are the 'contravariant components'.

The components of a contraction of $\tau(a, b, c)$ are

$$\tau_k^{ik} \equiv \tau(a^i, a^k, a_k) = \tau(a^i, \partial_b, b), \tag{10.12}$$

where we have used the 'summation convention' as well as the fact that $\partial_b = a^k a_k \cdot \partial_b$ and $a_k \cdot \partial_b b = a_k$. Thus, it is clear that our definition of contraction (10.4) conforms to the conventional one in tensor analysis. When formula (10.7) is expressed in terms of tensor components we get

$$\rho_k^i \equiv \rho(a^i, a_k) = \tau(a^i, a_j)\sigma(a^j, a_k) = \tau_j^i \sigma_k^j, \tag{10.13}$$

the familiar rule for 'multiplying' matrices.

Let us determine how different tensor components are related. Any frame $\{a'_k\}$ for $\mathscr{A}_n$ is related to the frame $\{a_k\}$ by a linear transformation $\underline{f}$ with components $f_k^j = a^j \cdot \underline{f}(a_k)$; thus,

$$a'_k = \underline{f}(a_k) = f_k^j a_j. \tag{10.14a}$$

The reciprocal frames are related by the adjoint transformation;

$$a^j = \overline{f}(a'^j) = f_k^j a'^k. \tag{10.14b}$$

The mixed components

$$\tau_{k'}^{i'j'} \equiv \tau(a'^i, a'^j, a'_k) \tag{10.15}$$

of the tensor $\tau(a, b, c)$ with respect to the frame $\{a'_k\}$ are therefore related to components τ_k^{ij} with respect to $\{a_k\}$ by the equations

$$\begin{aligned} \tau_k^{ij} f_r^k &= \tau(a^i, a^j, \underline{f}(a_r)) = \tau(\overline{f}(a'^i), \overline{f}(a'^j), \underline{f}(a_r)) \\ &= f_p^i f_q^j \tau(a'^p, a'^q, a'_r) = f_p^i f_q^j \tau_{r'}^{p'q'}. \end{aligned} \tag{10.16}$$

(The rather awkward use of primes here conforms with common practice in the literature.) This is the conventional rule for transforming components in tensor analysis.

In conventional covariant formulations of tensor analysis, tensors are defined by specifying their components with respect to some frames and assuming a transformation law of the form (10.16) for the components. But, as we have seen, the transformation law for tensor components obtains as an elementary theorem if tensors as defined as multilinear functions. Let us call these two different (but equivalent) definitions of tensors respectively the *covariant* and the *intrinsic* definitions. It seems obvious to us that the intrinsic definition is superior to the covariant one, because it specifies the essential linearity of tensors directly, without reference to an arbitrary frame. Yet the covariant definition is the one adopted in most physics, engineering and applied mathematics texts. Historical accident aside, the main reason for this practice is probably the lack of a method for performing computations with tensors without using indices. As we have shown how Geometric Calculus corrects this deficiency and provides other advantages, there are no remaining objections to the intrinsic definition.

Another point deserves clarification. Covariant tensor analysis makes it difficult to distinguish between a tensor's behavior under transformations and its linear properties. A distinction between *active* and *passive* transformations is commonly made. It is important to note that the covariant formalism itself provides no means to represent this distinction. The equations for both types of transformation are identical, and the distinction is made only in the interpretation of the equations. Textbooks are littered with evidence of muddles that arise from this practice. In contrast, the intrinsic approach avoids the problem completely. The transformation Eqns. (10.16) necessarily describe a passive transformation, in consequence of their derivation from the linear properties of a tensor. Thus, the passive transformations in covariant formulations merely express the fact that tensors are linear functions, whereas the intrinsic formulation expresses the fact directly without reference to transformations.

The transformation of a tensor induced by a linear transformation $\underline{f}$ of its domain $\mathscr{A}_n$ into another vector space $\mathscr{A}'_n$ is called an *active transformation*. Consider a rank two tensor $\tau(a, b)$ defined on $\mathscr{A}_n$. A tensor $\tau'(a', b')$ on $\mathscr{A}'_n$ is determined by the linear substitution

$$\tau'(a', b') = \tau(\underline{f}^{-1}(a'), \overline{f}(b')). \tag{10.17}$$

This transformation of τ into τ' is said to be *covariant* in the first place (or argument) and *contravariant* in the second place. In a similar way, the transformation of any tensor of grade zero is determined by specifying a substitution of covariant or contravariant type for each of its arguments. We wish to emphasize that such an assignment of transformation behavior to a tensor is quite independent of the specification of the tensor as a multilinear function. Indeed, the transformation (10.17) makes sense even if $\tau(a, b)$ is not a linear function. It is true that when tensors arise in physics and geometry, their transformation behaviors are usually determined automatically by their functions in the theory. This is probably why the fact that active and passive transformations describe distinct and independent properties of tensors is so often overlooked.

The main problem of tensor algebra is the classification of tensors by determining their canonical forms. We have already found canonical forms in preceding sections for alternating r-forms, for linear transformations and for some of the simpler multiforms of degree two. These special cases were complicated enough to show that a complete analysis of tensor canonical forms is too much to aim for here. But some general remarks will help put the problem in perspective. As we have explained for the special cases treated earlier, we understand the *canonical form* of a tensor to be the algebraic representation of the tensor which is simplest in some prescribed sense. And by 'algebraic representation' we mean, of course, a quantity composed only of multivectors combined by geometric sums and products.

One of the most important ways to classify a tensor $T(a_1, a_2, \ldots, a_r)$ is to decompose it into a sum of tensors with distinct symmetries in their arguments and then determine canonical forms for each of these. Without fully determining the classical solution to this problem, we would like to make some observations on the approach with Geometric Algebra. As determined by (2-3.10), the completely skewsymmetric part of T is easily obtained by operating on T with the skew-symmetrizer, and the result is a multiform of degree r, which can be put in canonical form by the method described in Section 3-9. By setting all the arguments of T equal, we get a *homogeneous function of degree r*

$$H_r(a) = T(a, a, \ldots, a). \tag{10.18}$$

Expressing the vector a as a function of its components α_k relative to a fixed frame in $\mathscr{A}_n$, we see that $H_r(a) = H(\alpha_1, \alpha_2, \ldots, \alpha_n)$ is a *homogeneous polynomial of degree r* in the n scalar variables α_k. This connects tensor theory to the classical theory of polynomials, and it justifies our choice of the word 'degree' for the number of tensor arguments. Now, by differentiating $H_r(a)$ we get the symmetric tensor

$$S(a_1, a_2, \ldots, a_r) = \frac{1}{r!}\, a_1 \cdot \partial_a a_2 \cdot \partial_a \ldots a_r \cdot \partial_a H(a). \tag{10.19}$$

This is easily seen to be the completely symmetric part of T. Other techniques of Geometric Calculus employed in this book are useful in tensor analysis, but we have said enough to establish a perspective on the subject.

Chapter 4

Calculus on Vector Manifolds

The modern approach to calculus on manifolds, as typified by ref. [La], begins with the general notion of a topological space, from which spaces of increasing complexity are built up by introducing a succession of structures such as differentiable maps, fiber bundles, differential forms, connections and metrics. Without disputing that there are good reasons for this approach, we wish to point out that it has some serious practical drawbacks. To begin with, each of the successive structures entails new assumptions, vocabulary and techniques which are difficult to formulate without duplicating, overlapping or modifying features of the structure to which it is added. Hence, the mathematical system tends to become increasingly redundant as it is built up. With care the redundancy can be minimized, but we doubt that it can be eliminated altogether, because a complex structure may admit to a formulation which is simpler than any which can be developed by building it up from standard structures. For example, standard formulations of linear and multilinear algebra as independent subjects reveal considerable redundancy when combined, redundancy which we claim to have eliminated in the preceding chapters by developing both subjects within the single corpus of Geometric Algebra.

Another drawback of the 'build-up approach' is that it defers proof of the most comprehensive theorems until each layer of the necessary mathematical apparatus has been successively constructed and analyzed in detail. The time required to master these details is so considerable that most physicists and applied mathematicians remain ignorant of many potentially useful results.

The issue we are raising can be set in a somewhat simpler context. It is certainly important to study the build-up of the real numbers from simpler structures. But few would dispute that, for applications at least, it is best to begin with a formulation of the real numbers as a single mathematical structure. Similarly, to establish the main results of differential and integral calculus on manifolds as quickly and easily as possible, it is necessary to forego the 'build-up approach' and construct at the outset a single compact mathematical structure from which all the desired results can be obtained without further assumptions. That is our objective in this chapter and those that follow.

Geometric Algebra supplies all the algebraic structure we need to define vector manifolds and develop a complete differential and integral calculus without introducing coordinates. It reduces the development of calculus almost completely to the study of a single differential operator, which we interpret as the ***derviative*** by a point in the manifold. From this derivative all other important differential operators (such as divergence and curl, exterior and covariant derivatives) can be obtained by simple algebraic operations. Besides the derivative, the key role of pseudoscalars and the closely related projection into the tangent algebra are unique features of Geometric Calculus. The discovery and exploitation of their properties throughout this chapter culminates in contributions to differential geometry in the chapter following.

The points of a vector manifold are vectors in the (infinite dimensional) Geometric Algebra which possesses an inner product. It might be thought, therefore, that our treatment of manifolds is limited to submanifolds of Euclidean space with an inherited Riemann structure. However, in developing the general theory we make no specific embedding assumptions, so no particular metrical structure is implied. Indeed, we deal with nonRiemanniam manifolds in Chapter 5. As explained in earlier chapters, we can use the inner produce as a grade-lowering operation without committing ourselves to a particular metric.

Whether or not every manifold can be represented as a vector manifold and so characterized by Geometric Calculus we leave as an open question (though we are convinced that the answer is in the affirmative). Generality is not our main interest here. Rather, with as little formal apparatus as possible, we aim to develop as much of manifold theory as is likely to be useful to physicists and applied mathematicians. The present chapter is concerned only with local properties of manifolds, but we think that it can contribute to improvements in the global theory in connection with the integration theory in Chapter 7.

This chapter has two main parts. The first four sections develop the properties of manifolds and the differential calculus of fields as far as possible without reference to differential geometry. Most of the results obtained are needed for efficient application of geometric calculus. The last three sections are concerned with the induced transformations of the calculus of fields from one manifold to another. Section 4-5 works out the general theory, but we especially want to direct attention to the examples worked out in Section 4-6. Conventional formulations must introduce coordinates and employ matrices to carry out computations, whereas our treatment is completely coordinate-free. The advantages of Geometric Calculus are most clearly revealed by comparing the way the alternative methods handle details of computations.

Section 4-7 shows that Geometric Calculus possesses the well-known virtues of complex function theory for describing conformal transformations of the plane. Of course, classical complex analysis is limited to two dimensions, whereas the present formalism applies to any finite dimensional manifold; indeed, it fully integrates complex analysis and real analysis into a single subject.

4-1. Vector Manifolds

A (differentiable) manifold is a set on which differential calculus can be carried out. Ordinarily an m-dimensional manifold $\mathcal{M}$ is defined by assuming that it can be locally parametrized by a set of (scalar) coordinates. The coordinates determine an invertible mapping of $\mathcal{M}$ into Euclidean m-space $\mathscr{E}_m$. Calculus on $\mathcal{M}$ is then developed by mapping quantities into $\mathscr{E}_m$ where the operations of calculus, being well-defined, can be carried out and the results mapped back to $\mathcal{M}$. In recent years, exterior differential forms have been widely employed to express calculus on manifolds in coordinate-free form (see, for instance, [La]). However, the resulting calculus is only nominally coordinate-free, for it is often necessary to introduce coordinates to carry out computations. Moreover, the calculus of differential forms is not sufficiently general to embrace all the important quantities defined on manifolds, so it must be supplemented by other mathematical structures such as tensors and fiber bundles.

In the usual approach to manifolds, coordinates supply the algebraic structure needed to support calculus. Being scalar quantities, coordinates submit to the algebraic operations of subtraction and division which are presupposed in the definition of a derivative as the limit of a difference quotient. But the necessary algebraic structure can be imposed on an *abstract manifold* more directly by assuming that its *points are vectors*, so points can be added and multiplied according to the rules of Geometric Algebra. Indeed, the employment of coordinates in the usual definition of a manifold can be regarded as an indirect means of assigning properties of vectors to the points of a manifold. As we will show, a fully developed coordinate-free differential calculus on *vector manifolds* can be achieved with only minor modifications and extensions of the apparatus developed in Section 2-1.

A *vector manifold* $\mathcal{M}$ is a set of vectors called *points* of $\mathcal{M}$ with certain properties to be described presently. If x and y are points of $\mathcal{M}$, then the vector $x - y$ is said to be a *chord* of $\mathcal{M}$, more specifically, the *chord from y to x*. The limit of a sequence of chords defines a tangent vector. A vector $a(x)$ is said to be *tangent to a point x in $\mathcal{M}$* if there is a curve $\{x(\tau);\, 0 < \tau < \epsilon\}$ in $\mathcal{M}$ extending from the point $x(0) = x$ such that

$$a(x) = a \cdot \partial x \equiv \frac{\mathrm{d}x(\tau)}{\mathrm{d}\tau}\bigg|_{\tau=0} = \lim_{\tau\to 0} \frac{x(\tau) - x}{\tau}. \tag{1.1}$$

If the chords of a curve are not null vectors, the curve can be parametrized by the magnitudes of its chords, $\sigma = \sigma(\tau) \equiv |x(\tau) - x|$, so

$$a = \frac{\mathrm{d}\sigma}{\mathrm{d}\tau}\frac{\mathrm{d}x}{\mathrm{d}\sigma}, \qquad |a| = \frac{\mathrm{d}\sigma}{\mathrm{d}\tau}, \qquad \hat{a} = \frac{\mathrm{d}x}{\mathrm{d}\sigma} = \frac{\mathrm{d}x}{|\mathrm{d}x|}. \tag{1.2}$$

The set $\mathscr{A}(x)$ of all vectors tangent to $\mathcal{M}$ at x is called the *tangent space* at x. At each *interior point* x of $\mathcal{M}$, the tangent space $\mathscr{A}(x)$ is an m-dimensional vector space, and $\mathcal{M}$ is said to be an *m-dimensional manifold*.

In the manner described in Section 1-2, the tangent space $\mathscr{A}(x)$ at an interior point generates a unique Geometric Algebra which we call *the tangent algebra* of $\mathscr{M}$ at x and denote by $\mathscr{G}(x) \equiv \mathscr{G}(\mathscr{A}(x))$. We assume that $\mathscr{A}(x)$ is *nonsingular* in the algebraic sense that it possesses a unit pseudoscalar $I = I(x)$, which we call the (unit) *pseudoscalar of* $\mathscr{M}$ *at* x. This allows for the possibility that $\mathscr{A}(x)$ is pseudo-Euclidean as defined in Section 1-5.

The unit pseudoscalar $I = I(x)$ is an m-blade valued function defined on the manifold $\mathscr{M}$. We wish to show how convenient it is to describe $\mathscr{M}$ by specifying the properties of the pseudoscalar function. We assume that $I = I(x)$ is defined at *every* point of $\mathscr{M}$. We say that $\mathscr{M}$ is *continuous* if $I = I(x)$ is a continuous function at every point x. A continuous pseudoscalar function $I = I(x)$ on $\mathscr{M}$ is said to assign an *orientation* to $\mathscr{M}$. There are two possible orientations corresponding to the two possible orientations $\pm I$ of a unit blade. A continuous manifold $\mathscr{M}$ is said to be *orientable* if its pseudoscalar function $I = I(x)$ is single-valued, and *nonorientable* if $I = I(x)$ is double-valued. We will be concerned only with orientable manifolds.

We say that a manifold $\mathscr{M}$ is *differentiable* if its pseudoscalar $I = I(x)$ is differentiable (in the sense specified below) at each point of $\mathscr{M}$. We say that $\mathscr{M}$ is *smooth* if $I = I(x)$ has derivatives of all orders. Although discontinuities of a manifold can be efficiently described as discontinuities of its pseudoscalar, we are concerned only with smooth manifolds in this book. To minimize repetition, we lay down the assumption of smoothness at once; we will repeat it only occasionally as a reminder.

A vector $a = a(x)$ associated with an interior point x of $\mathscr{M}$ is a tangent vector at x if and only if

$$a \wedge I = a(x) \wedge I(x) = 0. \tag{1.3}$$

According to Section 1-2, therefore, at an interior point the pseudoscalar $I = I(x)$ uniquely determines the tangent space $\mathscr{A}(x)$ and the tangent algebra $\mathscr{G}(x) = \mathscr{G}(\mathscr{A}(x)) = \mathscr{G}(I(x))$. At a *boundary point* of $\mathscr{M}$, every tangent vector satisfies (1.3), but some vectors which satisfy (1.3) are not tangent vectors in the sense of (1.1). Thus, at a boundary point x the tangent space $\mathscr{A}(x)$ is a subset of $\mathscr{G}^1(I(x))$. In this book we will be primarily concerned with the local properties of a manifold in neighborhoods of interior points, so we regard (1.1) and (1.3) as equivalent defining properties of a tangent vector unless otherwise stated.

The set of all linear combinations of *all chords* of a vector manifold $\mathscr{M}$ is a linear vector space which we call the *embedding space* of $\mathscr{M}$ and denote by $\overline{\mathscr{M}}$. The Geometric Algebra $\mathscr{G}(\overline{\mathscr{M}})$ is called the *embedding algebra* of $\mathscr{M}$. The tangent algebra $\mathscr{G}(x)$ at each point x of $\mathscr{M}$ is obviously a subalgebra of $\mathscr{G}(\overline{\mathscr{M}})$.

To describe a manifold $\mathscr{M}$ we often employ vectors in $\overline{\mathscr{M}}$ which are not tangent vectors of $\mathscr{M}$. Indeed, we did this in our first definition (1.1). However, our characterization of local properties of $\mathscr{M}$ will be completely indifferent to the dimension of $\overline{\mathscr{M}}$, which could be infinite though the dimension of $\mathscr{M}$ be finite. Determination of the possible dimension of $\overline{\mathscr{M}}$ from given conditions on the structure of $\mathscr{M}$ is called an *embedding problem*. For example, for the class of all manifolds with intrinsic geometries which are equivalent in some sense, we have the embedding problem of determining the minimum dimension for the embedding space of a manifold in the class. Embedding problems will not be studied in this book, but the apparatus we develop should be quite helpful in such studies.

According to Section 1-2, the pseudoscalar $I(x)$ determines the projection $P(x, A)$ of *any* multivector A into the tangent algebra $\mathcal{G}(x) = \mathcal{G}(I(x))$; specifically,

$$P(x, A) = [A \cdot I(x)] \cdot I^{-1}(x). \tag{1.4a}$$

We usually suppress the argument x and write

$$P(A) = (A \cdot I) \cdot I^{-1} = I^{-1} \cdot (I \cdot A). \tag{1.4b}$$

We do not repeat here the qualifications mentioned in connexion with Eqn. (1-2.9) which are necessary for (1.4) to provide the projections of scalars and m-vectors. If $A = A(x)$ is a function on $\mathcal{M}$, then we write (1.4b) with the understanding that at each point x the value of the function at x is projected into $\mathcal{G}(x)$, that is, $P(A) = P(x, A(x))$. A function $A = A(x)$ is said to be *tangent* to $\mathcal{M}$ at x if $P(x, A(x)) = A(x)$, that is if $A(x)$ is in $\mathcal{G}(x)$. If the function $A = A(x)$ is tangent to $\mathcal{M}$ at every point, we call it a *field* on $\mathcal{M}$ and write $P(A) = A$. The functional relation of the embedding algebra to the tangent algebras can be expressed by writing $P[\mathcal{G}(\overline{\mathcal{M}})] = \mathcal{G}(x)$.

The sum of fields A and B on $\mathcal{M}$ is the field $A + B$ with values $A(x) + B(x)$ in $\mathcal{G}(x)$ at each x of $\mathcal{M}$. Similarly the product AB is the field with values $A(x)B(x)$ in $\mathcal{G}(x)$. The set of all fields on $\mathcal{M}$ is therefore an algebra, which we call the *tangent algebra* of $\mathcal{M}$ and, with further abuse of our notation for Geometric Algebras, denote by $\mathcal{G}(\mathcal{M})$.

In Section 2-1 we developed the concepts of vector derivative, differential and adjoint on a *linear* vector manifold. With only minor modifications, which we discuss shortly, the powerful results which we established there are valid on any *smooth* vector manifold, and we will have many occasions to use them. We freely apply the nomenclature, notations and results of Section 2-1 to any vector manifold without repeating the explanations, qualifications and arguments made there.

For most of the results of Section 2-1 to hold on any smooth manifold $\mathcal{M}$, it is only necessary to generalize the definition (2-1.2) of the directional derivative as follows: Let $F = F(x)$ be a multivector valued function on $\mathcal{M}$, and let $a = a(x)$ be the tangent vector defined by (1.1). The *derivative* of F in the *direction* a is defined by

$$a \cdot \partial F = a(x) \cdot \partial_x F(x) \equiv \frac{\mathrm{d}F}{\mathrm{d}\tau}(x(\tau))\bigg|_{\tau=0} = \lim_{\tau \to 0} \frac{F(x(\tau)) - F(x)}{\tau}. \tag{1.5}$$

This definition differs from (2-1.2) in not assuming that $\mathcal{M}$ is a linear space. As before, we usually call $a \cdot \partial F$ the *directional derivative* when a is a specific vector, but when $a \cdot \partial F$ is to be regarded as a linear function of a we call it the *differential* of F and employ the notations

$$\underline{F}(a) \equiv F_a \equiv a \cdot \partial F. \tag{1.6a}$$

Unless otherwise stated, we assume $a = a(x)$ is a vector function defined on all of $\mathcal{M}$, so the differential is a linear function of $a(x)$ at each point x. When we wish to focus attention on the function at a particular point x, we may use one of the notations

$$\underline{F}(x, a(x)) \equiv F_a(x), \tag{1.6b}$$

and speak of the differential of F *at* x.

Although we develop the notion of differential before it, the central concept in our formulation of differential calculus on manifolds is the *derivative with respect to a point on a manifold*, denoted by the differential operator $\partial = \partial_x$. The operator $\partial = \partial_x$ is defined in terms of (1.5) in a manner described in Section 2-1. We will use all the properties of the vector derivative ∂ established in Section 2-1, but just now we recall only the fundamental fact that ∂ has the algebraic properties of a vector field, that is,

$$\partial = \langle\partial\rangle_1 = I^{-1}I\,\partial = I^{-1}I\cdot\partial = P(\partial). \tag{1.7}$$

It is most convenient to define *every* other differential operator on $\mathcal{M}$, for example, the directional derivative $a\cdot\partial$, as some function of the single fundamental operator ∂, the derivative with respect to a point.

When we deal with functions which are linear functions of other functions, it is convenient to generalize, or if you will, to modify our definition of the differential somewhat. Let $A_1 = A_1(x), \ldots, A_k = A_k(x)$ be multivector functions on $\mathcal{M}$, and consider the function

$$T(A_1, A_2, \ldots, A_k) = T(x, A_1(x), A_2(x), \ldots, A_k(x)). \tag{1.8a}$$

Recalling our definition of extensors in Section 3-10, we say that T is an *extensor* function of the functions $A_i = A_i(x)$, if it is a *linear* function of each A_i for each fixed x. Note that the differential of $F = F(x)$ defined by (1.6a) is an extensor function of $a = a(x)$.

We say that T is an *extensor field* if

$$PT(P(A_1), \ldots, P(A_k)) = T(A_1, \ldots, A_k). \tag{1.8b}$$

The simplest extensor field is, of course, the projection extensor defined by (1.4). We refer to T simply as an *extensor* when we do not wish to emphasize the distinction between (1.8a) and (1.8b). If the values of T are of homogeneous grade and the A_k are vector valued, then, for each fixed x, T is a tensor, as defined in Section 3-10. If in addition (1.8b) holds, then T is said to be a *tensor field*.

We define the *differential* T_a of an extensor function $T = T(A_1, \ldots, A_k)$ by

$$\begin{aligned} T_a(A_1, \ldots, A_k) &\equiv a\cdot\dot{\partial}\dot{T}(A_1, \ldots, A_k) \\ &= a\cdot\partial T(A_1, A_2, \ldots, A_k) - T(a\cdot\partial A_1, A_2, \ldots, A_k) - \\ &\quad - T(A_1, a\cdot\partial A_2, \ldots, A_k) - \cdots - T(A_1, A_2, \ldots, a\cdot\partial A_k). \end{aligned} \tag{1.9}$$

Like $T(A_1, \ldots, A_k)$, the function $T_b(A_1, \ldots, A_k)$ is a linear function of the A_i's. It is to preserve this linearity that terms like $T(a \cdot \partial A_1, A_2, \ldots, A_k)$ have been subtracted from the directional derivative $a \cdot \partial T$ in (1.9). We regard (1.6) as a special case of (1.9), obtaining when the function $T = T(x)$ is not regarded as an extensor. Whether or not a function is regarded as a function of other functions is, of course, to some extent a matter of convention, but this ambiguity causes no problems.

The subscript notation T_a for the differential has the advantage of conciseness, and it seldom leads to confusion in spite of many other uses we make of subscripts. However, sometimes it is desirable to consider the differential as an operator without reference to a particular extensor on which it acts. In that case, the operator notation d_a is convenient,* so then $\mathrm{d}_a T = T_a$. However, we will use the notation $a \cdot \partial$, as in (1.9), instead of the notation d_a, because it explicitly indicates its relation to the fundamental operator ∂, hence it is more amenable to algebraic manipulation.

In accordance with (1.9), we define and denote the second differential of a function $F = F(x)$ by

$$\underline{F}_b(a) \equiv F_{ab} \equiv b \cdot \partial \underline{F}(a) - \underline{F}(b \cdot \partial a). \tag{1.10}$$

This agrees with our definition of the second differential in Section 2-1, but here F_{ab} is an extensor function of vector functions $a = a(x)$ and $b = b(x)$. As we pointed out in Section 2-1, the symmetry $F_{ab} = F_{ba}$ is a consequence of the integrability condition. However, it is now necessary to add that this symmetry obtains only if a and b are tangent vectors. For vector manifolds in general, our operator expression (2-1.29) for the *integrability condition* must be written in the form

$$P(\partial \wedge \partial) = I^{-1} I \cdot (\partial \wedge \partial) = 0. \tag{1.11}$$

With the understanding that $P(a) = a$ and $P(b) = b$, we can write the *integrability condition* in the equivalent form

$$[a \cdot \partial, b \cdot \partial] = [a, b] \cdot \partial, \tag{1.12}$$

where

$$[a \cdot \partial, b \cdot \partial] \equiv a \cdot \partial b \cdot \partial - b \cdot \partial a \cdot \partial \tag{1.13}$$

is the commutator of directional derivatives and

$$[a, b] \equiv a \cdot \partial b - b \cdot \partial a \tag{1.14}$$

is the *Lie Bracket*.

The equivalence of (1.11) to (1.12) can be established with the help of the operator identity

$$(a \wedge b) \cdot (\partial \wedge \partial) = [a, b] \cdot \partial - [a \cdot \partial, b \cdot \partial], \tag{1.15}$$

* The subscript on d_a distinguishes it from the closely related operator d, the exterior differential defined by (1.20). In Sections 5-6 and 6-2 the notation d_a is used for different but related operators.

which itself can be established by the following steps

$$\begin{aligned}(a \wedge b) \cdot (\partial \wedge \partial) &= a \cdot (b \cdot \partial\partial) - b \cdot (a \cdot \partial\partial) \\ &= b \cdot \partial a \cdot \partial - (b \cdot \partial a) \cdot \partial - a \cdot \partial b \cdot \partial + (a \cdot \partial b) \cdot \partial \\ &= [a, b] \cdot \partial - [a \cdot \partial, b \cdot \partial]\,.\end{aligned}$$

Dotting (1.11) with $a \wedge b$ we have

$$(a \wedge b) \cdot P(\partial \wedge \partial) = P(a \wedge b) \cdot (\partial \wedge \partial) = (P(a) \wedge P(b)) \cdot (\partial \wedge \partial) = 0, \qquad (1.16)$$

which according to (1.15) is equivalent to (1.12) with the conditions $P(a) = a$ and $P(b) = b$. Conversely, beginning with (1.12) we can establish (1.16) which when differentiated by a and b gives us (1.11), thus, using (2-1.38), we have

$$\partial_b\, \partial_a (a \wedge b) \cdot P(\partial \wedge \partial) = \partial_b\, \partial_a a \cdot [b \cdot P(\partial \wedge \partial)] = \partial_b b \cdot P(\partial \wedge \partial) = 2P(\partial \wedge \partial).$$

Note that a and b, being arbitrary functions of x, can be regarded as independent variables in the tangent space at each point x. Since the tangent space is an n-dimensional linear space, we are justified in using the formulas for derivatives of linear functions such as (2-1.38). This is, in fact, just an application of the chain rule for differentiation. It is a device of great power which we shall often use in this way.

Equation (1.12) can be derived from the expression (1.5) for the directional derivative in terms of a limit. Having established the integrability condition, we have no further need to appeal to the limit process. Evidently, we could define a manifold and develop calculus without reference to limits by adopting certain basic formulas such as (1.7) and (1.11) as axioms.

As we noted in Section 2-1, the differential and the derivative of F are related by the identity

$$\partial F = \partial_a a \cdot \partial F = \partial_a F_a. \qquad (1.17)$$

The differential of ∂F is related to the second differential of F by the formula

$$b \cdot \partial(\partial F) = (\partial F)_b = P_b(\partial_a)F_a + \partial_a F_{ab}. \qquad (1.18)$$

We can establish (1.18) by expressing the variable a as the projection of a vector y in the embedding space $\overline{\mathscr{M}}$; thus $a = P(y)$, and $F_a = a \cdot \partial F = y \cdot \partial F = F_y$. It follows that

$$\partial_a F_a = P(\partial_y)F_y.$$

Since y and ∂_y are independent of x,

$$(\partial_a F_a)_b = P_b(\partial_y)F_y + P(\partial_y)F_{yb}.$$

This is equivalent to (1.18) by virtue of the linearity in y of F_{yb} and $F_y = F_{P(y)}$.

An obvious generalization of (1.18) will be useful. Let $T(B, A)$ be an extensor on $\mathscr{M}$ linear in multivector fields B and A. Suppose

$$Q(A) = \partial_B T(B, A), \tag{1.19a}$$

where $\partial_B = P(\partial_B)$ is the multivector derivative defined in Section 2-2. Then, the differential of $T(B, A)$ is related to the differential of $Q(A)$ by the equation

$$Q_b(A) = P_b(\partial_B) T(B, A) + \partial_B T_b(B, A). \tag{1.19b}$$

One other variant of the differential deserves discussion and a special notation. We define the *exterior differential* $\mathrm{d}T$ of an extensor function $T = T(A)$ by

$$\mathrm{d}T = \mathrm{d}T(B) \equiv \dot{T}(B \cdot \dot{\partial}). \tag{1.20}$$

The exterior differential plays a central role in Stokes' theorem (Section 7-3), where T is a *multiform** of degree k on the tangent algebra, that is, when

$$T(A) = T(P(A)) = T(\langle A \rangle_k). \tag{1.21}$$

Because of the linearity property, a discussion of (1.20) assuming (1.21) is sufficient.

The exterior differential $\mathrm{d}T$ can be obtained from the differential T_a by applying the skewsymmetrizer (2-3.10), as was done in writing (3-1.4c); thus

$$\mathrm{d}T(B) = \langle B\, \partial_a \wedge \partial_{a_k} \wedge \ldots \wedge \partial_{a_1} \rangle T_a(a_1 \wedge \ldots \wedge a_k), \tag{1.22a}$$

or, using the same notations as (3-1.4c),

$$\begin{aligned} \mathrm{d}T(B) &= k!\ (B \cdot \dot{\partial}) * \partial_{(k)} \dot{T}(a_{(k)}) = k!\ B * (\dot{\partial} \wedge \partial_{(k)}) \dot{T}(a_{(k)}) \\ &= (k+1)!\ B * \partial_{(k+1)} T_{a_{k+1}}(a_{(k)}). \end{aligned} \tag{1.22b}$$

Note that exterior differential differs from the skewsymmetrized differential by the factorial of $k + 1$. As an example, consider the case when T has degree 2. Using the identity

$$(a \wedge b \wedge c) \cdot \dot{\partial} = a \wedge b c \cdot \dot{\partial} - a \wedge c b \cdot \dot{\partial} + b \wedge c a \cdot \dot{\partial},$$

we see that

$$\begin{aligned} \mathrm{d}T(a \wedge b \wedge c) &= \dot{T}((a \wedge b \wedge c) \cdot \dot{\partial}) \\ &= c \cdot \dot{\partial} \dot{T}(a \wedge b) + b \cdot \dot{\partial} \dot{T}(c \wedge a) + a \cdot \dot{\partial} \dot{T}(b \wedge c) \\ &= T_c(a \wedge b) + T_b(c \wedge a) + T_a(b \wedge c). \end{aligned} \tag{1.23}$$

* The term 'multiform' was introduced in Section 3-9 to abbreviate 'multivector valued alternating form'. Our usage here is an obvious extension of the original terminology.

It should be obvious from these considerations that the exterior differential raises the degree of a multiform by one.

Now using (1.20) to compute the exterior derivative twice, we have

$$\mathrm{d}^2 T(C) = \mathrm{d}\dot{T}(C \cdot \dot{\partial}) = \dot{T}((C \cdot \dot{\partial}) \cdot \ddot{\partial}) = \dot{T}(C \cdot (\dot{\partial} \wedge \ddot{\partial})). \tag{1.24}$$

The notation here is awkward. The operator $\ddot{\partial}$ has two overdots to indicate that it is differentiated by $\dot{\partial}$, while both $\dot{\partial}$ and $\ddot{\partial}$ act on T. In Section 4-3 this derivative will be evaluated and shown to vanish when projected into the tangent algebra. If $C = P(C)$, we can apply (1.11) to get

$$\mathrm{d}^2 T = \dot{T}(C \cdot (\dot{\partial} \wedge \ddot{\partial})) = \dot{T}(C \cdot P(\dot{\partial} \wedge \ddot{\partial})) = 0. \tag{1.25a}$$

Thus the second exterior differential of a multiform vanishes identically. Like (1.11) and (1.12), the operator equation

$$\mathrm{d}^2 = 0 \tag{1.25b}$$

expresses the integrability condition.

When the multiform $T = T(A)$ is scalarvalued, it is called a *differential form*, and $\mathrm{d}T$ is exactly equivalent to Cartan's 'exterior derivative' of a differential form. For this reason, we have adopted his notation $\mathrm{d}T$, however, we use the term 'exterior differential' to be consistent with our general distinction between the terms 'differential' and 'derivative'.

According to Section 1-4, a k-form $T = T(A)$ can be expressed as the scalar product of some k-vector K with A. So for a differential k-form we have

$$T = T(x, A(x)) = \langle A(x)K(x)\rangle = \langle A\rangle_k \cdot K. \tag{1.26}$$

The exterior differential (1.20) can then be written

$$\begin{aligned} \mathrm{d}T &= \langle B \cdot \partial K\rangle = (\langle B\rangle_{k+1} \cdot \partial) \cdot K \\ &= \langle B\rangle_{k+1} \cdot (\partial \wedge K) = \langle BP(\partial \wedge K)\rangle. \end{aligned} \tag{1.27}$$

The quantity $\mathrm{d}T(B)$ is defined for all arguments $B = P(B)$ in the tangent algebra, so the scalar product $\langle B\, \partial \wedge K\rangle$ vanishes for any component of $\partial \wedge K$ not in the tangent algebra. To emphasize this, the projection operator P was included in the last term of (1.27). Equation (1.27) shows that the exterior differential of a differential k-form is exactly equivalent to the curl of some k-vector field projected into the tangent algebra. A more detailed comparison of Geometric Calculus with the conventional calculus of differential forms is carried out in Section 6-4.

We have introduced the notation $\mathrm{d}T$ to simplify comparison of Geometric Calculus with more conventional formalisms. But we shall use it sparingly, because we regard it as a mere abbreviation of the more explicit expression $\dot{T}(B \cdot \dot{\partial})$, and the explicit expression is needed for computations such as (1.27).

It should be mentioned that, besides the exterior differential, there is also an *interior differential*, which lowers degree by one, namely, the quantity $\dot{T}(B \wedge \dot{\partial})$

obtained from an extensor $T = T(A)$. We need not introduce a special notation for the interior differential, besides, we prefer the name '*tensor divergence*' to 'interior differential', because it agrees with conventional terminology in tensor analysis. A notation frequently used in the calculus of differential forms is explained in Section 6-4. The name 'divergence' is also appropriate for $\dot{\partial} \cdot \dot{T}(A)$, though this quantity is quite different from $\dot{T}(B \wedge \dot{\partial})$. The tensor divergence appears most frequently in connection with degree one tensors, when it has the form $\dot{T}(\dot{\partial})$. On the other hand, if $T = T(A) = A * K$ is a differential k-form, then, in parallel with (1.27), we have

$$\begin{aligned}\dot{T}(B \wedge \dot{\partial}) &= \langle B \wedge \partial K\rangle = (\langle B\rangle_{k-1} \wedge \partial) \cdot K \\ &= \langle B\rangle_{k-1} \cdot (\partial \cdot K) = \langle BP(\partial \cdot K)\rangle .\end{aligned} \tag{1.28}$$

Thus, the interior differential of a differential form is precisely equivalent to the projected divergence of some multivector field, just as the exterior differential is equivalent to the projected curl.

4-2. Projection, Shape and Curl

This section describes and derives the properties of $P_b(A)$, the first differential of the projection extensor $P(A)$. As the results of this section are used in subsequent sections, it will become clear that P_b completely characterizes the differential calculus of fields on a manifold. Indeed, it will be shown in the next chapter that P_b completely determines the intrinsic geometry of a manifold and that the second differential P_{ba} is needed only to describe extrinsic geometry.

The identity function on a manifold (which, of course, maps each point to itself) is not the trivial function one might at first suppose. The parenthesis ( ) is a universal symbol for the identity operator, so we can represent the identity function by writing $(x) = x$. The first differential of the identity function is precisely the projection of a vector into the tangent space at each point of the manifold, that is,

$$a \cdot \partial(x) = P(a) = (a \cdot I) \cdot I^{-1} = a \cdot II^{-1} . \tag{2.1}$$

This result is established for any differentiable vector manifold in the same way that we established it for a linear manifold in Section 2-1. In Section 1-2 we established the outermorphism property of projections, $P(A) \wedge P(B) = P(A \wedge B)$. Hence, the general multivector projection $P(A)$ defined by (1.4) is the outermorphism of the vector projection (2.1).

The second differential of the identity is, of course, the first differential of the projection, namely,

$$P_b(a) = b \cdot \dot{\partial}\dot{P}(a) = b \cdot \partial P(a) - P(b \cdot \partial a). \tag{2.2a}$$

More generally,

$$P_b(A) = b \cdot \dot{\partial}\dot{P}(A) = b \cdot \partial P(A) - P(b \cdot \partial A). \tag{2.2b}$$

This quantity is well defined at each point x of $\mathscr{M}$ for any multivector function $A = A(x)$ and vector function $b = b(x)$.

The properties of the projection differential P_b are derived from the general properties of the projection operator P which we determined in Section 1-2. Obviously, P_b is a linear, grade-preserving operator, that is,

$$P_b(\alpha A + \beta B) = \alpha P_b(A) + \beta P_b(B), \tag{2.3}$$

$$P_b(\langle A\rangle_k) = \langle P_b(A)\rangle_k, \tag{2.4a}$$

and, of course,

$$P_b(\langle A\rangle_0) = 0. \tag{2.4b}$$

Differentiating $A * P(B) = B * P(A)$, we get the symmetry property

$$A * P_b(B) = B * P_b(A). \tag{2.5}$$

Differentiating the outermorphism relation $P(A \wedge B) = P(A) \wedge P(B)$ we get

$$P_b(A \wedge B) = P_b(A) \wedge P(B) + P(A) \wedge P_b(B). \tag{2.6}$$

And differentiating $P^2(A) = P(A)$ we get

$$P_bP(A) + PP_b(A) = P_b(A). \tag{2.7}$$

We do not get a particularly useful result by differentiating the one remaining property of P, namely $P(P(A)B) = P(A)P(B)$, however, we make good use of this as well as other properties of P in computations that follow. Our list of general properties of P_b is completed by

$$P_b(A) = P_{P(b)}(A), \tag{2.8}$$

which follows from $b \cdot \partial = P(b) \cdot \partial$. The operator P_b can be interpreted as the rate of change of P in the direction $P(b)$, that is, that is, as a 'velocity of the projection operator'.

The above listed properties of P_b do not depend on the fact, expressed by (2.1), that $P(A)$ is the differential of the identity function. The integrability condition (1.12) applied to the identity function implies that $P_b(a) = P_a(b)$ if a and b are tangent vectors, as is readily proved by substituting (2.1) into (2.2a) and using (1.12). In view of (2.8), this property can be expressed in the form

$$P_bP(a) = P_aP(b), \tag{2.9}$$

which holds for any vector functions $a = a(x)$ and $b = b(x)$.

According to (2.8) only the tangential component of b contributes to $P_b(a)$. To ascertain how $P_b(A)$ depends on the tangency of A, we introduce the decomposition

$$A = A_\parallel + A_\perp, \tag{2.10a}$$

where

$$A_{\|} = P(A) \tag{2.10b}$$

is the *projection* of A into the tangent algebra, and

$$A_{\perp} = P_{\perp}(A) \equiv A - P(A) \tag{2.10c}$$

is the *rejection* of A from the tangent algebra. The operator P_b treats the components $A_{\|}$ and $A_{\perp}$ quite differently. Thus, the identity (2.7) admits the special cases

$$PP_b(A_{\|}) = PP_bP(A) = 0, \tag{2.11a}$$

$$PP_b(A_{\perp}) = P_b(A_{\perp}). \tag{2.11b}$$

From this it follows that identity (2.6) admits the special cases

$$P_b(A_{\|} \wedge B_{\|}) = P_b(A_{\|}) \wedge B_{\|} + A_{\|} \wedge P_b(B_{\|}), \tag{2.12a}$$

$$P_b(A_{\perp} \wedge B_{\|}) = P_b(A_{\perp}) \wedge B_{\|}, \tag{2.12b}$$

$$P_b(A_{\perp} \wedge B_{\perp}) = 0. \tag{2.12c}$$

From (2.12) it follows that if $A_{\perp}$ is a blade and $PP_b(A_{\perp}) = P_b(A_{\perp}) \neq 0$, then $A_{\perp}$ has a vector factor $a_{\perp}$ such that

$$P_b(A_{\perp}) = P_b(a_{\perp}) \wedge B_{\|} = P_b(a_{\perp})B_{\|}, \tag{2.13a}$$

where $B_{\|}$ is defined by the factorization

$$A_{\perp} = a_{\perp} \wedge B_{\|} = a_{\perp}B_{\|}. \tag{2.13b}$$

The principal object of our interest is the derivative of the projection operator

$$S(A) = \dot{\partial}\dot{P}(A) = \partial_b P_b(A). \tag{2.14}$$

We call $S(A)$ the *shape* of the function $A = A(x)$, and we refer to the operator S as the *shape operator* of the manifold $\mathcal{M}$. The algebraic properties of the shape operator can be derived from the above properties of the operator P_b. From (2.3) we immediately determine that the shape operator is linear, that is,

$$S(\alpha A + \beta B) = \alpha S(A) + \beta S(B). \tag{2.15}$$

However, before we present the remaining properties of shape in general, we examine the shape of a vector function in detail.

We first establish that, for a vector function $a = a(x)$,

$$S(a_{\|}) = \partial_b \wedge P_b(a), \tag{2.16a}$$

$$S(a_{\perp}) = \partial_b \cdot P_b(a). \tag{2.16b}$$

By (2.14) and (2.15), we can write

$$S(a) = S(a_{\parallel}) + S(a_{\perp}) = \partial_b \cdot P_b(a) + \partial_b \wedge P_b(a),$$

hence (2.16) can be proved by showing that

$$\partial_b \cdot P_b(a_{\parallel}) = 0, \tag{2.17a}$$

$$\partial_b \wedge P_b(a_{\perp}) = 0. \tag{2.17b}$$

We prove (2.17a) from (2.11a) as follows:

$$\partial_b \cdot P_b(a_{\parallel}) = P(\partial_b) \cdot P_b(a_{\parallel}) = \partial_b \cdot PP_b(a_{\parallel}) = 0.$$

To prove (2.17b), we first notice that (2.11b) implies that

$$P(\partial_b \wedge P_b(a_{\perp})) = \partial_b \wedge PP_b(a_{\perp}) = \partial_b \wedge P_b(a_{\perp}).$$

Hence if c and d are any vectors, we show, with the help of (2.5) and (2.9), that

$$\begin{aligned}(c \wedge d) \cdot (\partial_b \wedge P_b(a_{\perp})) &= (P(c) \wedge P(d)) \cdot (\partial_b \wedge P_b(a_{\perp}))\\ &= P(c) \cdot P_d(a_{\perp}) - P(d) \cdot P_c(a_{\perp})\\ &= a_{\perp} \cdot [P_d P(c) - P_c P(d)] = 0.\end{aligned}$$

This proves (2.17b). The reader may need to be reminded that we have used the formulas (2-1.46) and (2-1.19) in this proof. These and similar formulas from Chapters 1 and 2 will hereafter be taken for granted in proofs and computations.

Equations (2.16a, b) can be combined in the single equation

$$S(a) = S_a + N \cdot a, \tag{2.18}$$

where

$$S_a \equiv \dot{\partial} \wedge \dot{P}(a) = S(P(a)) \tag{2.19}$$

and

$$N \equiv \dot{P}(\dot{\partial}) = \partial_a \cdot S_a = \partial_a S_a. \tag{2.20}$$

We call the bivectorvalued function S_a the *curl tensor* or the *curl* of the manifold to emphasize the fact that it is the curl of the projection of a vector. We call N the spur of $\mathscr{M}$ to emphasize the fact that it is the spur of the tensor $P_b(a)$, thus,

$$\partial_b \cdot \partial_a P_b(a) = P_b(\partial_b) = \dot{P}(\dot{\partial}) = N.$$

To get (2.18) from (2.16b), we must show that the divergence of the projection is determined by the spur. This is easily done with the help of (2.5), thus,

$$\dot{\partial} \cdot \dot{P}(a) = a \cdot \dot{P}(\dot{\partial}) = a \cdot N. \tag{2.21}$$

We can also use (2.5) to put (2.17a) in the form

$$N \cdot P(a) = 0. \tag{2.22}$$

This says that N is everywhere orthogonal to tangent vectors of the manifold, which fits nicely with the connotation of 'spur' as something that 'sticks out' (of the manifold).

We have yet to complete our justification of (2.20). Equation (2.20) says that the spur is determined by the curl. This is proved by

$$\partial_a \cdot S_a = \partial_a \cdot (\partial_b \wedge P_b(a)) = \partial_a \cdot \partial_b P_b(a) - \partial_b \partial_a \cdot P_b(a) = N,$$

Equation (2.20) also says that

$$\partial_a \wedge S_a = 0. \tag{2.23}$$

This follows from the symmetry property (2.9), thus

$$\partial_a \wedge S_a = \partial_a \wedge \partial_b \wedge P_b P(a) = 0.$$

Equation (2.19) defines S_a in terms of $P_b(a)$. Conversely, we can express $P_b(a)$ in terms S_a. Dotting (2.19) with b, we have

$$b \cdot S_a = b \cdot (\partial_c \wedge P_c P(a)) = b \cdot \partial_c P_c P(a) - \partial_c b \cdot P_c P(a).$$

By virtue of (2.9) and (2.5), the last term can be written

$$\partial_c b \cdot P_c P(a) = \partial_c b \cdot P_a P(c) = \partial_c c \cdot PP_a(b) = PP_a(b),$$

while the first term can be re-expressed by (2.9), yielding

$$b \cdot S_a = P_a P(b) - PP_a(b). \tag{2.24}$$

Operating on (2.24) with P we get, because of (2.11a),

$$PP_a(b) = -P(b \cdot S_a). \tag{2.25}$$

On the other hand, if $b = P(b)$ in (2.24), we have

$$P_a P(b) = P(b) \cdot S_a. \tag{2.26}$$

Adding (2.25) and (2.26) and using (2.7), we get the desired expression

$$P_a(b) = P(b) \cdot S_a - P(b \cdot S_a). \tag{2.27}$$

Note that, by (2.26), (2.9) is equivalent to the relation

$$P(b) \cdot S_a = P(a) \cdot S_b. \tag{2.28}$$

This is just what one obtains by dotting (2.23) with $P(a \wedge b)$. Thus, (2.9), (2.23) and (2.28) are mutually equivalent expressions of the integrability condition applied to the identity function.

Note also that, by using (2.26), (2.11a) can be expressed as the condition

$$P(P(b)\cdot S_a)=0. \qquad (2.29)$$

Thus, the vector-valued tensor $P(b)\cdot S_a$ is normal to the manifold. The 'normality' of the spur expressed by (2.22) is a special consequence of (2.29), as is the relation

$$P(S_a)=0. \qquad (2.30)$$

Equation (2.30) can be more directly derived from (2.19) by using (2.11a).

We can generalize Eqn. (2.27) at once. Using (2.6) and the anticommutivity of the outer product, we get

$$P_b(a_1\wedge\ldots\wedge a_r)=\sum_{k=1}^{r}(-1)^{k+1}P_b(a_k)\wedge P(a_1)\wedge\ldots\check{P}(a_k)\ldots\wedge P(a_r), \qquad (2.31)$$

where the a_k's are vectors. Eliminating $P_b(a_k)$ with (2.27) and using (1-1.67), which applies because S_b is a bivector, we obtain

$$P_b(a_1\wedge\ldots\wedge a_r)=P(a_1\wedge\ldots\wedge a_r)\times S_b-P((a_1\wedge\ldots\wedge a_r)\times S_b). \qquad (2.32)$$

By virtue of linearity, (2.32) implies that for any multivector A,

$$P_b(A)=P(A)\times S_b-P(A\times S_b). \qquad (2.33)$$

With the help of (2.30), we easily get from (2.33) the special cases

$$P_b(A_\parallel)=P_b\dot{P}(A)=P(A)\times S_b, \qquad (2.34a)$$

$$P_b(A_\perp)=P_bP_\perp(A)=P(S_b\times A). \qquad (2.34b)$$

Thus, the differential of the projection is completely determined by the curl tensor.

Having examined the shape of a vector-valued function exhaustively, we are prepared to establish the general properties of the shape operator defined by (2.14). The most important properties of shape are expressed by

$$S(A_\parallel)=S(P(A))=\dot{\partial}\wedge\dot{P}(A), \qquad (2.35a)$$

$$S(A_\perp)=P(S(A))=\dot{\partial}\cdot\dot{P}(A). \qquad (2.35b)$$

Of course,

$$S(A)=S(A_\parallel)+S(A_\perp). \qquad (2.35c)$$

Moreover, because of (2.4), we have

$$\langle S(A)\rangle_{k+1}=SP(\langle A\rangle_k), \qquad (2.36a)$$

$$\langle S(A)\rangle_{k-1}=PS(\langle A\rangle_k). \qquad (2.36b)$$

Thus, the shape operator distinguishes $A_\parallel$ from $A_\perp$ by raising the grade of the former while lowering the grade of the latter.

Properties (2.35a, b) can be established by proving

$$\dot{\partial} \wedge \dot{P}(A_\perp) = 0, \tag{2.37a}$$

$$\dot{\partial} \cdot \dot{P}(A_\parallel) = 0. \tag{2.37b}$$

It suffices to consider $A = a_1 \wedge a_2 \wedge \ldots \wedge a_r = \langle A \rangle_r$. Using (2.31) and (2.19), we have

$$\begin{aligned} \dot{\partial} \wedge \dot{P}(a_1 \wedge \ldots \wedge a_r) &= \sum_{k=1}^{r} (-1)^{k+1} \dot{\partial} \wedge \dot{P}(a_k) \wedge P(a_1 \wedge \ldots \check{a}_k \ldots \wedge a_r) \\ &= \sum_{k=1}^{r} (-1)^{k+1} S(a_k) \wedge P(a_1 \wedge \ldots \check{a}_k \ldots \wedge a_r) \\ &= \dot{\partial} \wedge \dot{P}P(a_1 \wedge \ldots \wedge a_r), \end{aligned} \tag{2.38}$$

from which (2.37a) follows easily. The proof of (2.37b) is a little more trouble. Using (2.31) and (1-1.38), we have

$$\begin{aligned} \dot{\partial} \cdot \dot{P}(a_1 \wedge \ldots \wedge a_r) = \sum_{k=1}^{r} (-1)^{k+1} \{ &\dot{\partial} \cdot \dot{P}(a_k) P(a_1 \wedge \ldots \check{a}_k \ldots \wedge a_r) + \\ &+ P_{a_k}(a_1 \wedge \ldots \check{a}_k \ldots \wedge a_r) \}. \end{aligned}$$

We notice that, by (2.31) again,

$$\begin{aligned} &\sum_{k=1}^{r} (-1)^{k+1} P_{a_k}(a_1 \wedge \ldots \check{a}_k \ldots \wedge a_r) \\ &\quad = \sum_{j<k} (-1)^{j+k+1} [P_{a_j}(a_k) - P_{a_k}(a_j)] \wedge P(a_1 \wedge \ldots \check{a}_j \ldots \check{a}_k \ldots \wedge a_r), \end{aligned}$$

which vanishes if $P(a_k) = a_k$ for all k. Also, by (2.21) and (1-1.38)

$$\sum_{k=1}^{r} (-1)^{k+1} \dot{\partial} \cdot \dot{P}(a_k) P(a_1 \wedge \ldots \check{a}_k \ldots \wedge a_r) = P(N \cdot (a_1 \wedge \ldots \wedge a_r)).$$

Hence,

$$\begin{aligned} \dot{\partial} \cdot \dot{P}(a_1 \wedge \ldots \wedge a_r) = P(N \cdot (&a_1 \wedge \ldots \wedge a_r)) + \\ &+ \sum_{k=1}^{r} (-1)^{k+1} PP_{a_k}(a_1 \wedge \ldots \check{a}_k \ldots \wedge a_r). \end{aligned} \tag{2.39}$$

By virtue of (2.22) and (2.11a), the right side of (2.39) vanishes if $P(a_1 \wedge \ldots \wedge a_r) = a_1 \wedge \ldots \wedge a_r$, from which (2.37b) follows easily.

With (2.35) and (2.36) in hand, other properties of shape are readily established. For example, from (2.12a, b, c) we get

$$S(A_\| \wedge B_\|) = S(A_\|) \wedge B_\| + (-1)^r A_\| \wedge S(B_\|), \tag{2.40a}$$

$$S(A_\perp \wedge B_\|) = S(A_\perp) \wedge B_\| + (-1)^r \dot{P}(A_\perp) \wedge (\dot{\partial} \cdot B_\|), \tag{2.40b}$$

if $\langle A \rangle_r = A$, and

$$S(A_\perp \wedge B_\perp) = 0. \tag{2.40c}$$

From (2.38), we get

$$SP(a_1 \wedge \ldots \wedge a_r) = \sum_{k=1}^{r} (-1)^{k+1} S_{a_k} \wedge P(a_1 \wedge \ldots \check{a}_k \ldots \wedge a_r). \tag{2.41a}$$

From (2.39) and (2.34b), we get

$$PS(a_1 \wedge \ldots \wedge a_r) = P(N \cdot (a_1 \wedge \ldots \wedge a_r)) + \\ + \sum_{k=1}^{r} (-1)^{k+1} P(S(a_k) \times (a_1 \wedge \ldots \check{a}_k \ldots \wedge a_r)). \tag{2.41b}$$

The last two equations can be combined using (2.18), yielding

$$S(a_1 \wedge \ldots \wedge a_r) = \sum_{k=1}^{r} (-1)^{k+1} \{S(a_k) \wedge P(a_1 \wedge \ldots \check{a}_k \ldots \wedge a_r) + \\ + P(S(a_k) \times (a_1 \wedge \ldots \check{a}_k \ldots \wedge a_r))\}. \tag{2.41c}$$

The differential and shape of the pseudoscalar I are particularly important. Differentiating $II^{-1} = I \cdot I^{-1} = 1$, we get $(a \cdot \partial I) \cdot I^{-1} = 0$; hence,

$$P(a \cdot \partial I) = 0. \tag{2.42}$$

Differentiating $I = P(I)$ and using (2.42), we get

$$a \cdot \partial I = P_a(I). \tag{2.43}$$

Hence, by (2.35a),

$$\partial I = \partial \wedge I = S(I). \tag{2.44}$$

According to (2.36) the shape operator does not preserve grade. Nor does it preserve tangency, as is proved by applying (2.11a) to (2.35a) to get

$$PS(A_\|) = PSP(A) = 0. \tag{2.45}$$

However, the operator S^2 preserves both grade and tangency. Thus, from (2.35), we find that

$$S^2(A_\|) = PS^2(A_\|) = PS^2P(A), \tag{2.46a}$$

$$S^2(A_\perp) = P_\perp S^2(A) = SPS(A). \tag{2.46b}$$

And from (2.36), we ascertain

$$S^2(\langle A\rangle_r) = \langle S^2(A)\rangle_r. \tag{2.47}$$

The result of operating with S^2 on $a \wedge b = P(a \wedge b)$ is important in the study of curvature. With the help of (2.40) and (2.34b) we get

$$\begin{aligned} S^2(a \wedge b) &= S^2(a) \wedge b + P_b S(a) - P_a S(b) + a \wedge S^2(b) \\ &= S^2(a) \wedge b + 2P(S_b \times S_a) + a \wedge S^2(b). \end{aligned} \tag{2.48}$$

4-3. Intrinsic Derivatives and Lie Brackets

This section defines and studies the fundamental differential operators which preserve the tangency of fields on a manifold. Intrinsic differential identities are established which have wide applicability. The emphasis here is on results which are independent of the geometry of a manifold. The role of intrinsic derivatives in geometry will be discussed in Chapter 5.

The last half of this section concerns generalizations and variations of the Lie bracket concept. This material will not be used in the rest of the book, though the induced transformations of brackets will be determined in Section 4-5. The results are included for reference purposes, to use the Geometric Calculus for applications where Lie brackets play a prominent role.

Often we are concerned exclusively with *fields* on a vector manifold $\mathscr{M}$. Recall, that a field $A = A(x)$ on $\mathscr{M}$ has the property of tangency: $A = P(A)$. As a rule, the directional derivative $a \cdot \partial A$ of a field $A = A(x)$ is *not tangent* to $\mathscr{M}$. We can develop a differential calculus of fields by introducing the *directional coderivative* $a \cdot \nabla A$ of a defined by

$$a \cdot \nabla A \equiv P(a \cdot \partial A). \tag{3.1}$$

Similarly, we define the *coderivative* ∇A of A by

$$\nabla A \equiv P(\partial A) = \partial_a a \cdot \nabla A = \nabla_a a \cdot \nabla A. \tag{3.2}$$

Recall the definition (1.8b) of an extensor field $T = T(A_1, A_2, \ldots, A_k)$. Generalizing (3.1), we define the *codifferential* $\delta_a T$ of T by

$$\begin{aligned} \delta_a T \equiv PT_a(P(A_1), \ldots, P(A_k)) &= a \cdot \dot{\nabla}\dot{T}(P(A_1), \ldots, P(A_k)) \\ &= a \cdot \nabla T(A_1, \ldots, A_k) - T(a \cdot \nabla P(A_1), \ldots, A_k) - \\ &\quad - \cdots - T(A_1, \ldots, a \cdot \nabla P(A_k)). \end{aligned} \tag{3.3}$$

The prefix 'co' is used above to mean 'with', that is, 'with preservation of tangency'. Obviously, the projection operator P assures tangency of the coderivative by annihilating the component of the derivative which is not tangent.

For a tensor field T, the codifferential defined by (3.3) is equivalent to the usual 'covariant derivative' of tensor analysis. Of course, the meaning of the prefix 'co' in 'covariant' is quite different from its meaning in 'codifferential'.

Just as the codifferential $\delta_a T$ is defined as the projected differential of an extensor field by (3.3), it is natural to define the *exterior codifferential* δT as the projected exterior differential. Accordingly, for the extensor field

$$T = T(A) = PTP(A)\,, \tag{3.4a}$$

from (1.20) we have

$$\delta T = \delta T(B) \equiv P(\mathrm{d}T) = P(\dot{T}(B\cdot\dot{\partial})) \equiv \dot{T}(B\cdot\dot{\nabla}). \tag{3.4b}$$

As a rule, we shall use the notation δT sparingly as an abbreviation for the more explicit expression $\dot{T}(B\cdot\dot{\nabla})$. However, in the computations of Section 7-8, the notation δT is found to be a definite asset.

We wish to establish the general properties of the coderivative, especially so we can use it without reference to its definition (3.2) in terms of the derivative. For scalar fields, the derivative and coderivative are identical, that is, since $P(\partial) = \partial$,

$$\nabla\phi = \partial\phi \quad \text{if } \langle\phi\rangle = \phi.$$

The derivative $\nabla\phi$ is usually called the *gradient* of ϕ. Like the derivative, the coderivative of a general multivector valued function $A = A(x)$ can be decomposed into two parts

$$\nabla A = \nabla\cdot A + \nabla\wedge A, \tag{3.5a}$$

where

$$\nabla\cdot A \equiv P(\partial\cdot A), \tag{3.5b}$$

$$\nabla\wedge A \equiv P(\partial\wedge A). \tag{3.5c}$$

We call $\nabla\cdot A$ the *codivergence* of A and $\nabla\wedge A$ the *cocurl* of A. We may refer to ∇A as the *gradient* instead of the coderivative of A, because it generalizes the usual notion of gradient. Of course, $\nabla\cdot\phi = 0$ for scalar ϕ, by virtue of the algebraic property (1-1.21b), so the gradient of a scalar is the curl $\partial\wedge\phi = \nabla\wedge\phi = \nabla\phi$. In general, the coderivative is equivalent to the derivative only for fields on linear manifolds.

Considering once again Eqn. (1.27) and the attendant discussion, it is clear that the role of the cocurl in Geometric Calculus corresponds exactly to the role of the exterior derivative in the conventional calculus of differential forms. Similarly, Eqn. (1.28) shows that the codivergence corresponds to the 'interior derivative', To readers familiar with the subject, it will be clear that the concepts and terminology of de Rham's cohomology theory can be applied to multivector fields. Accordingly, we say that a multivector field $F = P(F)$ is *closed* if $\nabla\wedge F = 0$, and we say that it is *exact* if there exists another field $A = P(A)$ such that $F = \nabla\wedge A$ Equation

(3.11) below says that every exact field is closed. But, as standard accounts of cohomology theory show, not every closed field is exact. We shall not attempt a systematic reformulation of cohomology theory in the language of Geometric Calculus, but some results appear in Chapter 7.

Now let us examine the difference between derivative and coderivative in detail. Differentiating the tangency condition $A = P(A)$, we find that *for a field* $A = A(x)$ the derivative is related to the coderivative by

$$\partial A = \nabla A + S(A), \tag{3.6a}$$

where $S(A)$ is the shape of A defined by (2.14). Because of (2.35a), we have

$$\partial \wedge A = \nabla \wedge A + S(A), \tag{3.6b}$$

$$\partial \cdot A = \nabla \cdot A \quad \text{if } P(A) = A. \tag{3.6c}$$

Without the tangency condition on A and the notation for coderivative, Eqns. (3.6) take the form

$$\partial P(A) = P(\partial P(A)) + S(P(A)), \tag{3.7a}$$

$$\partial \wedge P(A) = P(\partial \wedge A) + S(P(A)), \tag{3.7b}$$

$$\partial \cdot P(A) = P(\partial \cdot P(A)). \tag{3.7c}$$

Operating on (3.7b) with P, we get, because of (2.45),

$$P(\partial \wedge P(A)) = P(\partial \wedge A) \tag{3.8a}$$

or, equivalently,

$$\nabla \wedge A = P(\partial \wedge A) \quad \text{if } A = P(A). \tag{3.8b}$$

The codivergence and cocurl of a field A are related by the duality theorem

$$\nabla \cdot A = [\nabla \wedge (AI)]I^{-1} \quad \text{if } P(A) = A, \tag{3.9}$$

where I is the pseudoscalar of the manifold. We can prove (3.9) as follows

$$\begin{aligned} P[\partial \wedge (AI)] &= P[\dot{\partial} \wedge (\dot{A}I)] + P[\dot{\partial} \wedge (A\dot{I})] = P[\dot{\partial} \wedge (P(\dot{A})I)] \\ &= P[\dot{\partial} \cdot (P(\dot{A}))I] = P[\dot{\partial} \cdot P(\dot{A})]I = \partial \cdot P(A)I. \end{aligned}$$

In the second step we used (2.24) and (2.45) to justify dropping the term

$$P[\dot{\partial} \wedge (A\dot{I})] = P[\dot{\partial} \wedge (A\dot{P}(I))] = P[\dot{\partial} \wedge (AP\dot{P}(I))] = 0.$$

The integrability condition comes in to play when we consider second derivatives. Taking $A = \partial$ in (3.7b), we get the operator equation

$$\partial \wedge \partial = \partial \wedge P(\partial) = P(\partial \wedge \partial) + S(\partial). \tag{3.10a}$$

But $P(\partial \wedge \partial) = 0$, according to (1.11), hence we have the general relation

$$\partial \wedge \partial A = S(\partial)A. \tag{3.10b}$$

From this we can prove that for any $A = A(x)$,

$$\nabla \wedge \nabla \wedge A = P(\partial \wedge \partial \wedge A) = 0. \tag{3.11}$$

Using (3.7b) we have

$$\partial \wedge P(\partial \wedge A) = P(\partial \wedge \partial \wedge A) + SP(\partial \wedge A).$$

Operating on this with P and using (2.45), (3.10b) and (2.30), we have

$$P(\partial \wedge P(\partial \wedge A)) = P(\partial \wedge \partial \wedge A) = P(S(\partial) \wedge A) = P(P(S(\partial)) \wedge A) = 0,$$

which establishes (3.11).

'Dual' to (3.11), we have the result

$$\partial \cdot (\partial \cdot P(A)) = 0, \tag{3.12a}$$

or, equivalently, because of (3.6c),

$$\nabla \cdot (\nabla \cdot A) = \partial \cdot (\partial \cdot A) = 0 \quad \text{if } P(A) = A. \tag{3.12b}$$

We can prove (3.12) by using (3.9) twice followed by (3.11), thus

$$\nabla \cdot (\nabla \cdot A) = \nabla \cdot \{[\nabla \wedge (AI)]I^{-1}\} = \nabla \wedge [\nabla \wedge (AI)]I^{-1} = 0.$$

The associative rule allows us to expand the *colaplacian* ∇^2 in two ways which, because of (3.11) and (3.12), yields

$$\nabla^2 A = \nabla \cdot \nabla A + \nabla \wedge \nabla A = \nabla \wedge (\nabla \cdot A) + \nabla \cdot (\nabla \wedge A) \quad \text{if } P(A) = A. \tag{3.13}$$

From this we can conclude that the operator $\nabla \wedge \nabla$ is grade-preserving, that is

$$\nabla \wedge \nabla \langle A \rangle_k = \langle \nabla \wedge \nabla A \rangle_k \quad \text{if } P(A) = A, \tag{3.14a}$$

or, equivalently,

$$\nabla \wedge \nabla A = (\nabla \wedge \nabla) \times A \quad \text{if } P(A) = A. \tag{3.14b}$$

This result can also be obtained by using (1-1.62b).

We can easily derive a large number of 'differential identities' for the coderivative in the same way we did for the derivative in Section 2-1. Identities engaging a simple derivative remain valid if ∂ is replaced by ∇. When two or more derivatives are engaged, the differences arising from the replacement of ∂ by ∇ are readily determined using the results above.

The properties of the directional coderviative can be derived from the definition

(3.1), but a more interesting and useful approach is to derive them from the properties of the coderviative by using differential identities. Let $a = a(x)$ and $b = b(x)$ be vector fields. According to (3.6b), since $P(a) = a$,

$$\partial \wedge a = \nabla \wedge a + S_a, \tag{3.15}$$

where S_a is the curl tensor defined by (2.19). Using the differential identity

$$b \cdot (\partial \wedge a) = b \cdot \partial a - \dot{\partial}\dot{a} \cdot b$$

and noting that $\dot{\partial}\dot{a} \cdot b = \dot{\nabla}\dot{a} \cdot b$, we obtain, by dotting (3.15) with b,

$$b \cdot \partial a = b \cdot \nabla a + b \cdot S_a. \tag{3.16}$$

Since $P(a \wedge b) = a \wedge b$, according to (2.28), we have, $a \cdot S_b = b \cdot S_a$, so from (3.16) we have

$$[a, b] \equiv a \cdot \partial b - b \cdot \partial a = a \cdot \nabla b - b \cdot \nabla a. \tag{3.17}$$

It follows that the Lie bracket $[a, b]$ preserves tangency, in spite of the fact that $a \cdot \partial b$ and $b \cdot \partial a$ do not; that is, regarding a and b as arbitrary vector-valued functions, we have

$$[P(a), P(b)] = P([a, b]). \tag{3.18}$$

This result can also be obtained directly from the integrability condition by operating on the identity function $(x) = x$ with (1.12). The tangency-preserving property (3.18) of the Lie bracket is evidently equivalent to the symmetry property (2.28) of the curl tensor, which was also derived from the integrability condition.

The Lie bracket satisfies the differential identity

$$[a, b] = \partial \cdot (a \wedge b) - b\partial \cdot a + a\partial \cdot b. \tag{3.19a}$$

Because of (3.6c), this reduces to

$$[a, b] = \nabla \cdot (a \wedge b) - b\nabla \cdot a + a\nabla \cdot b, \quad \text{if } P(a \wedge b) = a \wedge b. \tag{3.19b}$$

This shows that tangency of the Lie bracket is equivalent to tangency of the divergence.

We generalize the notation of Lie bracket to arbitrary functions $A = A(x)$ and $B = B(x)$ with the definition

$$[A, B] \equiv (A \cdot \partial) \wedge B - \dot{A} \wedge (\dot{\partial} \cdot B). \tag{3.20}$$

This *bracket* obviously has the distributive properties

$$[A, B + C] = [A, B] + [A, C], \tag{3.21a}$$

$$[A + B, C] = [A, C] + [B, C]. \tag{3.21b}$$

For $A_r \equiv \langle A \rangle_r$ and $B_s \equiv \langle B \rangle_s$ we easily prove the 'commutation rule'

$$[A_r, B_s]^\dagger = -[B_s^\dagger, A_r^\dagger] = (-1)^{(r-1)(s-1)}[B_s, A_r]. \tag{3.22a}$$

This admits the useful special case

$$[\dot{a}, B] = -[B, a] \quad \text{if } \langle a \rangle_1 = a. \tag{3.22b}$$

Using the algebraic identity (1-1.42), we easily find the generalization of (3.19a):

$$\partial \cdot (A_r \wedge B) = (\partial \cdot A_r) \wedge B + (-1)^r A_r \wedge (\partial \cdot B) + (-1)^{r+1}[A_r, B], \tag{3.23a}$$

which admits the special case

$$\partial \cdot (a \wedge B) = (\partial \cdot a)B - a \wedge (\partial \cdot B) + [a, B]. \tag{3.23b}$$

Since, according to (3.6c), the divergence preserves tangency, it follows at once from (3.23a) that the bracket $[A, B]$ preserves tangency, that is,

$$[P(A), P(B)] = P([A, B]). \tag{3.24}$$

With (3.12b) and (3.23a) it can be shown that the divergence of the bracket has the simple property

$$\nabla \cdot [A_r, B] = [\nabla \cdot A_r, B] + (-1)^{r+1}[A_r, \nabla \cdot B] \tag{3.25}$$

if $P(A_r) = A_r$ and $P(B) = B$. This is just what one would expect from a naive application of the product rule for differentiation. However, the curl of the bracket does not obey such a simple rule.

The bracket $[A_r, B]$ can be regarded as an operator on B. Using (1-1.23b) and (1-1.42), it can be shown that this operator has the property

$$[A_r, B_s \wedge C] = [A_r, B_s] \wedge C + (-1)^{s(r-1)} B_s \wedge [A_r, C]. \tag{3.26a}$$

This admits the special case

$$[A_r, b \wedge B_s] = [A_r, b] \wedge B_s + (-1)^s [A_r, B_s] \wedge b. \tag{3.26b}$$

By iterating (3.26b), or more easily by using (1-1.38), we get the useful expansions

$$[A_r, B_s] = \sum_{k=1}^{s} (-1)^{k+1}[A_r, b_k] \wedge b_1 \wedge \ldots \wedge \check{b}_k \wedge \ldots \wedge b_s, \tag{3.27a}$$

$$[a, B_s] = \sum_{k=1}^{s} b_1 \wedge \ldots \wedge b_{k-1} \wedge [a, b_k] \wedge b_{k+1} \wedge \ldots \wedge b_s, \tag{3.27b}$$

where $B_s = b_1 \wedge b_2 \wedge \ldots \wedge b_s$.

Just as the divergence and curl are dual to one another in the sense of Eqn. (3.9), we can introduce a *dual bracket* $\{A, B\}$ whose dual relation to the generalized Lie bracket is obvious from the definition

$$\{A, B\} \equiv [A, BI]I^{-1}. \tag{3.28}$$

The dual bracket obviously inherits some of the general properties of the bracket without modification. Thus, it has the linear property

$$\{A, B + C\} = \{A, B\} + \{A, C\}, \tag{3.29a}$$

$$\{A + B, C\} = \{A, C\} + \{B, C\}; \tag{3.29b}$$

and it preserves tangency,

$$P(\{A, B\}) = \{P(A), P(B)\}. \tag{3.30}$$

However, it does not possess any symmetry property comparable to (3.22). In fact, since the grade of the dual bracket is given by

$$\text{grade } \{A_r B_s\} = s - r + 1 > 0, \tag{3.31a}$$

we have

$$\{A_r, B_s\} = 0 \quad \text{if } r > s + 1. \tag{3.31b}$$

From (3.26a) we get, for $t + r \geqslant s + 1$,

$$C_t \cdot \{A_r, B_s\} = \{C_t \wedge A_r, B_s\} + (-1)^{r(s+1)}\{C_t, B_s\} \cdot A_r, \tag{3.32}$$

and from (3.25) we obtain, if A_r and B_s are tangents,

$$\nabla \wedge \{A_r, B\} = \{\nabla \cdot A_r, B\} + (-1)^{r+1}\{A_r, \nabla \wedge B\}. \tag{3.33}$$

The dual of the expression (3.23) for the Lie bracket in terms of the divergence is the following expression for the dual bracket in terms of the curl:

$$\{A_r, B_s\} = (-1)^{r+1}\nabla \wedge (A_r \cdot B_s) + (-1)^r(\nabla \cdot A_r) \cdot B_s + A_r \cdot \nabla \wedge B_s, \tag{3.34a}$$

which holds if A_r and B_s are fields and $s > 0$. We derive this from (3.23a); thus,

$$\begin{aligned}\{A_r, B_s\} &= [A_r, B_sI]I^{-1} \\ &= (-1)^{r+1}\nabla \cdot (A_r \wedge (B_sI))I^{-1} + (-1)^r(\nabla \cdot A_r) \wedge (B_sI)I^{-1} + \\ &\quad + A_r \wedge (\nabla \cdot (B_sI))I^{-1} \\ &= (-1)^{r+1}\nabla \cdot (A_r \cdot B_sI)I^{-1} + (-1)^r(\nabla \cdot A_r) \cdot (B_sII^{-1}) + \\ &\quad + A_r \wedge (\nabla \wedge B_sI)I^{-1} \\ &= (-1)^{r+1}\nabla \wedge (A_r \cdot B_s) + (-1)^r(\nabla \cdot A_r) \cdot B_s + A_r \cdot (\nabla \wedge B_s).\end{aligned}$$

In view of (3.31b), the only cases excluded from (3.34a) are the easily established

$$\{a, \beta\} = -\beta\nabla \cdot a + a \cdot \nabla\beta, \tag{3.34b}$$

and

$$\{\alpha, \beta\} = -\beta\nabla\alpha, \tag{3.34c}$$

where a is a vector field and α and β are scalar fields.

Finally, we have a relation between the two kinds of brackets expressed by the identity, for $s > 0$,

$$\begin{aligned} \langle C_t\{A_r, B_s\}\rangle = (-1)^{r+1}\langle [C_t, A_r]B_s\rangle + \langle A_r \cdot \nabla B_s \cdot C_t\rangle + \\ + (-1)^r\langle B_s \cdot C_t\nabla \cdot A_r\rangle. \end{aligned} \tag{3.35a}$$

This vanishes unless $s + 1 = r + t$, in which case it can be written in the equivalent form

$$\begin{aligned} C_t \cdot \{A_r, B_s\} = (-1)^{r+1}[C_t, A_r] \cdot B_s + (A_r \cdot \nabla) \cdot (B_s \cdot C_t) + \\ + (-1)^r(B_s \cdot C_t) \cdot (\nabla \cdot A_r). \end{aligned} \tag{3.35b}$$

To prove this, we note that (3.34a) gives immediately

$$\begin{aligned} C_t \cdot \{A_r, B_s\} = (-1)^{r+1}C_t \cdot (\nabla \wedge (A_r \cdot B_s)) + (-1)^r C_t \cdot ((\nabla \cdot A_r) \cdot B_s) + \\ + C_t \cdot (A_r \cdot (\nabla \wedge B_s)). \end{aligned}$$

With the help of (1-1.42) and (3.20) we see that

$$\begin{aligned} C_t \cdot (\nabla \wedge (A_r \cdot B_s)) &= (C_t \cdot \nabla) \cdot (A_r \cdot B_s) \\ &= [(C_t \cdot \nabla) \wedge A_r] \cdot B_s + (-1)^{r+1}[C_t \wedge (A_r \cdot \nabla)] \cdot B_s + \\ &\quad + (-1)^r(C_t \wedge A_r) \cdot (\nabla \wedge B_s) \\ &= [C_t, A_r] \cdot B_s + (-1)^{r+1}(A_r \cdot \nabla) \cdot (B_s \cdot C_t) + \\ &\quad + (-1)^r(C_t \wedge A_r) \cdot (\nabla \wedge B_s). \end{aligned}$$

On substituting this into the first term on the right of (3.36), we get (3.35b) as required.

The Lie bracket has many important properties and applications not mentioned here, but we have achieved our purpose of showing how it fits into the general system of differential identities in Geometric Calculus.

4-4. Curl and Pseudoscalar

This section considers the results of Section 4-2 from a different point of view and derives some important general properties of pseudoscalars. The approach here is easier to interpret geometrically than the one in Section 4-2, but it is not as directly applicable to the generalization considered in Section 4-5.

Section 4-2 studied differentials and derivatives of the projection operator P without making use of the fact that P can be expressed in terms of the pseudoscalar by

$$P(A) = (A \cdot I) \cdot I^{-1}. \tag{4.1}$$

Equation (2.43) expressed the differential of I in terms of the differential of P. Obviously, Eqn. (4.1) makes the converse possible, to determine the differential P_a from $a \cdot \partial I$. This will be done by relating the differentials and derivatives of I to the fundamental curl tensor $S_a = \dot{\partial} \wedge \dot{P}(a)$, which was introduced in Section 4-2.

According to (2.43) and (2.34a),

$$a \cdot \partial I = P_a(I) = I \times S_a. \tag{4.2}$$

This shows that the bivector-valued tensor S_a can be interpreted as the 'angular velocity' of the unit pseudoscalar I as it 'slides' along the manifold in the direction $P(a)$.

According to (2.30), $P(S_a) = 0$, which, by (4.1), can be written

$$S_a \cdot I = 0 \tag{4.3a}$$

if we exclude one-dimensional manifolds, so grade $I \geqslant 2$.

On the other hand, from definition (2.19), it follows that

$$S_a \wedge I = 0. \tag{4.3b}$$

According to (1-1.63), we can write

$$IS_a = I \cdot S_a + I \times S_a + I \wedge S_a,$$

so (4.3a, b) is equivalent to the relation

$$IS_a = I \times S_a = -S_a I. \tag{4.4}$$

Using (4.4), we can solve (4.2) for the curl tensor, obtaining

$$S_a = I^{-1} a \cdot \partial I = I^{-1} P_a(I) = -(a \cdot \partial I) I^{-1} = -P_a(I) I^{-1}. \tag{4.5}$$

We have seen that all the results in Sections 4-2 and 4-3 can be derived from the properties of the projection operator P and the curl tensor defined by $S_a \equiv \dot{\partial} \wedge \dot{P}(a)$. On the other hand, Eqns. (4.1) and (4.5) show that we could as well have obtained the same results beginning with the pseudoscalar I and the curl tensor defined by $S_a \equiv I^{-1} a \cdot \partial I$. This approach has the advantage of beginning with the obvious geometrical interpretation of S_a as the rate of rotation of I. We have chosen the former approach, however, because it can be directly generalized, as will be seen in the next section.

Recalling the definition (2.20) of the spur N, we obtain from (4.5)

$$N = \partial_a S_a = -(\partial I)I^{-1}. \tag{4.6}$$

So from (2.44) we have, for a pseudoscalar of grade m,

$$\partial I = \partial \wedge I = S(I) = -NI = (-1)^{m+1} IN. \tag{4.7}$$

This entails that

$$\partial \cdot I = N \cdot I = 0, \tag{4.8}$$

and if $I^{-1} = I^\dagger$,

$$N^2 = |\partial I|^2 = |\partial \wedge I|^2. \tag{4.9}$$

Differentiating (4.7) and using the fact that $N \cdot \partial = 0$, we obtain

$$\partial^2 I = -(N^2 + \partial N)I. \tag{4.10}$$

Hence

$$\partial \wedge \partial \wedge I = -(\partial \wedge N) \wedge I = -I \wedge \partial \wedge N = 0, \tag{4.11}$$

since $I \wedge \partial = 0$ always. Also,

$$(\partial \wedge N) \cdot I = -(\partial \wedge \partial) \cdot I = -\partial \cdot (\partial \cdot I) = 0. \tag{4.12}$$

So (4.10) can be written

$$\partial^2 I = -(N^2 + \partial \cdot N)I + I \times (\partial \wedge N). \tag{4.13}$$

Multiplying (4.13) by I^{-1}, we see that

$$-I^{-1} \cdot (\partial^2 I) = N^2 + \partial \cdot N. \tag{4.14}$$

Evidently the spur tells us some fundamental things about the manifold – exactly what is discussed in Chapter 7.

It is of interest to examine the consequences of applying the integrability condition to the pseudoscalar. Substituting (4.4) into (4.2) and differentiating we get

$$b \cdot \partial P_a(I) = b \cdot \partial a \cdot \partial I = I(b \cdot \partial S_a + S_b S_a),$$

or, equivalently,

$$P_{ab}(I) + P_a P_b(I) = b \cdot \partial a \cdot \partial I - (b \cdot \partial a) \cdot \partial I = I(S_{ab} + S_b S_a), \tag{4.15}$$

where

$$S_{ab} \equiv b \cdot \partial S_a - S_{b \cdot \partial a}$$

is the differential of the curl tensor. So, by the integrability condition (1.16) we have

$$(P_a P_b - P_b P_a)(I) = (a \wedge b) \cdot (\partial \wedge \partial) I = 0, \tag{4.16}$$

or, equivalently,

$$S_{ab} - S_{ba} = 2S_a \times S_b. \tag{4.17}$$

Thus, as a consequence of the integrability condition applied to the pseudoscalar, we find that the skew part of S_{ab} can be determined from S_a without differentiation.

4-5. Transformations of Vector Manifolds

We reserve the word *transformation* to refer to a function which preserves the defining properties of a manifold almost everywhere. In particular, a transformation f of a vector manifold $\mathscr{M}$ maps each point x of $\mathscr{M}$ to a point $x' = f(x)$ of some vector manifold $\mathscr{M}'$. This section determines how the 'differential calculus' of functions and fields on $\mathscr{M}$, as formulated in preceding sections of this chapter, relates to the differential calculus on $\mathscr{M}'$.

The differential $\underline{f}(a) = a \cdot \partial f$ of the transformation f can be regarded as a linear transformation of each vector field $a = a(x)$ on $\mathscr{M}$ into a vector field $\underline{f}(a) = a' = a'(x')$ on $\mathscr{M}'$; more precisely,

$$\underline{f} : a(x) \to a'(x') = \underline{f}(x, a(x)) = \underline{f}(f^{-1}(x'), a(f^{-1}(x'))), \tag{5.1}$$

where, for clarity, we have made the dependence of $\underline{f}$ on x explicit and assumed the existence of f^{-1} to exhibit the dependence on x'.

At each point x in $\mathscr{M}$ the differential $\underline{f}$ is a linear transformation of the tangent space $\mathscr{A}(x)$ to the tangent space $\mathscr{A}'(x')$ at x' in $\mathscr{M}'$. We extend this immediately to a linear transformation of the tangent algebra $\mathscr{G}(x) = \mathscr{G}(\mathscr{A}(x))$ to the tangent algebra $\mathscr{G}(x') = \mathscr{G}(\mathscr{A}'(x'))$ by employing the concept of *outermorphism* developed in Section 3-1. The definitions, notations and results of Section 3-1 apply without modification to the differential transformations of tangent spaces and tangent albebras of interest here. Accordingly, we understand that $\underline{f}$ transforms any multivector field $A = A(x)$ on $\mathscr{M}$ into a field $\underline{f}(A) = A' = A'(x')$ on $\mathscr{M}'$; more precisely,

$$\underline{f} : A(x) \to A'(x') = \underline{f}(A(x)) = \underline{f}(A(f^{-1}(x'))). \tag{5.2}$$

It is convenient to refer to $\underline{f}$ in this extended sense as *the differential* of f whenever f is a vector transformation.

Let $I = I(x)$ and $I' = I'(x')$ be the unit pseudoscalar fields on $\mathscr{M}$ and $\mathscr{M}'$ respectively. If $\underline{f}(I(x)) \neq 0$, we say that f is *nonsingular* at x, and (5.2) gives

$$\underline{f} : I(x) \to \underline{f}(I(x)) = J_f I'(x'). \tag{5.3}$$

The 'scale factor' $J_f = J_f(x') = J_f(f(x))$ is called the *Jacobian* of f. We will restrict our attention in this book to transformations which are everywhere nonsingular. If f is everywhere nonsingular, then $\underline{f}$ is invertible everywhere and $\mathscr{M}'$ has the same dimension as $\mathscr{M}$. It follows also that f itself is invertible. This is a consequence of the fundamental *inverse function theorem* to be studied in Section 7-6. In the meantime we assume always that f^{-1} exists.

Following Section 3-1, we introduce the adjoint $\overline{f}$ of the differential $\underline{f}$ defined by

$$\overline{f}(A') \equiv \partial_B \underline{f}(B) * A', \tag{5.4}$$

where $\partial_B = P(\partial_B)$ is the multivector derivative defined on the tangent algebra of $\mathscr{M}$. Since $\underline{f}$ is uniquely determined by f, so also is $\overline{f}$. Therefore, it is quite appropriate to refer to $\overline{f}$ as *the adjoint* of f. The adjoint $\overline{f}$ can be regarded as a linear transformation of the tangent algebra of fields on $\mathscr{M}'$ onto the tangent algebra of fields on $\mathscr{M}$, specifically, a field $A' = A'(x')$ on $\mathscr{M}'$ undergoes the transformation

$$\overline{f} : A'(x') \to \overline{f}(A'(x')) = \overline{f}(A'(f(x))). \tag{5.5}$$

To express the differential and adjoint transformation directly and succinctly in terms of f, we need to re-examine the process of simplicial differentiation used in Section 3-1. We use the abbreviated notation

$$\underline{f}_{(r)} \equiv \underline{f}(a_1) \wedge \underline{f}(a_2) \wedge \ldots \wedge \underline{f}(a_r), \tag{5.6a}$$

where the $a_k = a_k(x)$ are now vector fields on $\mathscr{M}$. Similarly, we write

$$f_{(r)} \equiv f(x_1) \wedge f(x_2) \wedge \ldots \wedge f(x_r), \tag{5.6b}$$

which is a function of r points x_k on $\mathscr{M}$. Since

$$\underline{f}(a_k) = a_k(x) \cdot \partial_{x_k} f(x_k) \big|_{x_k = x},$$

we can express (5.6a) in terms of (5.6b) by

$$\underline{f}_{(r)} = a_1 \cdot \partial_1 a_2 \cdot \partial_2 \ldots a_r \cdot \partial_r f_{(r)}, \tag{5.6c}$$

where we have used the abbreviation $\partial_k \equiv \partial_{x_k}$, and it is understood that the derivatives on the right side of (5.6c) are evaluated at $x_1 = x_2 = \ldots = x_r = x$. In accordance with (2-3.6b), the simplicial derivative on the tangent algebra is

$$\underline{\partial}_{(r)} \equiv (r!)^{-1} \underline{\partial}_r \wedge \ldots \wedge \underline{\partial}_2 \wedge \underline{\partial}_1, \tag{5.7a}$$

where the underbar notation serves to distinguish differentiation by tangent vectors $\underline{\partial}_k \equiv \partial_{a_k}$ from differentiation by points. Since,

$$\partial_k = \underline{\partial}_k a_k \cdot \partial_k,$$

we have

$$\partial_{(r)} \equiv (r!)^{-1} \partial_r \wedge \ldots \wedge \partial_2 \wedge \partial_1$$
$$= \underline{\partial}_{(r)} a_1 \cdot \partial_1 a_2 \cdot \partial_2 \ldots a_k \cdot \partial_k. \tag{5.7b}$$

Now, using (5.7) and (5.6) in connection with (3-1.4), we have the following expression for the differential transformation of a field $A = A(x)$ on $\mathscr{M}$:

$$\underline{f}(\langle A \rangle_r) = A * \underline{\partial}_{(r)} \underline{f}_{(r)} = A * \partial_{(r)} f_{(r)}. \tag{5.8}$$

Similarly, for the adjoint transformation of a field $A' = A'(x') = A'(f(x))$ on $\mathscr{M}'$, (3-1.11) yields

$$\overline{f}(\langle A' \rangle_r) = \underline{\partial}_{(r)} \underline{f}_{(r)} * A' = \partial_{(r)} f_{(r)} * A'. \tag{5.9}$$

The differential and adjoint of a given transformation can be computed directly from (5.8) and (5.9) or (5.4).

It is evident from (5.8) and (5.9) that the differential and adjoint are well-defined not only as transformations of fields but of any multivector functions on $\mathscr{M}$ and $\mathscr{M}'$. However, it follows easily from (5.8) and (5.9) that

$$\underline{f} = \underline{f} P = P' \underline{f} = P' \underline{f} P, \tag{5.10a}$$
$$\overline{f} = \overline{f} P' = P \overline{f} = P \overline{f} P', \tag{5.10b}$$

where P and P' are projections into the tangent algebras of $\mathscr{M}$ and $\mathscr{M}'$ respectively. Thus $\underline{f}$ and $\overline{f}$ automatically project functions into the tangent algebra of the one manifold before transforming them into the tangent algebra of the other manifold.

Of course, functions on $\mathscr{M}$ can be related to functions on $\mathscr{M}'$ by direct substitution as well as by differential and adjoint transformations. Indeed, since the differential $\underline{f}$ is obtained by differentiation from the function $f(x)$ defined on $\mathscr{M}$, substitution of $x = f^{-1}(x')$ is necessary to express $\underline{f}$ as a function on $\mathscr{M}'$. Direct substitution also relates differentiation on $\mathscr{M}$ to differentiation on $\mathscr{M}'$. From a function $F' = F'(x')$ defined on $\mathscr{M}'$, we obtain by direct substitution another function

$$F(x) = F'(f(x)) = F'(x') \tag{5.11}$$

defined on $\mathscr{M}$. The directional derivative of F on $\mathscr{M}$ is related to the directional derivative of F' on $\mathscr{M}'$ by the chain rule

$$a \cdot \partial F = a \cdot \partial_x F(x) = a \cdot \partial_x F'(f(x))$$
$$= (a \cdot \partial_x f(x)) \cdot \partial_{x'} F'(x') = \underline{f}(a) \cdot \partial' F'. \tag{5.12a}$$

where $\partial' = \partial_{x'}$ is the derivative on $\mathscr{M}'$. This relation is summarized by the operator identity

$$a \cdot \partial = \underline{f}(a) \cdot \partial' = a' \cdot \partial'. \tag{5.12b}$$

There is ambiguity here and throughout our formulation of transformation theory as to whether a quantity such as $a' = \underline{f}(a)$ is to be regarded as a function on $\mathscr{M}$ or or $\mathscr{M}'$. Actually either interpretation may be preferred over the other depending on how the quantity in question is to be used. We depend on the context to resolve any ambiguities.

Using the symmetry property of differential and adjoint, from (5.12b) we get

$$a \cdot \partial = \underline{f}(a) \cdot \partial' = a \cdot \overline{f}(\partial'). \tag{5.12c}$$

Differentiating (5.12c) by the tangent vector a, we get

$$\partial_x = \partial_a \underline{f}(a) \cdot \partial_{x'} = \partial_x f(x) \cdot \partial_{x'} = \overline{f}(\partial_{x'}) \tag{5.13a}$$

or, more succinctly,

$$\partial = \overline{f}(\partial'). \tag{5.13b}$$

This is a result we found in Section 2-1, but now its significance is more apparent. Equation (5.13) formulates the chain rule as a rule for transforming the derivative on $\mathscr{M}'$ to the derivative on $\mathscr{M}$. Of course $\partial_{x'} \neq \underline{f}(\partial_x)$; rather, the inverse of (5.13b) is found from (5.3) and (3-1.21a) to be

$$\partial_{x'} = \overline{f}^{-1}(\partial_x) = [\underline{f}(I)]^{-1}\underline{f}(I\,\partial_x) = (J_f I')^{-1}\underline{f}(I\,\partial_x). \tag{5.14}$$

It should be noted that in the coordinate-free formulations of the chain rule (5.12) and (5.13) the multiplication of ∂ by the point $x' = f(x)$ is essential. Though $f = f(x)$ is a function on $\mathscr{M}$, it is not generally a field. This is one of a number of places where restricting the calculus on a manifold to fields alone, as is done in the theory of differential forms (see Section 6-4), creates unnecessary complications. The consideration of functions with values outside the tangent algebra is essential to the coordinate-free theory developed here.

To relate the calculus on $\mathscr{M}$ to the calculus on $\mathscr{M}'$, we must examine the differentials of the differential and adjoint transformations defined, respectively, by

$$\begin{aligned} \underline{f}_b(A) &\equiv b \cdot \partial \underline{f}(A) - \underline{f}(b \cdot \partial A) \\ &= b' \cdot \partial' \underline{f}(A) - \underline{f}(b' \cdot \partial' A) \equiv \underline{f}_{b'}(A), \end{aligned} \tag{5.15a}$$

and

$$\begin{aligned} \overline{f}_b(A') &\equiv b \cdot \partial \overline{f}(A') - \overline{f}(b \cdot \partial A') \\ &= b' \cdot \partial' \overline{f}(A') - \overline{f}(b' \cdot \partial' A') \equiv \overline{f}_{b'}(A'), \end{aligned} \tag{5.15b}$$

where $b' = \underline{f}(b)$, and we have used (5.12b) to exploit the ambiguity in our choice of independent variables.

We hope it is clear by now why we have introduced and used so many different notations for differentials. Geometric functions play many different roles, and the various notations for differentials help us identify and distinguish these roles. Generally we prefer the underbar notation $\underline{f}$ for the differential of a function f given the role of a transformation, and we extend it automatically to denote the differential outermorphism. But if the function f is to be regarded as a field or some other kind of function, we usually indicate its differentials by subscripts. Sometimes it is convenient to employ both notations at once, as we have in writing $\underline{f}_b$ above.

Each of the properties of P_b ascertained in Section 2 generalizes to some property of $\underline{f}_b$ and $\overline{f}_b$. Indeed, for the identity transformation $f(x) = (x) = x$,

$$\underline{f}(A) = P(A) = P'(A) = \overline{f}(A)$$

and

$$\underline{f}_b(A) = P_b(A) = \overline{f}_b(A).$$

Thus, $\underline{f}$ and $\underline{f}_b$ are direct generalizations of P and P_b.

From the outermorphism property of differential and adjoint, we determine immediately that

$$\underline{f}_b(A \wedge B) = \underline{f}_b(A) \wedge \underline{f}(B) + \underline{f}(A) \wedge \underline{f}_b(B), \tag{5.16a}$$

$$\overline{f}_b(A' \wedge B') = \overline{f}_b(A') \wedge \overline{f}(B') + \overline{f}(A') \wedge \overline{f}_b(B'). \tag{5.16b}$$

This generalizes Eqn. (2.6). Differentiating (5.10a, b), we get the operator identities

$$\underline{f}_b = P'_b \underline{f} + P' \underline{f}_b = \underline{f}_b P + \underline{f} P_b, \tag{5.17a}$$

$$\overline{f}_b = P_b \overline{f} + P \overline{f}_b = \overline{f}_b P' + \overline{f} P'_b. \tag{5.17b}$$

This generalizes Eqn. (2.7). The choice $A = \langle A \rangle_1 = a$ reduces (5.15a) to the second differential of f, which by virtue of the integrability condition is symmetric, that is,

$$\underline{f}_b(P(a)) = f_{ab} = f_{ba} = \underline{f}_a(P(b)). \tag{5.18}$$

This, of course, generalizes (2.9). We need not list other obvious properties of $\underline{f}_b$ and $\overline{f}_b$ such as the fact that they are grade-preserving operators.

The relation of $\overline{f}_b$ to $\underline{f}_b$ can be determined by differentiating (5.4). With the help of (1.19), we find

$$\overline{f}_b(A') = \partial_B \underline{f}_b(B) * A' + P_b(\overline{f}(A')). \tag{5.19}$$

Because of (5.10) and (2.11a), (5.19) implies

$$P\overline{f}_b(A') = \partial_B \underline{f}_b(B) * A'. \tag{5.20}$$

This result can also be obtained directly by differentiating

$$B * \overline{f}_b(A') = A' * \underline{f}_b(B), \tag{5.21}$$

which is the obvious generalization of (2.5).

With the help of (5.20) we can derive the important result

$$\nabla_b \wedge \bar{f}_b(A') \equiv P(\partial_b \wedge \bar{f}_b(A')) = 0. \tag{5.22}$$

Indeed, for vector arguments b, c and a', (5.20) and (5.18) give us

$$P(\partial_b \wedge \bar{f}_b(a')) = \partial_b \wedge \partial_c f_{bc} \cdot a' = 0,$$

whence (5.22) follows for arbitrary A' by induction with (5.16b) as follows:

$$\begin{aligned} P\{\partial_b \wedge \bar{f}_b(a' \wedge B')\} &= P\{\partial_b \wedge \bar{f}_b(a') \wedge \bar{f}(B') + \partial_b \wedge \bar{f}(a') \wedge \bar{f}_b(B')\} \\ &= -\bar{f}(a') \wedge P\{\partial_b \wedge \bar{f}_b(B')\}. \end{aligned}$$

Equation (5.22) is equivalent to the condition that the exterior differential of $\underline{f}$ vanish identically; according to the definition (1.20),

$$\mathrm{d}\underline{f}(A) \equiv \dot{\underline{f}}(A \cdot \dot{\partial}) = 0, \tag{5.23a}$$

or simply,

$$\mathrm{d}\underline{f} = 0. \tag{5.23b}$$

Equation (5.23) can be proved by relating it to (5.22), but instead we prove it directly from the integrability condition. First note that, for a vector field a,

$$\underline{f}(a) = a \cdot \partial f = \mathrm{d}f.$$

So, by (1.25)

$$\mathrm{d}[\underline{f}(a)] = \mathrm{d}^2 f = 0.$$

Hence, the exterior derivative of $\underline{f}(a \wedge b) = \underline{f}(a) \wedge \underline{f}(b)$ also vanishes;

$$\mathrm{d}[\underline{f}(a \wedge b)] = \mathrm{d}[\underline{f}(a)] \wedge \underline{f}(b) + \underline{f}(a) \wedge \mathrm{d}[\underline{f}(b)] = 0.$$

From these first steps, the proof of (5.23) is easily completed. Note that by taking f to be the identity transformation in (5.23), one gets

$$\mathrm{d}P = 0. \tag{5.24}$$

Thus the exterior differential of the projection operator vanishes identically.

Since the curl tensor S_b determines P_b completely, its transformation properties are of the utmost interest. The relations of P_b to $\bar{f}_b$ and P'_b are expressed by (5.17b), which, because of (2.11a), can be put in the form

$$\bar{f}_b(A') = P_b(A_\parallel) + P\bar{f}_b(A') = \bar{f}_b(A'_\parallel) + \bar{f}P'_b(A'_\perp), \tag{5.25}$$

where $A'_\parallel = P'(A')$ and $A_\parallel = \bar{f}(A')$. Using $\partial_b = P(\partial_b) = \bar{f}(\partial_{b'})$, we get

$$\begin{aligned} \partial_b \wedge \bar{f}_b(A') &= \partial_b \wedge P_b(A'_\parallel) + P(\partial_b \wedge \bar{f}_b(A')) \\ &= \partial_b \wedge \bar{f}_b(A'_\parallel) + \bar{f}(\partial_{b'} \wedge P'_{b'}(A'_\perp)). \end{aligned}$$

The last terms on these lines vanish by virtue of (5.22) and (2.37a) respectively. So, recalling the definition (2.35a) of the shape tensor, we have

$$S(A_{\|}) = \partial_b \wedge \bar{f}_b(A'_{\|}) = \partial_b \wedge \bar{f}_b(A') \quad \text{for } A_{\|} = \bar{f}(A'). \tag{5.26}$$

As a corollary of (5.26), we have

$$\partial_b \wedge \bar{f}_b(A'_{\perp}) = 0. \tag{5.27}$$

For vector arguments, (5.26) reduces to the desired expression for the curl tensor

$$S_a = S(\bar{f}(a')) = \partial_b \wedge \bar{f}_b(a') = \partial_b \wedge \bar{f}_b(P'(a')). \tag{5.28}$$

The remarkable feature of (5.28) is that the curl tensor on $\mathscr{M}$ is completely determined by $\bar{f}$ and $\bar{f}_b$ without reference to the curl tensor on $\mathscr{M}'$.

We wish to know how intrinsic derivatives on $\mathscr{M}'$ are related to intrinsic derivatives on $\mathscr{M}$. Differentiating $A = \bar{f}(A')$ and using $\partial = \bar{f}(\partial')$ and (5.26), we get

$$\partial \wedge A = \partial \wedge \bar{f}(A') = S(A_{\|}) + \bar{f}(\partial' \wedge A'). \tag{5.29}$$

Because of (2.45), this yields

$$\nabla \wedge A = P(\partial \wedge A) = \bar{f}(\partial' \wedge A') = \bar{f}(\nabla' \wedge A') \tag{5.30a}$$

or

$$\nabla \wedge \bar{f}(A') = \bar{f}(\nabla') \wedge \bar{f}(A') = \bar{f}(\nabla' \wedge A'). \tag{5.30b}$$

Thus, the cocurl operation 'commutes' with the adjoint transformation.

The rule for transforming the codivergence can be derived from the rule (5.30) for transforming the cocurl because of the 'duality relation'

$$(\nabla' \cdot A')I' = \nabla' \wedge (A'I'),$$

which, according to (3.9), holds if $P'(A') = A'$. For this purpose, recall the relation (3-1.20b), which, with the help of (5.3), can be put in the form

$$\bar{f}(\underline{f}(A)I') = J_f AI. \tag{5.31}$$

Using these relations, we find

$$\begin{aligned}\bar{f}[(\nabla' \cdot \underline{f}(A))I'] &= \bar{f}[\nabla' \wedge (\underline{f}(A)I')] = \nabla \wedge \bar{f}[\underline{f}(A)I'] \\ &= \nabla \wedge (J_f AI) = \nabla \cdot (J_f A)I.\end{aligned}$$

This can be 'inverted' by using (5.31) in the form $\bar{f}(B'I') = J_f\underline{f}^{-1}(B')I$ to get the desired expression for transformation of the codivergence:

$$\nabla' \cdot \underline{f}(A) = \underline{f}\{J_f^{-1}(\nabla \cdot (J_f A))\} = \underline{f}(\nabla \cdot A + (\nabla \log J_f) \cdot A), \tag{5.32a}$$

where $A = P(A)$. Because of (3.6c), (5.32a) can be expressed in a more general form as an equation for the transformation of the divergence:

$$\partial' \cdot \underline{f}(A) = \underline{f}(\partial \cdot P(A) + (\partial \log J_f) \cdot A). \tag{5.32b}$$

The 'dual roles' played by $\overline{f}$ and $\underline{f}$ in (5.30) and (5.32) are evident.

According to (3-1.14b), for a vector b',

$$b' \cdot \underline{f}(A) = \underline{f}(\overline{f}(b') \cdot A). \tag{5.33}$$

Using this in connection with (5.32), we find

$$\dot{\nabla}' \cdot \underline{f}(\dot{A}) = \underline{f}(\nabla \cdot A), \tag{5.34}$$

and

$$\dot{\partial}' \cdot \underline{\dot{f}}(A) = \partial_{b'} \cdot \underline{f}_{b'}(A) = \underline{f}((\partial \log J_f) \cdot A) = (\partial' \log J_f) \cdot \underline{f}(A). \tag{5.35}$$

Thus, the last term in (5.32) arises from the divergence of $\underline{f}$.

Equation (5.33) also gives us the operator equation

$$\underline{f}(A) \cdot \partial' = \underline{f}(A \cdot \partial), \tag{5.36}$$

which generalizes (5.12). Obviously (5.32) differs from (5.36) only by the action of ∂' on $\underline{f}$.

From the transormation (5.32) of the divergence, we easily derive the rule for transforming the generalized Lie bracket. Noting that derivatives of the Jacobian vanish when (5.32) is applied to (3.23a), we obtain for fields A and B

$$\underline{f}([A, B]) = [\underline{f}(A), \underline{f}(B)]. \tag{5.37}$$

Thus, the bracket operation applied to fields commutes with the differential transformation.

The induced transformation of the dual brackets can be obtained by transforming (3.34a). Recalling that (3-1.14a) says that for $r \leqslant s$,

$$\overline{f}[\underline{f}(A_r) \cdot B'_s] = A_r \cdot \overline{f}(B'_s),$$

and using (5.30), we find the transformation of the first and third terms on the right side of Eqn. (3.34a);

$$\overline{f}[\nabla' \wedge (\underline{f}(A_r) \cdot B'_s)] = \nabla \wedge \overline{f}(\underline{f}(A_r) \cdot B'_s) = \nabla \wedge (A_r \cdot \overline{f}(B_s)), \tag{5.38a}$$

$$\overline{f}[\underline{f}(A_r) \cdot (\nabla' \wedge B'_s)] = A_r \cdot \overline{f}(\nabla' \wedge B'_s) = A_r \cdot (\nabla \wedge \overline{f}(B'_s)). \tag{5.38b}$$

Using (5.32) we find the transformation of the second term in (3.34a);

$$\overline{f}([\nabla' \cdot \underline{f}(A_r)) \cdot B'_s] = (\nabla \cdot A_r + (\nabla \log J_f) \cdot A_r) \cdot \overline{f}(B'_s). \tag{5.38c}$$

Applying Eqns. (5.38) to the terms of (3.34a) we finally get the desired transformation formula;

$$\overline{f}[\{\underline{f}(A_r), B_s'\}] = \{A_r, \overline{f}(B_s')\} - [A_r \cdot (\nabla \log J_f)] \cdot \overline{f}(B_s'). \tag{5.39}$$

Our derivation shows that the Jacobian on the right side of (5.39) arises from the transformation of the term in (5.38c) containing the divergence of A_r. So we can get a more symmetrical transformation by subtracting that term from the dual bracket; thus,

$$\overline{f}[\{\underline{f}(A_r), B_s'\} + (\dot{\underline{f}}(\dot{A}_r) \cdot \dot{\nabla}') \cdot B_s'] = \{A_r, \overline{f}(B_s')\} + (\dot{A}_r \cdot \dot{\nabla}) \cdot \overline{f}(B_s'). \tag{5.40}$$

Alternatively, given a nonvanishing scalar field $\alpha = \alpha(x)$ we can define an *α-dual bracket* by

$$\{A_r, B_s\}_\alpha \equiv [A_r, B_s p]p^{-1} = \{A_r, B_s\} + [A_r \cdot (\nabla \log \alpha)] \wedge B_s \tag{5.41}$$

where $p \equiv \alpha I$. We readily show that the α-dual bracket obeys the simple transformation formula

$$\overline{f}[\{\underline{f}(A_r), B_s'\}_{\alpha J_f}] = \{A_r, \overline{f}(B_s')\}_\alpha, \tag{5.42}$$

where, as before, J_f is the Jacobian of the transformation. Equation (5.42) differs from (5.39) only because the change in scale of the pseudoscalar induced by the transformation has been incorporated in the definition of the dual bracket.

4-6. Computation of Induced Transformations

In the preceding section we have shown how the general theory of induced transformations of fields and their derivatives can be formulated in simple coordinate-free form with Geometric Algebra. Here we wish to demonstrate that Geometric Algebra is equally valuable as a practical computational tool. For three important general classes of transformations, we compute the induced transformations without using coordinates and formulate the results as simple relations which are often awkward to express without Geometric Algebra. We carry out a few elementary computations to the very end to show that Geometric Algebra is helpful all the way. The examples examined here also show that application of the algebra is by no means automatic; like any tool, it must be skillfully wielded to be most effective.

Example 1. Consider the class of transformation $f\colon \mathscr{M} \to \mathscr{M}'$ of the form

$$f : x \to x' = f(x) = \lambda x, \tag{6.1}$$

where $\lambda = \lambda(x)$ is a scalar field. The differential of f can be computed from (5.7) and (5.8) with the techniques for differentiation developed in Chapter 2. For an arbitrary r-vector A_r, we have

$$\begin{aligned} \underline{f}(A_r) &\equiv A_r \cdot \partial_{(r)} f_{(r)} = A_r \cdot \partial_{(r)} (\lambda_1 x_1) \wedge (\lambda_2 x_2) \wedge \ldots \wedge (\lambda_r x_r) \\ &= (r!)^{-1} A_r \cdot [(\partial_r \lambda_r + \lambda \partial_r) \wedge \ldots \wedge (\partial_1 \lambda_1 + \lambda \partial_1)] x_{(r)} \\ &= \lambda^r A_r \cdot \partial_{(r)} x_{(r)} = \lambda^{r-1} A_r \cdot [(\partial \lambda) \wedge \partial_{(r-1)}] x_{(r-1)} \wedge x \\ &= \lambda^r P(A_r) + \lambda^{r-1} [A_r \cdot (\partial \lambda)] \cdot \partial_{(r-1)} x_{(r-1)} \wedge x \\ &= \lambda^r P(A_r) + \lambda^{r-1} [P(A_r) \cdot \partial \lambda] \wedge x. \end{aligned}$$

Hence, if A_r is a field on $\mathscr{M}$,

$$\underline{f}(A_r) = \lambda^r [A_r + (A_r \cdot \partial \log \lambda) \wedge x]. \tag{6.2}$$

In a similar way we compute the adjoint transformation:

$$\begin{aligned} \overline{f}(B'_r) &= \partial_{(r)} f_{(r)} \cdot B'_r = \partial_{(r)} [(\lambda_1 x_1) \wedge \ldots \wedge (\lambda_r x_r)] \cdot B'_r \\ &= (r!)^{-1} (\partial_r \lambda_r + \lambda \partial_r) \wedge \ldots \wedge (\partial_1 \lambda_1 + \lambda \partial_1) x_{(r)} \cdot B'_r \\ &= \lambda^r \partial_{(r)} x_{(r)} \cdot B'_r + \lambda^{r-1} [(\partial \lambda) \wedge \partial_{(r-1)}] (x_{(r-1)} \wedge x) \cdot B'_r \\ &= \lambda^r P(B'_r) + \lambda^{r-1} [(\partial \lambda) \wedge P(x \cdot B'_r)]. \end{aligned}$$

Hence,

$$\overline{f}(B'_r) = \lambda^r P[B'_r + (\partial \log \lambda) \wedge (x \cdot B'_r)]. \tag{6.3}$$

To determine further properties of the transformation, properties of λ and of the range or domain of f must be specified. For example, if $\mathscr{M}'$ is the unit sphere in an $(m + 1)$-dimensional Euclidean space $\mathscr{E}_{m+1}$, then f is the *central projection* of an arbitrary differentiable hypersurface $\mathscr{M}$ in $\mathscr{E}_{m+1}$ onto the sphere; of course $\mathscr{M}$ is completely determined by specifying λ. The tangent I' to the unit sphere can be expressed as an explicit function of x;

$$I' = I'(x') = \hat{x}' i = \hat{x} i = I'(x), \tag{6.4a}$$

where $\hat{x} \equiv |x|^{-1} x$ is the unit normal to the sphere and i is the unit pseudoscalar of $\mathscr{E}_{m+1}$. Similarly, the tangent I is related to the normal n of $\mathscr{M}$ by

$$I = I(x) = n(x) i = ni = (-1)^m in. \tag{6.4b}$$

Recalling (5.3), (6.2) is seen to relate I' to I;

$$\underline{f}(I) = J_f I' = \lambda^m [I + (I \, \partial \log \lambda) \wedge x]. \tag{6.5a}$$

This can be expressed as a relation between normals by using (6.4);

$$\underline{f}(I)i^\dagger = J_f\hat{x} = \lambda^m [n + (n \wedge \partial \log \lambda) \cdot x]. \tag{6.5b}$$

From the symmetry of (6.1), it is clear that (6.5b) can be inverted by interchanging n and x, replacing x by x', λ by λ^{-1} and J_f and $J_{\bar{f}}^{-1}$. Hence,

$$n = \lambda^{-m} J_f [\hat{x} + (\hat{x} \wedge \partial' \log \lambda^{-1}) \cdot x'] = \lambda^{-m} J_f [\hat{x} - \lambda^2 |x| \partial' \lambda^{-1}]. \tag{6.6}$$

Thus we have an explicit expression for the normal of $\mathscr{M}$ in terms of quantities defined on the sphere. In a similar way other properties of $\mathscr{M}$ are related to properties of the sphere. Note that the Jacobian J_f is just a normalization factor for (6.6), so we get an explicit expression for the magnitude of J_f by squaring (6.6); thus

$$J_f = \pm\lambda^m [1 + x^2 \lambda^4 (\partial' \lambda^{-1})^2]^{-1/2}. \tag{6.7a}$$

The sign of the Jacobian is easily determined by noting that (6.6) also implies

$$J_f = \lambda^m n \cdot \hat{x}. \tag{6.7b}$$

This, by the way, is just what one obtains by using (6 .5a).

As another important special case of (6.1), we consider *inversions* of Euclidean m-space $\mathscr{E} = \mathscr{M} = \mathscr{M}'$. With $\lambda = x^{-2}$, (6.1) becomes

$$f: x \to x' = x^{-1} = x^{-2}x. \tag{6.8}$$

Since $\partial |x| = \hat{x} = x/|x|$, $\partial \log \lambda = -\partial \log x^2 = -x^{-2} 2x$, and $(-1)^r \hat{x} A_r = A_r \hat{x} + 2A_r \cdot \hat{x}$, (6.2) and (6.3) reduce to

$$\underline{f}(A_r) = x^{-2r} [A_r - 2(A_r \cdot \hat{x})\hat{x}] = (-1)^r x^{-2r} \hat{x} A_r \hat{x} = \bar{f}(A_r). \tag{6.9a}$$

It should be noted that the equivalence of $\underline{f}$ and $\bar{f}$ is a consequence of $\partial_a \wedge \underline{f}(a) = \partial \wedge f = 0$. For a tangent vector, (6.9a) can be written

$$\underline{f}(a) = -x^{-1} a x^{-1} = -x' a x' = \bar{f}(a). \tag{6.9b}$$

Hence the transformation of the derivative is given by

$$\partial = \bar{f}(\partial') = -x^{-1}\, \partial' x^{-1}, \tag{6.10a}$$

or

$$\partial' = -x\, \partial x = x^2(\partial - 2\hat{x}\hat{x} \cdot \partial), \tag{6.10b}$$

where, of course, ∂ is understood not to differentiate the explicit x. From (6.10) we find immediately that the transformation of the Laplacian is given by

$$\partial^2 = x^{-4} \partial'^2 = x'^4 \partial'^2. \tag{6.11}$$

The Jacobian of the transformation is read off immediately from (6.9a);

$$J_f = I^\dagger \underline{f}(I) = -x^{-2m} = -x'^{2m}. \tag{6.12}$$

Example 2. Consider the class of transformations of the form

$$f: x \to x' = f(x) = x + n\beta(x), \tag{6.13}$$

where n is a fixed vector and $\beta = \beta(x)$ is a scalar field. Computation of the differential of (6.13) is quite similar to that of (6.1);

$$\begin{aligned}\underline{f}(A_r) &= A_r \cdot \partial_{(r)} f_{(r)} \\ &= A_r \cdot \partial_{(r)}(x_1 + \beta_1 n) \wedge \ldots \wedge (x_r + \beta_r n) \\ &= A_r \cdot \partial_{(r)} x_{(r)} + A_r \cdot [(\partial\beta) \wedge \partial_{(r-1)}] x_{(r-1)} \wedge n \\ &= P(A_r) + P[A_r \cdot (\partial\beta)] \wedge n.\end{aligned}$$

Hence, if A_r is a field

$$\underline{f}(A_r) = A_r + [A_r \cdot (\partial\beta)] \wedge n. \tag{6.14}$$

Similarly,

$$\overline{f}(B_r') = P[B_r' + (\partial\beta) \wedge (n \cdot B_r')]. \tag{6.15}$$

Noting that $(I \wedge n)a = (I \wedge n) \cdot a = In \cdot a - (Ia) \wedge n$ for any tangent vector a, and defining $P_\perp(n) \equiv I^\dagger (I \wedge n) = n - P(n)$, we get from (6.14)

$$K_f \equiv I^\dagger \underline{f}(I) = 1 + n \cdot \partial\beta + (\partial\beta) P_\perp(n), \tag{6.16}$$

$$|J_f|^2 = (1 + n \cdot \partial\beta)^2 + (\nabla\beta)^2 |I \wedge n|^2. \tag{6.17}$$

From (6.15),

$$\partial = \overline{f}(\partial') = P[\partial' + (\partial\beta) n \cdot \partial']. \tag{6.18}$$

If the domain of f in (6.13) is a hyperplane $\mathscr{M}$ in $\mathscr{E}_{m+1}$, then f is the transformation of the hyperplane into an arbitrary hypersurface $\mathscr{M}'$ in $\mathscr{E}_{m+1}$. We suppose that n is the normal to $\mathscr{M}$ and that the origin $x = 0$ is in $\mathscr{M}$ so that

$$x \cdot n = 0 \quad \text{and} \quad n \cdot \partial = 0. \tag{6.19}$$

This minor simplification incurs no loss of generality since a 'hypertransformation' can always be reduced to this form by utilizing a transformation of the type (6.13). If i is the tangent to $\mathscr{E}_{m+1}$, then the tangent I of $\mathscr{M}$ is related to the normal by

$$I = ni. \tag{6.20a}$$

Similarly, the tangent I' is related to the normal n' of $\mathscr{M}'$ by

$$I' = n'i. \tag{6.20b}$$

Because of (6.19), (6.14) reduces to

$$\underline{f}(A_r) = A_r + (A_r \cdot \partial\beta)n. \tag{6.21}$$

Hence,

$$\underline{f}(I) = J_f I' = I(1 + (\partial\beta)n),$$

or, using (6.20)

$$K_f \equiv I^\dagger \underline{f}(I) = J_f nn' = 1 + (\partial\beta)n = 1 - n\,\partial\beta. \tag{6.22}$$

Solving this for n',

$$n' = J_f^{-1}(n - \partial\beta), \tag{6.23}$$

where

$$J_f = |n - \partial\beta| = (1 + (\partial\beta)^2)^{1/2}. \tag{6.24}$$

Equation (6.18) reduces to

$$\partial = \overline{f}(\partial') = P(\partial') + (\partial\beta)n \cdot \partial' = \partial' + (\partial\beta - n)n \cdot \partial', \tag{6.25}$$

since $P(\partial') = I^\dagger I \cdot \partial' = n(n \wedge \partial') = \partial' - nn \cdot \partial'$. But ordinarily one is more interested in ∂' as a function of ∂, since it can be used to reduce differentiation on the curved surface $\mathscr{M}'$ to differentiation on the flat surface $\mathscr{M}$. This is more easily accomplished by using the inverse of (6.13). First note that, because of (6.19),

$$\beta = \beta(x) = n \cdot x'; \tag{6.26}$$

thus, whatever its x-dependence, the x'-dependence of β is always linear. Hence

$$f^{-1} : x' \to x = f^{-1}(x') = x' - nn \cdot x' = P(x'), \tag{6.27}$$

which displays x as simply the *orthogonal projection* of x' into the hyperplane. Computation of the differential and adjoint of f^{-1} is similar to that of f. From (6.21), since $B'_r \cdot \partial'\beta = B'_r \cdot \partial' x' \cdot n = B'_r \cdot n$, we can write down immediately

$$\underline{f}^{-1}(B'_r) = B'_r - (B'_r \cdot n)n = P(B'_r), \tag{6.28}$$

and from (6.15), since $n \cdot A_r = 0$ for any tangent field A_r on $\mathscr{M}$,

$$\overline{f}^{-1}(A_r) = P'(A_r) = A_r - n'(n' \cdot A_r), \tag{6.29}$$

where P' is the projection into the tangent algebra of $\mathscr{M}'$. From (6.29) and (6.23) we get the desired expression

$$\begin{aligned} \partial' &= \overline{f}^{-1}(\partial) = P'(\partial) = \partial - n'n' \cdot \partial \\ &= \partial + J_f^{-2}(n - \partial\beta)(\partial\beta) \cdot \partial. \end{aligned} \tag{6.30}$$

In a similar manner we can easily evaluate the expressions of Section 4-5 for the transformation of curl, divergence and Lie brackets.

For the sake of concreteness, we consider a couple of specific hypersurfaces. With the choice $\beta(x) = x^2$, $f(x) = x + x^2 n$ transforms a hyperplane into a hyperparaboloid. Since $\partial\beta = \partial x^2 = 2x$ in this case, the details of the induced transformations are described by the simple algebraic equations

$$\begin{aligned}
\underline{f}(A_r) &= A_r + 2A_r \cdot xn, \\
\overline{f}(B_r') &= P[B_r' + 2x \wedge (n \cdot B_r')] \\
&= n[n \wedge B_r' + 2(nx) \wedge (n \cdot B_r')], \\
J_f &= (1 + 4x^2)^{1/2}, \\
\partial' &= \partial + 2(1 + 4x^2)^{-1}(n - 2x)x \cdot \partial.
\end{aligned}$$

The transformation of the unit ball in $\mathscr{M} = \mathscr{E}_m$ into a hemisphere $\mathscr{M}'$ in $\mathscr{E}_{m+1}$ is obtained by the choice $\beta(x) = (1 - x^2)^{1/2}$,

$$x' = f(x) = x + (1 - x^2)^{1/2} n, \quad x^2 \leqslant 1.$$

Since $\partial\beta = -(1 - x^2)^{-1/2} x$, we find, for instance,

$$\begin{aligned}
\underline{f}(A_r) &= A_r - (1 - x^2)^{-1/2} A_r \cdot xn, \\
J_f &= (1 - x^2)^{-1/2} = -x^{-1}\,\partial\beta, \\
\partial' &= \partial - [(1 - x^2)^{-1/2} n + x]\, x \cdot \partial.
\end{aligned}$$

Example 3. As a final example, we consider the *stereographic projection* of a hyperplane $\mathscr{E}_m$ onto the unit sphere in $\mathscr{E}_{m+1}$. As in the last example, we write n for the unit normal to $\mathscr{E}_m$ and take $n \cdot x = 0$. The transformation f of a point x in $\mathscr{E}_m$ to a point x' on the sphere can be written in the following three significant forms:

$$x' = f(x) = n + \lambda^{-1}(x - n) \tag{6.31a}$$

$$= -(x - n)^{-1} n (x - n) \tag{6.31b}$$

$$= (x - n)^{-1} x^{-1} n x (x - n), \tag{6.13c}$$

where

$$\lambda = \tfrac{1}{2}|x - n|^2 = \tfrac{1}{2}(1 + x^2) = (1 - n \cdot x')^{-1}. \tag{6.32}$$

Form (6.31a) expresses x' as a 'point of division' of the line segment connecting n and x. Form (6.31b) expresses x' as obtained from n by a *reflection* in the hyperplane with normal $x - n$. Form (6.31c) expresses x' as obtained from n by a *rotation* trough twice the angle between the vectors x and $x - n$. The equivalence of forms (6.31b) and (6.31c) is obvious since $n \cdot x = 0$ implies

$$xn = -nx. \tag{6.33}$$

We can find the differential transformation of a tangent vector a by evaluating the directional derivative of

$$f(x) = -\frac{(x-n)n(x-n)}{|x-n|^2}.$$

Since $a \cdot \partial x = a$ and $a \cdot \partial |x-n|^2 = a \cdot \partial x^2 = 2a \cdot x$, we have

$$\underline{f}(a) = a \cdot \partial f(x) = -\frac{an(x-n) - (x-n)na - 2a \cdot xf(x)}{|x-n|^2}$$

$$= -\frac{(x-n)}{|x-n|^4}\{(x-n)an + na(x-n) - 2a \cdot xn\}(x-n)$$

$$= -\frac{(x-n)}{|x-n|^4} 2a(x-n) = \frac{2}{|x-n|^2}(x-n)^{-1}nan(x-n).$$

Hence,

$$\underline{f}(a) = -\lambda^{-1}(x-n)^{-1}a(x-n) = \lambda^{-1}\phi^{-1}a\phi, \tag{6.34a}$$

where λ is defined by (6.32) and

$$\phi = n(x-n). \tag{6.34b}$$

Thus, the vector $a' = \underline{f}(a)$ tangent to the sphere at a' is obtained from $a = a(x)$ by a rotation specified by the spinor ϕ and a dilation by λ^{-1}.

From (6.34) we easily determine all the significant properties of the induced transformations of tangent multivector fields. Since $\phi\phi^{-1} = 1$, for vector fields a and b (6.34) implies

$$\underline{f}(a)\underline{f}(b) = \lambda^{-2}\phi^{-1}ab\phi. \tag{6.35}$$

The scalar part of (6.35) is

$$\underline{f}(a) \cdot \underline{f}(b) = \lambda^{-2}a \cdot b. \tag{6.36}$$

This proves that the stereographic projection is a *conformal* transformation. The bivector part of (6.35) is

$$\underline{f}(a \wedge b) = \underline{f}(a) \wedge \underline{f}(b) = \lambda^{-2}\phi^{-1}a \wedge b\phi. \tag{6.37}$$

This obviously generalizes to

$$\underline{f}(A_r) = \lambda^{-r}\phi^{-1}A_r\phi \tag{6.38}$$

for a tangent r-vector field A_r. Obviously,

$$\underline{f}(A_r)\underline{f}(B_s) = \lambda^{-r-s}\phi^{-1}A_rB_s\phi. \tag{6.39}$$

Note that this is not equal to $\underline{f}(A_rB_s)$ unless $A_rB_s = A_r \wedge B_s$.

The inverse of (6.38) is obviously

$$\underline{f}^{-1}(B_r') = \lambda^r \phi B_r' \phi^{-1}. \tag{6.40}$$

And, since the adjoint of a rotation is equal to its inverse, we can immediately write down

$$\overline{f}(B_r') = \lambda^{-r} \phi B_r' \phi^{-1} = \lambda^{-2r} \underline{f}^{-1}(B_r'). \tag{6.41}$$

But, as a check, we compute $\overline{f}(a')$ directly;

$$\begin{aligned}\overline{f}(a') &= \nabla a' \cdot f(x) = \nabla 2a' \cdot (x-n)^{-1} \\ &= 2|x-n|^{-2}[a' - nn \cdot a' - 2xa' \cdot (x-n)^{-1}] \\ &= \lambda^{-1}[a' - 2(x-n)a' \cdot (x-n)^{-1}] \\ &= -\lambda^{-1}(x-n)a'(x-n)^{-1} \\ &= \lambda^{-1} n(x-n)a'(x-n)^{-1} n = \lambda^{-1}\phi a' \phi^{-1}.\end{aligned}$$

4-7. Complex Numbers and Conformal Transformations

The theory of functions defined on the two-dimensional *real plane* $\mathscr{R}_2$ made great progress, especially during the last century, when points of $\mathscr{R}_2$ were represented by complex numbers. The algebra of complex numbers greatly simplified the construction and geometrical interpretation of many important functions with range and domain in a plane. But complex algebra is incapable of describing geometrical features of functions in more than two dimensions. This inherent limitation to two dimensions has contributed to the development of complex function theory as an independent discipline, distinguished from the theory of 'real functions' on m-dimensional real manifolds by differences in definitions and notations which obscure their common basis. The artificial separation of complex analysis from real analysis persists today in textbooks, in curricula and in mathematical practice.

This section shows how Geometric Algebra unites the real and complex planes algebraically. All the results of complex function theory are thus readily obtained as a special case of the Geometric Calculus on vector manifolds, and appropriate generalizations to higher dimensions become apparent. This approach reinterprets the role of complex analysis in mathematics, eschewing the common convention of regarding complex numbers as scalars. But none of the valuable algebraic features of complex analysis are thereby sacrificed; indeed, they are enhanced by integrating them into the calculus of vectors.

After establishing notations and conventions which make it easy to relate the conventional language of complex analysis to the language of Geometric Calculus, this section shows how the notion of analytic function can be defined and generalized without complex analysis, and furthermore, that conformal transformations of the plane are as efficiently described by the general transformation theory developed in Section 4-5, as by the usual method of complex analysis. Chapter 7 shows that the integral theorems of analytic function theory are also included as special cases of more general theorems in Geometric Calculus.

Let $\mathscr{G}(\mathbf{i})$ be the Geometric Algebra with a two-blade **i** as unit pseudoscalar. To emphasize relations to traditional concepts, we call the even subalgebra of $\mathscr{G}(\mathbf{i})$ the *complex plane* and designate it by $\mathscr{C} = \mathscr{G}^+(\mathbf{i})$. Every element z of $\mathscr{C}$ can be written in the form $z = x + \mathbf{i}y$, where x and y are scalars. We may call z a 'complex number' because $\mathbf{i}^2 = -1$. Since $z^\dagger = x - \mathbf{i}y$, reversion reduces to 'complex conjugation in $\mathscr{C}$'. The real and imaginary parts of a complex number correspond to its scalar and bivector parts, that is,

$$x = \operatorname{Re}\{z\} = \langle z\rangle = \frac{z + z^\dagger}{2}, \tag{7.1a}$$

$$y = \operatorname{Im}\{z\} = -\mathbf{i}\langle z\rangle_2 = \frac{z - z^\dagger}{2\mathbf{i}}. \tag{7.1b}$$

The odd elements of $\mathscr{G}(\mathbf{i})$ comprise a two-dimensional vector space $\mathscr{R}_2 = \mathscr{G}^-(\mathbf{i}) = \mathscr{G}^1(\mathbf{i})$, called the *real plane* here to emphasize that it is a linear space 'over the reals'. Every vector **x** in $\mathscr{R}_2$ satisfies the equation

$$\mathbf{x} \wedge \mathbf{i} = 0, \tag{7.2a}$$

or, equivalently,

$$\mathbf{x}\mathbf{i} = -\mathbf{i}\mathbf{x}. \tag{7.2b}$$

We have noted this fact before, namely that the vectors 'in' a plane anticommute with the bivectors, the pseudoscalars of the plane. In the rest of this section it will be understood that all vectors are elements of $\mathscr{R}_2$. And perhaps it is well to explicitly mention the assumption that $\mathbf{x}^2 > 0$ if **x** is a nonzero vector.

In this section we use boldface letters to denote vectors and simple bivectors. This allows us to use the conventional symbol x for the 'real part' of z, while distinguishing it from the vector **x**. Moreover, the boldface **i** emphasizes that $\sqrt{-1}$ here is not to be interpreted as a scalar.

A linear mapping of the real plane onto the complex plane is given by the equation

$$z = \mathbf{a}^{-1}\mathbf{x} = x + \mathbf{i}y, \tag{7.3}$$

where **a** is fixed nonzero vector. Of course

$$x = \mathbf{a}^{-1} \cdot \mathbf{x} = \frac{\mathbf{a} \cdot \mathbf{x}}{\mathbf{a}^2}, \tag{7.4a}$$

$$\mathbf{i}y = \mathbf{a}^{-1} \wedge \mathbf{x} = \frac{\mathbf{a} \wedge \mathbf{x}}{\mathbf{a}^2}. \tag{7.4b}$$

To each point $\mathbf{x}$ in $\mathscr{R}_2$, Eqn. (7.3) corresponds a unique point z in $\mathscr{C}$. According to (7.3), every complex number z can be uniquely expressed as the *ratio* of some vector $\mathbf{x}$ to an arbitrarily chosen vector $\mathbf{a}$. Vectors collinear to $\mathbf{a}$ correspond to points on the real axis, the subspace of scalars in $\mathscr{C}$. Complex conjugation in $\mathscr{C}$ corresponds to a reflection in $\mathscr{R}_2$ 'through' the vector $\mathbf{a}$; thus,

$$z^\dagger = \mathbf{x}\mathbf{a}^{-1} = \mathbf{a}^{-1}(\mathbf{a}^{-1}\mathbf{x}\mathbf{a}) = \mathbf{a}^{-1}z\mathbf{a}. \tag{7.5a}$$

Multiplying by $\mathbf{a}$, we get

$$\mathbf{a}z^\dagger = z\mathbf{a}, \tag{7.5b}$$

which is also an obvious consequence of (7.2b).

Solving (7.3) for $\mathbf{x}$, we get

$$\mathbf{x} = \mathbf{a}z. \tag{7.6}$$

Equation (7.6) provides complex numbers with a geometric interpretation. It exhibits z as an operator which rotates and dilates a vector $\mathbf{a}$ into a vector $\mathbf{x}$. Since $\mathbf{a}$ can be any vector in $\mathscr{R}_2$ and, given $\mathbf{a}$, $\mathbf{x}$ is unique, by (7.6), $z \neq 0$ determines a linear transformation, more specifically, a homothetic transformation of $\mathscr{R}_2$ onto $\mathscr{R}_2$. Each homothetic endomorphism of the real plane is uniquely represented by a point in the complex plane; points on the positive real axis represent dilatations, while points on the unit circle in $\mathscr{C}$ represent rotations. Multiplication of complex numbers represents composition of homothetic transformations. Indeed, complex numbers are just the simplest kind of *spinors*, as they were defined in Section 3-8.

Having determined how the real and complex planes are related algebraically and geometrically, we are prepared to study how the complex derivative is related to differentiation in the real plane.

Let $z' = F(z)$ be a mapping of $\mathscr{C}$ into $\mathscr{C}$. According to (7.3),

$$F(z) = F(\mathbf{a}^{-1}\mathbf{x}) = F(x + \mathrm{i}y), \tag{7.7}$$

so F determines a mapping of $\mathscr{R}_2$ into $\mathscr{C}$. For functions with values in $\mathscr{G}(\mathrm{i})$, the derivative is equal to the coderivative, that is,

$$\nabla \equiv \nabla_{\mathbf{x}} = \partial_{\mathbf{x}}. \tag{7.8}$$

Complex derivatives can be defined in terms of scalar derivatives by

$$\frac{\mathrm{d}}{\mathrm{d}z} \equiv \tfrac{1}{2}(\partial_x - \mathrm{i}\,\partial_y), \tag{7.9a}$$

$$\frac{\mathrm{d}}{\mathrm{d}z^\dagger} \equiv \tfrac{1}{2}(\partial_x + \mathrm{i}\,\partial_y). \tag{7.9b}$$

The vector derivative is related to the scalar derivatives by

$$\mathbf{a} \cdot \nabla = \partial_x, \tag{7.10a}$$

$$\mathbf{a} \wedge \nabla = \mathrm{i}\,\partial_y. \tag{7.10b}$$

Hence the complex and vector derivatives are related by

$$\mathbf{a}\nabla = \mathbf{a}\cdot\nabla + \mathbf{a}\wedge\nabla = 2\frac{\mathrm{d}}{\mathrm{d}z^\dagger}, \tag{7.11a}$$

$$\nabla\mathbf{a} = \mathbf{a}\cdot\nabla - \mathbf{a}\wedge\nabla = 2\frac{\mathrm{d}}{\mathrm{d}z}. \tag{7.11b}$$

The function $F(z)$ is said to be *analytic* at a point z if

$$\frac{\mathrm{d}F}{\mathrm{d}z^\dagger}(z) = \partial_x F + \mathrm{i}\,\partial_y F = 0. \tag{7.12}$$

The real and imaginary parts of (7.12) are the so-called Cauchy–Riemann equations. By virtue of (7.11a), (7.12) is equivalent to the equation

$$\nabla F = 0. \tag{7.13}$$

In contrast to (7.12), Eqn. (7.13) obviously generalizes immediately to functions with any multivector values defined on a vector manifold of any dimension. So we regard (7.13) to be more fundamental than (7.12) and take it to be the defining property of analytic functions in general. In Chapter 7 we shall see that this leads to a generalization of Cauchy's integral formula.

If $F(z)$ is analytic, then, because of (7.12), the complex derivative (7.9a) reduces to

$$\frac{\mathrm{d}F}{\mathrm{d}z} = \partial_x F = \mathbf{a}\cdot\nabla F. \tag{7.14}$$

Thus, the complex derivative of an analytic function is precisely equivalent to the directional derivative in the real plane in the direction corresponding to the real axis. In most books on complex analysis, the complex derivative is defined by

$$\frac{\mathrm{d}F}{\mathrm{d}z}(z) \equiv \lim_{h\to 0} \frac{F(z+h) - F(z)}{h}, \tag{7.15}$$

where h is complex, and $F(z)$ is said to be analytic if the limit exists and is independent of h. We object to this definition, because it confuses two distinct concepts, the directional derivative (7.14) and the vanishing of the vector derivative (7.13); in so doing, it obscures the way to generalize the concept of analytic function to manifolds of higher dimension.

Let us consider some simple analytic functions. The simplest such function is evidently $F(z) = z = \mathbf{a}^{-1}\mathbf{x}$. According to (2-1.40), we have indeed,

$$\nabla z = \nabla\mathbf{a}^{-1}\mathbf{x} = 0. \tag{7.16}$$

It is instructive to derive this result from (2-1.34) and (2-1.38) instead; thus,

$$\nabla(\mathbf{a}^{-1}\mathbf{x}) = \nabla(-\mathbf{x}\mathbf{a}^{-1} + 2\mathbf{x}\cdot\mathbf{a}^{-1}) = -2\mathbf{a}^{-1} + 2\mathbf{a}^{-1} = 0.$$

According to (2-1.36), we have

$$\nabla\mathbf{x}^{-1} = 0 \quad \text{at } \mathbf{x} \neq 0, \tag{7.17a}$$

hence

$$\nabla z^{-1} = \nabla(\mathbf{a}^{-1}\mathbf{x})^{-1} = \nabla\mathbf{x}^{-1}\mathbf{a} = 0. \tag{7.17b}$$

Obviously, $\mathbf{z} = \mathbf{a}^{-1}\mathbf{x}$ and $z^{-1} = \mathbf{x}^{-1}\mathbf{a}$ are analytic functions of **a** as well as **x**. More generally, we see that any power of z is an analytic function of **a** and **x**; that is,

$$\nabla z^k = \nabla_{\mathbf{x}}(\mathbf{a}^{-1}\mathbf{x})^k = 0, \tag{7.18a}$$

or, differentiating by **a** instead of **x**,

$$\nabla_{\mathbf{a}} z^k = \nabla_{\mathbf{a}}(\mathbf{a}^{-1}\mathbf{x})^k = 0. \tag{7.18b}$$

The fact that z^k is an analytic function of both vector variables **a** and **x** has significant advantages in applications which are not available when complex numbers are used without relating them to vectors.

The complex function $z' = F(z)$ corresponds to a transformation $\mathbf{x}' = f(\mathbf{x})$ of $\mathscr{R}_2$ into $\mathscr{R}_2$ according to the relation

$$z' = \mathbf{a}^{-1}\mathbf{x}' = F(z) = F(\mathbf{a}^{-1}\mathbf{x}) = \mathbf{a}^{-1}f(\mathbf{x}). \tag{7.19}$$

By virtue of (7.11b),

$$\frac{\mathrm{d}F}{\mathrm{d}z} = \tfrac{1}{2}(\nabla\mathbf{a})(\mathbf{a}^{-1}f) = \tfrac{1}{2}\nabla f. \tag{7.20}$$

In particular $\mathrm{d}z/\mathrm{d}z = \frac{1}{2}\nabla\mathbf{x} = 1$. The correspondence of complex derivative to vector derivative exhibited by (7.20) holds for differentiable functions in general. But if F is analytic, then, by comparing (7.14) with (7.20), we get the special relation

$$\frac{\mathrm{d}F}{\mathrm{d}z} = \mathbf{a}\cdot\nabla F = \mathbf{a}^{-1}\mathbf{a}\cdot\nabla f = \tfrac{1}{2}\nabla f. \tag{7.21}$$

Thus $\nabla F = 0$ does not imply $\nabla f = 0$, but, according to (7.21) it does imply that ∇f is completely determined by the directional derivative $\mathbf{a}\cdot\nabla f$ in *any* direction **a**.

Since $\mathscr{R}_2$ is a vector manifold, we can use the general method of Section 4-5 to analyze the transformation $x' = f(x)$. We wish to compare conformal transformations of the real plane with analytic functions. The transformation f is said to be conformal if

$$\underline{f}(\mathbf{a})\cdot\underline{f}(b) = \lambda^2\mathbf{a}\cdot\mathbf{b}, \tag{7.22a}$$

that is, if the angle between any pair of tangent vectors is preserved by the induced transformation. From our study of orthogonal transformations in Section 3-5, we know that (7.22a) implies that there exists a rotor U such that

$$\underline{f}(\mathbf{a}) = \mathbf{a}\cdot\nabla f = \lambda U^\dagger\mathbf{a}U. \tag{7.22b}$$

Equations (7.22a, b) hold for any conformal transformation of a vector manifold. For the two-dimensional case of interest here, (7.5b) permits us to write (7.22b) in the simple form

$$\underline{f}(\mathbf{a}) = \mathbf{a} \cdot \nabla f = \mathbf{a}\psi = \psi^\dagger \mathbf{a}, \tag{7.22c}$$

where $\psi = \psi(\mathbf{x}) = \lambda U^2$ is a complex-valued or, better, a spinor field on $\mathscr{R}_2$. It will be noted that $|\psi|^2 = \psi\psi^\dagger = \psi^\dagger\psi = \lambda^2$.

All the general properties of conformal transformations of the plane are consequences of (7.22). The adjoint of f is given by

$$\bar{f}(\mathbf{b}) = \nabla_{\mathbf{a}}\mathbf{b} \cdot \underline{f}(\mathbf{a}) = \psi\mathbf{b} = \mathbf{b}\psi^\dagger. \tag{7.23}$$

This is easily derived from (7.22) by noting that $\mathbf{b} \cdot \underline{f}(\mathbf{a}) = \langle \mathbf{ba}\psi \rangle = \langle \mathbf{a}\psi\mathbf{b} \rangle = \mathbf{a} \cdot (\psi\mathbf{b})$.

We can solve (7.22) for ψ in two ways: by differentiating by $\mathbf{a}$ and by dividing by $\mathbf{a}$. Thus, we obtain

$$\nabla_{\mathbf{a}}\underline{f}(\mathbf{a}) = \dot{\nabla}\dot{f} = 2\psi = 2\mathbf{a}^{-1}\underline{f}(\mathbf{a}). \tag{7.24}$$

But this is identical to the condition on f specified by (7.21). Therefore, the analyticity condition $\nabla F = 0$ on the function $F(\mathbf{a}^{-1}\mathbf{x}) = \mathbf{a}^{-1}f(x)$ is precisely equivalent to the condition that $f(\mathbf{x}) = \mathbf{a}F(\mathbf{a}^{-1}\mathbf{x})$ be a conformal transformation of the real plane. By differentiating (7.24) we discover that

$$\nabla\psi = \tfrac{1}{2}\nabla^2 f = 0, \tag{7.25}$$

since $\nabla(\mathbf{a}^{-1}\mathbf{a} \cdot \nabla f) = \mathbf{a} \cdot \nabla(\nabla\mathbf{a}^{-1}f) = \mathbf{a} \cdot \nabla(\nabla F) = 0$. Of course, we are not regarding $\mathbf{a}$ as a function of $\mathbf{x}$ here.

It seems that the analyticity of ψ expressed by (7.25) is more significant than the analyticity of F, for ψ directly determines the induced transformation of vector fields according to (7.22) and (7.23), whereas F merely re-expresses the transformation f of the real plane as a transformation of the complex plane. Equations (7.22) through (7.25) express the fundamental facts about conformal transformation of the real plane, so our earlier references to the complex plane and complex derivatives are really quite unnecessary for applications, although they are needed to translate results from the literature of complex variable theory into the language of Geometric Calculus.

Conformal transformations of the plane have some other properties which we should consider. From (7.22) we see that $\underline{f}(\mathbf{a})\underline{f}(\mathbf{b}) = \lambda^2\mathbf{ab}$, hence

$$\underline{f}(\mathbf{a} \wedge \mathbf{b}) = \underline{f}(\mathbf{a}) \wedge \underline{f}(\mathbf{b}) = \lambda^2 \mathbf{a} \wedge \mathbf{b}. \tag{7.26}$$

This tells us immediately that $J_f = \lambda^2$ is the Jacobian of f, and, using (7.24), we see that

$$J_f = \lambda^2 = \psi\psi^\dagger = \tfrac{1}{4}|\nabla f|^2 = \tfrac{1}{4}[(\nabla \cdot f)^2 - (\nabla \wedge f)^2]. \tag{7.27}$$

Using (7.25) now, we find that

$$\nabla J_f = \nabla\lambda^2 = \psi^\dagger \nabla \psi^\dagger = (\nabla\psi^\dagger)\psi. \tag{7.28}$$

We are most interested in the induced transformations of vector fields and their derivatives. Of course, from (5.13) and (7.23) we get

$$\nabla = \bar{f}(\nabla') = \psi\nabla', \tag{7.29}$$

where $\nabla' \equiv \nabla_{\mathbf{x}'} = \partial_{\mathbf{x}'}$. Now, if $\mathbf{E}' = \mathbf{E}'(\mathbf{x}')$ is a vector field on $\mathscr{R}_2$, then the adjoint transformation gives us a new vector field

$$\mathbf{E}(\mathbf{x}) = \bar{f}(\mathbf{E}'(f(\mathbf{x})) = \psi(\mathbf{x}')\mathbf{E}'(\mathbf{f}(\mathbf{x})), \tag{7.30a}$$

or, more briefly,

$$\mathbf{E} = \bar{f}(\mathbf{E}') = \psi\mathbf{E}'. \tag{7.30b}$$

Operating on (7.30) with (7.29) and using (7.25), we find the induced transformation

$$\nabla\mathbf{E} = \nabla\bar{f}(\mathbf{E}') = \lambda^2\nabla'\mathbf{E}'. \tag{7.31}$$

Taking $\mathbf{E}' = \nabla'$ in (7.31), we get the rule for transforming the Laplacian:

$$\nabla^2 = \lambda^2\nabla'^2. \tag{7.32}$$

We have used the symbol **E** here to emphasize applicability of the present theory to physical problems in which **E** is an electric field. Indeed, much of analytic function theory was developed during the nineteenth century precisely to solve such problems. A static electric field **E** satisfies the equation

$$\nabla\mathbf{E} = \rho, \tag{7.33}$$

where ρ is the charge density. In general, Eqn. (7.33) defines fields in a three-dimensional space, but for certain charge distributions, it reduces essentially to an equation in two dimensions, and the results of this section can be applied to its study. From (7.31) we find that the change of charge distribution induced by a conformal transformation of the plane is

$$\rho = \lambda^2\rho'. \tag{7.34}$$

As (7.33) says, in charge-free regions a static electric field is analytic. And (7.31) shows that $\nabla'\mathbf{E}' = 0$ implies $\nabla\mathbf{E} = 0$, that is, *analyticity of fields is preserved by conformal transformations of the plane*. This is the property of conformal transformations of most significance in physical applications. Beside it, the conformal property (7.22a) is hardly more than incidental. Actually, the essential equivalence of 'conformality' to 'analyticity' is a special feature of two dimensions. The class of analytic functions is quite distinct from the class of conformal transformations in higher dimensions.

For completeness we determine the differential transformation of Eqn. (7.33), to be compared with the adjoint transformation (7.31). Using (7.23), (7.29) and (7.28), we find

$$\nabla' \underline{f}(\mathbf{E}) = \psi^{-1} \nabla(\psi^{\dagger} E) = \psi^{-1} (\nabla \psi^{\dagger})\mathbf{E} + \nabla E = \lambda^{-2} (\nabla \lambda^2)\mathbf{E} + \nabla \mathbf{E}.$$

Hence,

$$\nabla' \underline{f}(\mathbf{E}) = \lambda^{-2} \nabla(\lambda^2 \mathbf{E}). \tag{7.35}$$

Chapter 5

Differential Geometry of Vector Manifolds

This chapter continues the study of calculus on vector manifolds begun in Chapter 4. The emphasis here is on the central object of classical differential geometry, the curvature tensor. We have endeavored to supply simple and systematic derivations of all properties of the curvature tensor including relations to extrinsic geometry, behavior under transformations, and generalization to nonRiemannian curvature. We believe that some of our results are new, but our main objective is to demonstrate the unique advantages of the method and to develop the calculus to the point where application to any problem in differential geometry is straightforward.

The advantages of formulating differential geometry in terms of Geometric Calculus are best seen by examining the details developed in this chapter, but some general features deserve mention here. In the first place, the formulation of differential geometry in this chapter is completely independent of coordinates. Other formulations in the literature are at best only *nominally coordinate-free*; by this we mean that they provide coordinate-free formulations of the general theorems, but they must resort to coordinates for many proofs and most computations. We have been able to achieve a coordinate-free formulation of differential geometry, because in the preceding chapter we avoided coordinates in our definition of a manifold and in our formulation of the chain rule and induced transformations.

Another advantage of our approach is its unification of intrinsic and extrinsic geometry. This unification simplifies the study of intrinsic geometry itself and shows how extrinsic geometry is reflected in the intrinsic structure.

It should be pointed out that our formalism has special advantages for undergraduate courses in classical differential geometry of surfaces in Euclidean three-space, as set forth, for example, by Struik [St]. Unlike tensor analysis and the calculus of differential forms, Geometric Calculus contains the vector algebra in the form used by Struik, including the vector cross product. References [H4] and [H1] explain how the cross product fits into Geometric Algebra. So the geometry in Struik is already expressed in terms of Geometric Calculus. Thus, our approach retains all the virtues of Struik's vector formulation while fitting it into the general theory of manifolds without unnecessary alterations of method or notation. Translation of the classical geometry in Struik into our formalism is easily accomplished with the help of Section 5-2. The resulting material is, of course, appropriate for a first course in differential geometry.

Our approach to differential geometry generalizes a special method which has long been used for hypersurfaces, as can be seen by comparing our Section 5-2 with the account by Hicks [Hi] as well as by Struik [St]. The classical method characterizes the geometry of a hypersurface by equations for its normal as it "slides" over the surface. Our method characterizes the geometry of every manifold by the curl tensor, which describes how the pseudoscalar 'slides' over the manifold. Conventional formulations of intrinsic geometry employ only tensors with values in the tangent algebra, so they overlook the curl tensor, which does not belong to this class. This impoverishes and complicates the theory unnecessarily.

5-1. Curl and Curvature

In Section 4-4, we learned that the curl tensor S_a can be interpreted as the angular velocity of a pseudoscalar as it slides along a curve in the manifold with tangent a. In our approach to geometry the curl tensor takes over the function of the 'Christoffel symbols' or 'coefficients of connection' in classical formulations of Riemannian geometry. But the curl tensor contains more information about the manifold than do the coefficients of connexion, and it is defined without reference to any coordinate system. The Riemann curvature tensor depends on derivatives of the coefficients of connexion, but it is a simple algebraic function of the curl tensor. Indeed, as we show below, the commutator $S_a \times S_b$ determines both intrinsic and extrinsic curvatures. We have good reason, therefore, to regard the curl, rather than the curvature, as the fundamental object in different geometry.

In spite of the importance of the curl, the curvature tensor can be formulated and some of its properties are more easily derived without reference to the curl. So in this section we keep our attention centered on the curvature tensor and seek to establish each of its properties with the most direct argument available.

The curvature tensor is commonly introduced as a measure of the failure of commutivity of codifferentials. According to (4-3.3) and (4-3.1), the codifferential of a function $A = A(x)$ is defined by

$$\delta_a A = a \cdot \nabla A \equiv P(a \cdot \partial A) = a \cdot \partial P(A) - P_a(A). \tag{1.1}$$

The commutator of codifferentials $[\delta_a, \delta_b] \equiv \delta_a \delta_b - \delta_b \delta_a$ is related to the coderivative by the identity

$$[\delta_a, \delta_b] A = [a \cdot \nabla, b \cdot \nabla] A - [a, b] \cdot \nabla A = (b \wedge a) \cdot (\nabla \wedge \nabla) A. \tag{1.2}$$

The proof of (1.2) is similar to the argument establishing (4-1.15), so it can be omitted.

We determine the curvature tensor by using the integrability condition to evaluate the commutator (1.2). We are primarily concerned with the codifferentials of fields, so we assume $P(A) = A$ and $P(a \wedge b) = a \wedge b$ throughout this section. Differentiating (4-3.1) and using (4-2.33) we get

$$\begin{aligned} b \cdot \nabla a \cdot \nabla A &= P(b \cdot \partial a \cdot \partial A) + P_b P_a(A) \\ &= P(b \cdot \partial a \cdot \partial A) + P(S_b \times (A \times S_a)). \end{aligned} \tag{1.3}$$

Recalling the integrability condition (4-1.12), we find that

$$[\delta_a, \delta_b]A = [P_a, P_b]A = A \times R(a \wedge b), \tag{1.4}$$

where $[P_a, P_b] \equiv P_a P_b - P_b P_a$ and

$$R(a \wedge b) \equiv P(S_a \times S_b) \tag{1.5a}$$

$$= P_a(S_b) = \partial_v \wedge P_a P_b(v) \tag{1.5b}$$

$$= b \cdot \nabla S_a = \delta_b S_a \tag{1.5c}$$

$$= \partial_u \wedge \partial_v P_u(a) \cdot P_v(b). \tag{1.5d}$$

To get the right side of (1.4) from (1.3), we used the Jacobi identity

$$S_a \times (A \times S_b) - S_b \times (A \times S_a) = A \times (S_a \times S_b).$$

Since $P(S_a \times S_b)$ is linear and skewsymmetric in a and b, we know from Section 1-4 that it is a function of the bivector $a \wedge b$; accordingly, we have adopted the notation $R(a \wedge b)$. The first term in (1.5b) follows from (1.5a) by (4-2.16a), and the second term then follows from (4-2.12b), because of (4-2.16a) and (4-2.30). We get (1.5c) from (1.5b) by using (1.1) and $P(S_b) = 0$. Finally, to get (1.5d), we use (1-1.68) to get the identity

$$\begin{aligned} S_a \times S_b &= [\partial_v \wedge P_v(a)] \times [\partial_u \wedge P_u(b)] \\ &= P_v(a) \cdot \partial_u \, \partial_v \wedge P_u(b) - \partial_v \wedge \partial_u P_v(a) \cdot P_u(b) \\ &\quad + P_u(b) \cdot \partial_v P_v(a) \wedge \partial_u - \partial_v \cdot \partial_u P_v(a) \wedge P_u(b). \end{aligned} \tag{1.6}$$

According to (4-2.11a), $c \cdot P_v(a) = 0$ if c, v, a are vector fields, so the first and third terms on the right of (1.6) vanish, and besides (1.5d), we see that

$$P_\perp(S_a \times S_b) = P_b(\partial_v) \wedge P_v(a) \tag{1.7}$$

and

$$c \cdot P_\perp(S_a \times S_b) = 0. \tag{1.8}$$

Because of (1.8), the operator P is unnecessary when (1.5a) is substituted into (1.4), that is

$$[\delta_a, \delta_b]A = [P_a, P_b]A = A \times (S_a \times S_b). \tag{1.9}$$

We call the bivector valued function $R(a \wedge b)$ defined by (1.5) the *curvature (tensor)* of the manifold $\mathscr{M}$. We sometimes refer to it as the '*Riemann* curvature' if the tangent algebra of $\mathscr{M}$ is assumed to be Euclidean in the sense of Section 1-4. The Riemann curvature of classical differential geometry is the (scalar-valued) tensor of rank four $R(a, b, c, d) \equiv (c \wedge d) \cdot R(a \wedge b)$; a proof that this is equivalent to the conventional covariant tensor form is given in Section 6-2. Comparison with Cartan's 'curvature form' is made in Section 6-4. Equations (1.4) and (1.5) display important properties of the curvature tensor which are not apparent in conventional formulations. Equation (1.4) *evaluates* the commutator of codifferentials in terms of differentials of the projection and in terms of the curvature tensor. Equation (1.5a) reveals the remarkable fact that the curvature tensor is a simple *algebraic* function of the curl tensor. Thus, the curl tensor completely characterizes the intrinsic geometry of a manifold.

The general properties of the curvature can easily be established from its sundry equivalent expressions given by (1.5), as we now show. At each point of $\mathscr{M}$, the curvature tensor defines a linear transformation of tangent bivectors into tangent bivectors; that is, to every bivector field B on $\mathscr{M}$ there corresponds another bivector field $R(B)$ defined by

$$R(B) \equiv \tfrac{1}{2} B \cdot (\partial_b \wedge \partial_a) R(a \wedge b). \tag{1.10}$$

To use the terminology introduced in Sections 3-9 and 4-1, the curvature $R = R(B)$ is a multiform field of degree two and grade two. Indeed, the curvature is a protractionless multiform in the sense of Section 3-9; for, by (1.5b) and (4-2.37a), we have

$$\partial_a \wedge R(a \wedge b) = \partial_a \wedge P_a(S_b) = 0. \tag{1.11}$$

As shown in Section 3-9, Eqn. (1.11) is equivalent to two weaker conditions on R, namely, that R is *symmetric* in the sense that for bivector fields A and B

$$A \cdot R(B) = B \cdot R(A), \tag{1.12a}$$

and R satisfies the *Ricci identity*

$$a \cdot R(b \wedge c) + b \cdot R(c \wedge a) + c \cdot R(a \wedge b) = 0, \tag{1.12b}$$

for vector fields a, b, c. It is not difficult to show that the Ricci identity is equivalent to the property (4-3.11) that $\nabla \wedge \nabla \wedge A = 0$.

The curvature tensor satisfies the famous *Bianchi identity*, which can be cast in either of the equivalent forms

$$\dot{\nabla} \wedge \dot{R}(a \wedge b) = 0, \tag{1.13a}$$

or

$$\delta_c R(a \wedge b) + \delta_a R(b \wedge c) + \delta_b R(c \wedge a) = 0. \tag{1.13b}$$

Equation (1.13b) can be further abbreviated by using the exterior codifferential (4-3.4b) and recalling (4-1.23), thus,

$$\delta R = 0, \tag{1.13c}$$

that is, the Bianchi identity says that the curvature tensor has a vanishing exterior differential. We can prove (1.13) easily by differentiating (1.5d) with the help of (4-1.19) and using (4-2.11a) to get

$$\begin{aligned}\delta_c R(a \wedge b) &= \partial_u \wedge \partial_v \delta_c(P_u(a) \cdot P_v(b)) \\ &= \partial_u \wedge \partial_v(P_{uc}(b) \cdot P_v(a) - P_{uc}(a) \cdot P_v(b)).\end{aligned} \tag{1.14}$$

Equation (1.13b) follows trivially from (1.14), and

$$\dot{\nabla} \wedge \dot{R}(a \wedge b) = \partial_c \delta_c R(a \wedge b) = \partial_c \wedge \partial_u \wedge \partial_v(P_{uc}(b) \cdot P_v(a) - P_{uc}(a) \cdot P_v(b)),$$

which vanishes to give (1.13a) because $P_{uc} = P_{cu}$.

It is interesting to express the fundamental Eqn. (1.4) in terms of the exterior codifferential (4-3.4b); the result is simply

$$\delta^2 A = A \times R = -R \times A. \tag{1.15}$$

This should be compared with the second exterior differential $d^2 A = 0$. Since $\delta R = 0$, all 'powers' of the exterior differential can be immediately computed from (1.15), thus

$$\delta^3 A = \delta A \times R, \tag{1.16a}$$

$$\delta^4 A = (A \times R) \times R \equiv A \times R^2, \tag{1.16b}$$

and, in general,

$$\delta^{2k} A = A \times R^k, \tag{1.16c}$$

$$\delta^{2k+1} A = \delta A \times R^k, \tag{1.16d}$$

where R^k indicates the k-fold commutator with R, as shown explicitly in (1.16b) for $k = 2$. The multiform $\delta^k A$ has degree k, so, of course,

$$\delta^{m+1} A = 0, \tag{1.16e}$$

where m is the dimension of $\mathscr{M}$. The k-vector argument of $\delta^k A$ is suppressed in (1.15) and (1.16), giving the formulas a somewhat misleading appearance of simplicity, which rapidly vanishes when (4-1.22) is used to make the argument explicit. We will not have occasion to use Eqns. (1.16), but (1.15) along with (1.13c) find important applications in Section 7-8.

Now let us consider some of the classical tensors constructed from the curvature tensor.

The contraction of the curvature tensor is the *Ricci tensor*

$$R(b) \equiv \partial_a \cdot R(a \wedge b) = \partial_a R(a \wedge b), \tag{1.17}$$

where we have used (1.11). Also, by contracting (1.11) we get

$$\partial_b \wedge R(b) = 0, \tag{1.18a}$$

which as we saw in Section 3-4 is equivalent to the condition

$$a \cdot R(b) = b \cdot R(a). \tag{1.18b}$$

Thus the Ricci tensor is a symmetric linear transformation of vector fields into vector fields.

From (1.5b) and (1.5c) we find easily

$$R(b) = S^2(b) = -\nabla S_b = -\nabla \cdot S_b. \tag{1.19}$$

Hence, the shape operator on vector fields is a square root of the Ricci tensor.

The contraction of the Ricci tensor is commonly called the *scalar curvature* and denoted by R. (We will not use the scalar curvature often enough to allow it to be confused with the Riemann curvature which we sometimes also denote by the single symbol R as in Eqns. (1.15) and (1.16).) From (1.17) and (1.18a) we have

$$R \equiv \partial_b \cdot R(b) = \partial_b R(b) = \partial_b\, \partial_a R(a \wedge b). \tag{1.20}$$

Contraction of (1.19) gives, with the help of (4-2.19) and (4-2.11a), $R = -\partial_b \cdot (\nabla \cdot S_b) = -(\partial_b \wedge \nabla) \cdot S_b = \nabla \cdot (\partial_b \cdot S_b)$. So, recalling the definition (4-2.20) of the spur N, we get

$$R = \partial_b \cdot S^2(b) = \partial \cdot N. \tag{1.21}$$

By contracting the Bianchi identity we get relations of special interest in Einstein's geometric theory of gravitation. From (1.13a) we get

$$\partial_a \cdot (\dot{\nabla} \wedge \dot{R}(a \wedge b)) = \partial_a \cdot \dot{\nabla} \dot{R}(a \wedge b) - \dot{\nabla} \wedge (\partial_a \cdot \dot{R}(a \wedge b)) = 0.$$

By virtue of (4-1.19),

$$\dot{\nabla} \dot{R}(b) = P(\partial_v P_v(\partial_a)) R(a \wedge b) + \dot{\nabla} \partial_a \dot{R}(a \wedge b).$$

But $P(\partial_v P_v(\partial_a)) = \partial_v P P_v P(\partial_a) = 0$. Hence the contracted Bianchi identity can be written

$$\dot{\nabla} \wedge \dot{R}(b) = \dot{R}(\dot{\nabla} \wedge b), \tag{1.22}$$

where $\dot{R}(\dot{\nabla} \wedge b) \equiv P(\dot{R}(\dot{\partial} \wedge b))$. Contracting (1.22), we have

$$\partial_b \cdot (\dot{\nabla} \wedge \dot{R}(b)) = \partial_b \cdot \dot{\nabla} \dot{R}(b) - \dot{\nabla} \partial_b \cdot \dot{R}(b) = \partial_b \cdot \dot{R}(\dot{\nabla} \wedge b)$$

or

$$2\dot{R}(\dot{\nabla}) = \nabla R, \tag{1.23}$$

where $\dot{R}(\dot{\nabla}) \equiv P(\dot{R}(\dot{\partial}))$. If we define the *Einstein tensor* $G(a)$ by

$$G(a) \equiv R(a) - \tfrac{1}{2}aR, \tag{1.24}$$

then (1.23) takes the form

$$\dot{G}(\dot{\nabla}) = 0. \tag{1.25}$$

Recalling the relation of the simplicial derivative to the multivector derivative, we can write (1.20) in the form

$$\partial_B R(B) = \tfrac{1}{2}\,\partial_b \wedge \partial_a R(a \wedge b) = \tfrac{1}{2}R, \tag{1.26}$$

where ∂_B is the derivative by the bivector variable B. Of course, (1.26) is equivalent to the equations

$$\partial_B \cdot R(B) = \tfrac{1}{2}R, \tag{1.27a}$$

$$\partial_B \times R(B) = 0, \tag{1.27b}$$

$$\partial_B \wedge R(B) = 0. \tag{1.27c}$$

Now we return to the fundamental Eqn. (1.4), and show that it can be put in the alternative form

$$\nabla \wedge \nabla A = S^2(A) = \partial_B R(B) \times A. \tag{1.28}$$

First we use (1.2) and (2-2.41) followed by (4-3.14b) to prove

$$\tfrac{1}{2}\,\partial_a\,\partial_b\,[\delta_a, \delta_b]A = \nabla \wedge \nabla A = (\nabla \wedge \nabla) \times A. \tag{1.29}$$

To prove

$$\tfrac{1}{2}\,\partial_a\,\partial_b\,[P_a, P_b]A = S^2(A), \tag{1.30}$$

we use the grade-preserving property of the operator established by (1.29) along with several results from Section 4-2 which are evident in the following computation:

$$\begin{aligned}\partial_a\,\partial_b P_b P_a(A) &= \partial_a\,\partial_b \cdot P_b P_a(A) = \partial_a \wedge [\partial_b \cdot P_b P_a(A)]\\ &= -\partial_b \cdot (\partial_a \wedge P_b P_a(A)) + \partial_a \cdot \partial_b P_b P_a(A)\\ &= -S^2(A) + \partial_a \cdot \partial_b P_a P_b(A).\end{aligned}$$

Combining this with

$$\partial_a\,\partial_b P_a P_b(A) = -\,\partial_a\,\partial_b P_b P_a(A) + 2\,\partial_a \cdot \partial_b P_a P_b(A)$$

we get (1.30). Of course the right side of (1.28) is obtained by simply replacing the simplicial derivative by the bivector derivative.

Noting the grade-preserving property (1.29), from (1.28) we get

$$\partial_B \cdot (R(B) \times A) = 0 \quad \text{if } \langle A \rangle_1 = 0, \tag{1.31a}$$

$$\partial_B R(B) \times A = \partial_B \times (R(B) \times A), \tag{1.31b}$$

$$\partial_B \wedge (R(B) \times A) = 0. \tag{1.31c}$$

Setting $A = b$ and dotting (1.31c) with a, we get using (1.21a),

$$(a \cdot \partial_B) \wedge (R(B) \cdot b) = R(a \wedge b). \tag{1.32}$$

Equation (1.28) gives us some other interesting relations. Applying (1.28) to a vector field b and using (1.19), we get

$$S^2(b) = \nabla \wedge \nabla b = \partial_B R(B) \cdot b = R(b). \tag{1.33}$$

Again, applying (1.28) to the bivector field $a \wedge b$ and using (4-2.38) we get

$$S^2(a \wedge b) = \nabla \wedge \nabla a \wedge b = R(a) \wedge b + a \wedge R(b) - 2R(a \wedge b). \tag{1.34}$$

Clearly, the operator S^2 contains all the information in the curvature tensor, but it has the advantage of applying to any field, whereas the curvature tensor operates directly only on bivector fields and, by contraction, on vector fields.

So far in this section we have been concerned only with intrinsic geometrical properties of a manifold. At the end of Section 4-4 we found that the differential of the curl tensor S_{ab} satisfies the relation

$$S_{ab} - S_{ba} = b \cdot \partial S_a - a \cdot \partial S_b + S_{[a,b]} = 2S_a \times S_b. \tag{1.35}$$

This result can also be established by computation from (1.6) or (1.5a, d) and (1.7). The tangential component of (1.35) expresses a fact about curvature which we have already expressed by (1.5a, c). But the normal component of (1.35) gives us the important new result

$$P_\perp(S_{ab} - S_{ba}) = P_\perp(b \cdot \partial S_a - a \cdot \partial S_b) + S_{[a,b]} = 2P_\perp(S_a \times S_b). \tag{1.36}$$

This equation describes extrinsic geometric properties of the manifold. As we show in the next section, Eqn. (1.36) generalizes the so-called Codazzi–Mainardi equations for hypersurfaces in Euclidean space. Note that, by virtue of (1.8), we have from (1.36)

$$c \cdot P_\perp(S_{ab} - S_{ba}) = c \cdot P_\perp(b \cdot \partial S_a - a \cdot \partial S_b) + c \cdot S_{[a,b]} = 0. \tag{1.37}$$

Since the tensor

$$S_a \times S_b = P(S_a \times S_b) + P_\perp(S_a \times S_b) \tag{1.38}$$

unifies in a simple way the *'intrinsic' curvature* $P(S_a \times S_b) = R(a \wedge b)$ with the tensor $P_\perp(S_a \times S_b)$, which can fairly be called the *extrinsic curvature*, it seems appropriate to refer to $S_a \times S_b$ as the *total curvature*.

The total curvature satisfies the *generalized Bianchi identity*

$$(S_a \times S_b)_c + (S_b \times S_c)_a + (S_c \times S_a)_b = 0, \tag{1.39}$$

where

$$(S_a \times S_b)_c \equiv c \cdot \partial(S_a \times S_b) - S_{c \cdot \partial a} \times S_b - S_a \times S_{c \cdot \partial b} \tag{1.40}$$

is the differential of the total curvature. To prove (1.39) we take the differential of (1.35) to get

$$2(S_a \times S_b)_c = S_{abc} - S_{bac} = S_{abc} - S_{bca}, \tag{1.41}$$

where the last term was obtained by using the symmetry $S_{bac} = S_{bca}$ of the second differential of the curl tensor S_b. Adding three copies of (1.41) differing only by a cyclic permutation of the arguments *a, b, c*, we get (1.39) immediately. Projection of (1.39) into the tangent algebra gives us another proof of the Bianchi identity (1.13b) since

$$\delta_c R(a \wedge b) = P((S_a \times S_b)_c) + PP_c(S_a \times S_b),$$

and the last term cancels other similar terms from (1.39) because of the Jacobi identity

$$S_c \times (S_a \times S_b) + S_a \times (S_b \times S_c) + S_b \times (S_c \times S_a) = 0.$$

5.2. Hypersurfaces in Euclidean Space

For a hypersurface in Euclidean space, the role of its pseudoscalar can be taken over by its normal. This section shows how easily our general formalism is adapted to this special point of view. The results facilitate comparison of our formalism with the classical approach to hypersurfaces. The effectiveness of our calculus in geometrical computations is demonstrated by several examples.

Let $\mathscr{M}$ be an m-dimensional hypersurface in $\mathscr{E}_{m+1}$. If i denotes the (constant) unit pseudoscalar of $\mathscr{E}_{m+1}$, then at a generic point x of $\mathscr{M}$, the unit tangent $I = I(x)$ can be written

$$I = ni, \tag{2.1}$$

where $n = n(x)$ is the unit normal to $\mathscr{M}$. In the last chapter and the preceding section, the geometry of an arbitrary vector manifold has been expressed in terms of I and its derivatives. But, by virtue of (2.1), for hypersurfaces in Euclidean space the role of the tangent I can be taken over by the normal n. Thus, the curl tensor for $\mathscr{M}$ can be written

$$S_a = Ia \cdot \partial I^{-1} = na \cdot \partial n = n\underline{n}(a) = n \wedge \underline{n}(a), \tag{2.2}$$

where $\underline{n}(a) \equiv a \cdot \partial n$ is the differential of n. The function $n = n(x)$ maps $\mathscr{M}$ into the unit sphere $\mathscr{S}_m$ in $\mathscr{E}_{m+1}$. The differential $\underline{n}$ transforms tangent fields on $\mathscr{M}$ linearly into tangent fields on $\mathscr{S}_m$. Indeed, the condition $n^2 = 1$ implies that $n \cdot (a \cdot \partial n) = n \cdot \underline{n}(a) = 0$; hence the vector field $\underline{n}(a)$ is tangent to $\mathscr{M}$ at each point x as well as to $\mathscr{S}_m$ at each corresponding point $n = n(x)$. Thus the tangent space at x in $\mathscr{M}$ is identical to the tangent space at $n(x)$ in $\mathscr{S}_m$. It follows that

$$a \cdot \nabla n = a \cdot \partial n \equiv \underline{n}(a), \tag{2.3}$$

and, from (2.2) used in (4-2.28), that

$$a \cdot \underline{n}(b) = b \cdot \underline{n}(a). \tag{2.4}$$

Therefore, $\underline{n}$ is a symmetric linear transformation of vector fields on $\mathscr{M}$ to fields on $\mathscr{S}_m$.

While (2.1) allows the normal n to take over the role of I, Eqn. (2.2) allows $\underline{n}(a)$ to take over the role of S_a on $\mathscr{M}$. Thus, substitution of (2.2) into (4-3.16) gives

$$b \cdot \nabla a - b \cdot \partial a = na \cdot \underline{n}(b). \tag{2.5}$$

And, again using (2.2), the intrinsic curvature can be written

$$R(a \wedge b) = P(S_a \times S_b) = \underline{n}(b) \wedge \underline{n}(a) = \underline{n}(b \wedge a). \tag{2.6}$$

On the right of (2.6) we are again using the notation for outermorphism from Section 3-1. Equation (2.6) shows that the transformation of bivector fields on $\mathscr{M}$ under the differential outermorphism $\underline{n}$ is equivalent to the Riemann curvature. Therefore the intrinsic geometry of $\mathscr{M}$ is completely described by algebraic properties of the symmetric linear transformation $\underline{n}$, with the set of all tangent multivector fields on $\mathscr{M}$ taken as its domain. The properties of $\underline{n}$ can be formulated and analyzed by the method of Chapter 3.

The concept of mean curvature plays an important role in standard discussions of hypersurfaces. The *mean curvature* $H(x)$ at each point x of $\mathscr{M}$ is given by the trace of $\underline{n}$, which by (4-1.17) is equal to the divergence of the normal; thus,

$$H = \frac{1}{m}\,\partial_a \underline{n}(a) = \frac{1}{m}\,\partial \cdot n = \frac{-n \cdot N}{m} = \frac{-nN}{m}, \tag{2.7}$$

where we have also related H to the spur N by Eqn. (4-2.20). Evidently the spur is the appropriate generalization of mean curvature to arbitrary vector manifolds.

Recall the definition of characteristic multivector from Section 3-2. The scalar curvature R can be expressed as the second characteristic multivector of $\underline{n}$:

$$\begin{aligned} R &\equiv (\partial_a \wedge \partial_b) \cdot R(b \wedge a) = (\partial_a \wedge \partial_b) \cdot (\underline{n}(b) \wedge \underline{n}(a)) \equiv 2\underline{\partial}_{(2)} \cdot \underline{n}_{(2)} \\ &= (\partial_2 \wedge \partial_1) \cdot (n_1 \wedge n_2) \equiv 2\partial_{(2)} \cdot n_{(2)}, \end{aligned} \tag{2.8}$$

where $n_k \equiv n(x_k)$, and we have adopted the notation of (4-5.7a) and used (4-5.7b) to replace derivatives by tangent vectors with derivatives by points.

The transformation $n = n(x)$ of $\mathscr{M}$ into the sphere $\mathscr{S}_m$ is commonly called the *Gauss map* of the hypersurface $\mathscr{M}$, and its Jacobian κ is called the *Gaussian curvature*. Of course, the Jacobian of n is the determinant of its differential $\underline{n}$. Thus, using (4-5.7) again, we have

$$\kappa = \underline{\partial}_{(m)} \cdot \underline{n}_{(m)} \equiv \frac{1}{m!} (\underline{\partial}_m \wedge \ldots \wedge \underline{\partial}_1) \cdot (\underline{n}_1 \wedge \ldots \wedge \underline{n}_m)$$

$$= \partial_{(m)} \cdot n_{(m)} \equiv \frac{1}{m!} (\partial_m \wedge \ldots \wedge \partial_1) \cdot (n_1 \wedge \ldots \wedge n_m)$$

$$= I^{-1}\underline{n}(I). \tag{2.9a}$$

We can calculate κ from whichever of these forms is most convenient. Since $\underline{n}$ is a symmetric transformation, the dots in (2.9) can be included or omitted at will.

When m is even, the Gaussian curvature is simply related to the curvature tensor. From (2.6) and (2.9) we find that, for $I = e_1 \wedge e_2 \wedge \ldots \wedge e_m$,

$$R(e_1 \wedge e_2) \wedge \ldots \wedge R(e_{m-1} \wedge e_m) = \underline{n}(e_1) \wedge \underline{n}(e_2) \wedge \ldots \wedge \underline{n}(e_m)$$

$$= \underline{n}(I) = \kappa I. \tag{2.9b}$$

This result is obviously independent of the choice of vector factors e_k for I.

Though the local intrinsic geometry of $\mathscr{M}$ is fully characterized by the algebraic properties of $\underline{n}$, derivatives of $\underline{n}$ are needed to describe the extrinsic geometry of $\mathscr{M}$. 'Dotting' Eqn. (1.35) or (1.36) by n, we find that

$$n \cdot (b \cdot \partial S_a - a \cdot \partial S_b) = b \cdot \partial \underline{n}(a) - a \cdot \partial \underline{n}(b)$$

$$= b \cdot \nabla \underline{n}(a) - a \cdot \nabla \underline{n}(b) = \underline{n}([b, a]). \tag{2.10}$$

This result may be recognized as equivalent to the well-known Codazzi–Mainardi equations by comparison with Eqn. 10 in Chapter 2 of ref. [Hi].

With the above results in hand, the entire discussion of hypersurfaces in Chapter 2 of ref. [Hi] can easily be translated into the language of geometric calculus, so we need not pursue the subject further. But an example is called for to illustrate computations.

Let $\mathscr{M}$ be the m-sphere of radius $|x|$ in $\mathscr{E}_{m+1}$. Then at a generic point x of $\mathscr{M}$ the outward unit normal is given by the explicit function $n = n(x) = x/|x|$. Since $|x|$ is constant on $\mathscr{M}$, we have $\partial|x| = \partial_x|x| = 0$. But for any m-dimensional manifold we have $\partial x = \partial_x x = m$ and $a \cdot \partial x = a$ if $a = a(x)$ is a field. Hence,

$$\underline{n}(a) = \frac{a}{|x|} \tag{2.11}$$

and

$$\partial_a \underline{n}(a) = \partial n = \partial \cdot n = \frac{m}{|x|}. \tag{2.12}$$

Hence the intrinsic curvature is

$$R_{ab} = \underline{n}(b) \wedge \underline{n}(a) = \frac{b \wedge a}{|x|^2}, \tag{2.13}$$

the mean curvature is $H = |x|^{-1}$, and the Gaussian curvature is, by (4-6.7a),

$$\kappa = \underline{\partial}_{(m)} \underline{n}_{(m)} = |x|^{-m} \partial_{(m)} x_{(m)} = |x|^{-m}. \tag{2.14}$$

Computations of other geometric properties of a sphere are equally trivial.

We now discuss the computation of the shape and curvature of an arbitrary hypersurface $\mathscr{M}'$ in $\mathscr{E}_{m+1}$ which is parametrized (at least locally) by its projection into a hyperplane $\mathscr{M}$ in $\mathscr{E}_{m+1}$. We are given a differentiable function f which maps a point x in $\mathscr{M}$ to a point $x' = f(x)$ in $\mathscr{M}'$. The properties of such a function were discussed in detail in Example 2 of Section 4-6 (see Eqns. (4-6.13) through (4-6.30)). The definitions and results obtained there will be used here with little commentary.

As in Eqn. (2.1), $I' = I'(x') = n'i$ denotes the unit tangent to $\mathscr{M}'$ at x', where $n' = n'(x')$ is the unit normal; the (constant) unit normal to $\mathscr{M}$ is denoted by n. To every tangent vector field $a' = a'(x')$ on $\mathscr{M}'$ there corresponds a unique tangent vector field $a = a(x)$ determined by the differential function: $a' = \underline{f}(a) = a \cdot \partial f$. We will express the geometry of $\mathscr{M}'$ in terms of tensor fields on $\mathscr{M}$. Differentiation on $\mathscr{M}'$ is simply related to differentiation on $\mathscr{M}$ by the operator equation $a' \cdot \partial' = a \cdot \partial$.

We are now in position to get a simple explicit expression for the curl tensor $S_{a'}$ of $\mathscr{M}'$. Recalling the results given by (4-6.23) and (4-6.24), namely

$$n' = J_f^{-1}(n - \partial\beta) \qquad \text{where } J_f = |n - \partial\beta| = (1 + (\partial\beta)^2)^{1/2}, \tag{2.15}$$

we find

$$\underline{n}'(a') = a' \cdot \partial' n' = -J_f^{-1} a' \cdot \partial' \partial\beta - n'a' \cdot \partial' \ln J_f. \tag{2.16}$$

Defining a tensor field $h(a)$ by

$$h(a) \equiv J_f^{-1} a \cdot \partial\, \partial\beta = J_f^{-1} a' \cdot \partial' \partial\beta, \tag{2.17}$$

and using (2.2), we get the shape tensor in the form

$$S_{a'} = h(a) \wedge n'. \tag{2.18}$$

Dotting this on the left with n', we get

$$\underline{n}'(a') = n' \cdot (h(a) \wedge n') = -h(a) + n'n' \cdot h(a). \tag{2.19}$$

Comparing (2.19) with (2.16), we find

$$n' \cdot h(a) = -a' \cdot \partial' \ln J_f = -a \cdot \partial \ln J_f, \tag{2.20}$$

which could have been obtained by differentiating J_f directly.

Just as the geometry of the hypersurface $\mathscr{M}'_m$ can be characterized by the tangent field $\underline{n}'(a')$ on the unit hypersphere, so it can be alternatively characterized by the tangent field $h(a) = h(\underline{f}^{-1}(a'))$ on the hyperplane $\mathscr{M}$. The tangency of $h(a)$ to $\mathscr{M}$, that is,

$$n \cdot h(a) = 0 \tag{2.21}$$

follows immediately from (2.17). It is also important to note that h is a *symmetric* linear transformation of tangent fields on $\mathscr{M}$ into tangent fields on $\mathscr{M}$, that is,

$$a \cdot h(b) = b \cdot h(a). \tag{2.22}$$

Using (2.19) in (2.16), we find the expression for the intrinsic curvature in terms of h;

$$R_{a'b'} = h(a) \wedge h(b) - n' \cdot h(a) n' \wedge h(b) + n' \cdot h(b) n' \wedge h(a). \tag{2.23}$$

The mean curvature is computed with the help of (2.15) and (4-6.30) which expresses ∂' in terms of ∂;

$$H = \frac{1}{m} \partial' \cdot n' = -\frac{\partial' \cdot \partial\beta}{mJ_f} = \frac{1}{mJ_f} [J_f^{-2} (\partial\beta \cdot \partial)^2 \beta - (\partial\beta)^2]. \tag{2.24}$$

The Gaussian curvature can be computed in several instructive ways. For example, if $\{a_k\}$ is an orthonormal frame on $\mathscr{M}$, we have $I = a_1 a_2 \ldots a_m$, and

$$\underline{f}(I) = J_f I' = \underline{f}(a_1) \wedge \ldots \wedge \underline{f}(a_m) = a'_1 \wedge \ldots \wedge a'_m .$$

Hence,

$$\begin{aligned}
\kappa &= I'^{\dagger} \underline{n}'(I') = J_f^{-1} I'^{\dagger} \underline{n}'(a'_1 \wedge \ldots \wedge a'_m) \\
&= J_f^{-1} i^{\dagger} n' \wedge \underline{n}'(a'_1) \wedge \ldots \wedge \underline{n}'(a'_m) \\
&= J_f^{-1} i^{\dagger} n' \wedge h(a_1) \wedge \ldots \wedge h(a_m) \\
&= J_f^{-2} I^{\dagger} h(a_1) \wedge \ldots \wedge h(a_m) = J_f^{-2} I^{\dagger} \underline{h}(I) \\
&= J_f^{-2} \det h = (-1)^m J_f^{-m-2} \partial_{(m)} (\partial\beta)_{(m)} \\
&= (-1)^m J_f^{-m-2} \det (\partial_j \partial_k \beta).
\end{aligned} \tag{2.25}$$

The computations are now easily completed given any explicit form of the function $\beta = \beta(x)$.

5.3. Related Geometries

In Section 4-5 we determined how fields and their derivatives can be transformed from one manifold to another. This section applies and extends the results of Section 4-5 to describe the induced transformation of the intrinsic geometry on a manifold. Accordingly, the definitions and results of Section 4-5 are taken for granted here.

The main result of this section is Eqn. (3.11) describing the induced transformation of the curvature tensor. It is needed for applications in the next two sections.

Let $h = h(A) = h(x, A(x))$ be an extensor function defined on the vector manifold $\mathscr{M}$. The substitution $x = f^{-1}(x')$ expresses h as an extensor function on the manifold $\mathscr{M}'$. According to (4-5.12) we have $a \cdot \partial = a' \cdot \partial'$ if $a' = \underline{f}(a)$. Hence

$$h_a(A) \equiv a \cdot \partial h(A) - h(a \cdot \partial A)$$

$$= a' \cdot \partial' h(A) - h(a' \cdot \partial' A) \equiv h_{a'}(A), \tag{3.1a}$$

or, more briefly,

$$h_a = h_{a'}. \tag{3.1b}$$

Thus, the differential of h has the same values, whether regarded as a function on $\mathscr{M}$ or on $\mathscr{M}'$. To relate the second differentials of h on $\mathscr{M}$ and $\mathscr{M}'$, we note that

$$(b' \cdot \partial' a') \cdot \partial' = (\underline{f}_b(a) + \underline{f}(b \cdot \partial a)) \cdot \partial'$$

$$= \underline{f}_b(a) \cdot \partial' + (b \cdot \partial a) \cdot \partial.$$

So

$$h_{ab} = b \cdot \partial h_a - h_{b \cdot \partial a}$$

$$= b' \cdot \partial' h_{a'} - h_{(b' \cdot \partial' a' - \underline{f}_b(a))}$$

or

$$h_{ab} = h_{a'b'} + h_{\underline{f}_b(a)}. \tag{3.2}$$

Since $\underline{f}_b(a) = \underline{f}_a(b)$, we have

$$h_{ab} - h_{ba} = h_{a'b'} - h_{b'a'}. \tag{3.3}$$

This result is independent of the fact that h_a is a differential, which, of course, implies that

$$h_{ab} = h_{ba}. \tag{3.4}$$

Now suppose that $h = h(A)$ is an extensor *field* on $\mathscr{M}'$ when A is a field on $\mathscr{M}$, so h transforms fields on $\mathscr{M}$ into fields on $\mathscr{M}'$. We can express this by the operator equation

$$h = P'h = hP = P'hP. \tag{3.5}$$

This generalizes the definition of extensor field given by (4-1.8b). To ascertain the implications of (3.5) for the differential of h, consider

$$h_b = (hP)_b = h_bP + hP_b.$$

Operating with this on P_aP and using (4-2.11a) expressed in the form $PP_aP = 0$, we get

$$h_bP_aP = hP_bP_aP. \tag{3.6a}$$

Similarly, considering $h_b = (P'h)_b$, we get

$$P'P'_ah_b = P'P'_aP'_bh. \tag{3.6b}$$

The *codifferential* of an extensor h satisfying (3.5) is appropriately defined by

$$\delta_a h \equiv P'h_aP = P'h_{a'}P \equiv \delta_{a'}h. \tag{3.7}$$

This generalizes (4-3.3), our previous definition of 'codifferential'. Using (3.6a, b) we find that the *second codifferential* of h satisfies

$$\begin{aligned}\delta_b\delta_a h &= P'(\delta_a h)_b P \\ &= P'(P'_bh_a + h_{ab} + h_aP_b)P \\ &= P'(P'_bP'_ah + h_{ab} + hP_aP_b)P.\end{aligned}$$

We can eliminate projection operators from this expression by using (3.5) and

$$P_bP_aP = PP_bP_aP = PP_bP_a,$$

which is easily derived from (4-2.11a) and (4-2.11b). Hence,

$$\delta_b\,\delta_a h = P'_bP'_ah + P'h_{ab}P + hP_aP_b. \tag{3.8}$$

Now using (3.3) and (3.4) we find that the commutator of codifferentials satisfies

$$\begin{aligned}[\delta_a, \delta_b]h &= [P'_a, P'_b]h - h[P_a, P_b] \\ &= [P'_{a'}, P'_{b'}]h - h[P_{a'}, P_{b'}] = [\delta_{a'}, \delta_{b'}]h.\end{aligned} \tag{3.9}$$

We can express the fact that this tensor is a function of a and b only through the product $a \wedge b$ by introducing the notation

$$\delta^2_{a \wedge b} h \equiv [\delta_a, \delta_b] h = [\delta_{a'}, \delta_{b'}] h = \delta^2_{a' \wedge b'} h. \tag{3.10}$$

Recalling Eqn. (1.4), we see that (3.9) relates the curvatures on the two manifolds $\mathscr{M}$ and $\mathscr{M}'$. Thus, for $A = P(A)$ and $A' = h(A) = P'(A')$ with $a' = \underline{f}(a)$ and $b' = \underline{f}(b)$, we have the fundamental relations

$$\begin{aligned}[\delta_a, \delta_b] h(A) &= [P'_{a'}, P'_{b'}] A' - h([P_{a'} P_b] A)\\ &= [\delta_{a'}, \delta_{b'}] A' - h([\delta_{a'} \delta_b] A)\\ &= R'(a' \wedge b') \times A' - h(R(a \wedge b) \times A)\\ &= [\delta_{a'}, \delta_{b'}] h(A).\end{aligned} \tag{3.11}$$

In particular, this holds if $h = \underline{f}$. If we introduce a bivector field $B' = h(B)$ and use the notation (3.10), we have the more succinct expression

$$\begin{aligned}\delta^2_B \underline{f}(A) &= R'(\underline{f}(B)) \times \underline{f}(A) - \underline{f}(R(B) \times A)\\ &= R'(B') \times A' - \underline{f}(R(B) \times A) = \delta^2_{B'} \underline{f}(A).\end{aligned} \tag{3.12}$$

By interchanging P and P' in our discussion of h, we get the analog of (3.12) involving the adjoint transformation, namely,

$$\begin{aligned}\delta^2_B \overline{h}(A') &= R(B) \times h(A') - \overline{h}(R'(h(B')) \times A')\\ &= R(B) \times A - \overline{h}(R'(B') \times A') = \delta^2_{B'} \overline{h}(A'),\end{aligned} \tag{3.13}$$

where $B' = h(B)$, but we have $A = \overline{h}(A')$ instead of the relation $A' = h(A)$ required in (3.11) and (3.12).

5-4. Parallelism and Projectively Related Geometries

In this section we review some well-known properties of parallel vector fields and geodesics to show how they can be expressed with Geometric Calculus. Then we study projective transformations of manifolds, defined as transformations which preserve geodesics. Although the main result of this study was discovered by Weyl, our method is distinguished by its complete independence of coordinates and the new form it gives to basic relations.

Let the function $x = x(\tau)$ be a curve $\mathscr{C}$ in some region $\mathscr{R}$ of a vector manifold $\mathscr{M}$. The 'velocity' v of the curve is

$$v = \frac{\mathrm{d}x}{\mathrm{d}\tau} = P(v), \tag{4.1}$$

where P is the projection into the tangent algebra of $\mathcal{M}$. Let $a = a(x)$ be a vector field defined at points x in $\mathcal{R}$. On the curve $\mathcal{C}$, $a = a(x(\tau))$ is a function of τ. According to (2-2.27), the *derivative* by τ is related to the directional derivative by x by the chain rule as follows:

$$\frac{\mathrm{d}a}{\mathrm{d}\tau} = \left(\frac{\mathrm{d}x}{\mathrm{d}\tau}\right) \cdot \partial_x a(x) = v \cdot \partial a. \tag{4.2}$$

We define the *coderivative* by τ to be the projection of the derivative into the tangent algebra of $\mathcal{M}$, namely

$$\frac{\delta a}{\delta \tau} \equiv P\left(\frac{\mathrm{d}a}{\mathrm{d}\tau}\right) = v \cdot \nabla a = \delta_v a. \tag{4.3}$$

Thus, the *coderivative* of a vector field on a curve is identical to the *codifferential* of the field evaluated at the velocity of the curve.

As we have seen in Chapter 4, the differential is related to the codifferential of the vector field by the equation

$$\begin{aligned} v \cdot \partial a &= P(v \cdot \partial a) + v \cdot \dot{\partial} \dot{P}(a) \\ &= v \cdot \nabla a + P_v(a) = \delta_v a + a \cdot S_v. \end{aligned} \tag{4.4}$$

Since $P(a \cdot S_v) = PP_v(a) = PP_vP(a) = 0$, we identify

$$P_\perp\left(\frac{\mathrm{d}a}{\mathrm{d}\tau}\right) = P_v(a) = a \cdot S_v \tag{4.5}$$

as the component of $\mathrm{d}a/\mathrm{d}\tau$ normal to the manifold, while (4.3) is the tangential component.

We say that the *direction* of the vector field $a = a(x)$ is *uniform* on the curve $\mathcal{C}$ if $\delta a/\delta\tau$ is proportional to a, that is, if

$$a \wedge \frac{\delta a}{\delta \tau} = a \wedge (\delta_v a) = 0. \tag{4.6}$$

When (4.6) is satisfied, tangent vectors $a(x)$ and $a(y)$ at points x and y on $\mathcal{C}$ are said to be *parallel* to one another *with respect to the curve*, and either one is said to be obtainable from the other by *parallel displacement*.

Equation (4.6) is independent of the magnitude of a, for it remains valid if a is replaced by λa where $\lambda = \lambda(x)$ is a scalar field. Some authors say that the direction of a is 'constant' if (4.6) is satisfied; however, (4.5) shows that the normal component of $\mathrm{d}a/\mathrm{d}\tau$ cannot vanish in general unless the curl tensor vanishes, so we have adopted the weaker adjective 'uniform'.

We say that a curve $\mathscr{C}$ is a *geodesic* if the direction of its velocity $v = \mathrm{d}x/\mathrm{d}\tau$ is uniform on the curve, that is, if

$$v \wedge \frac{\delta v}{\delta \tau} = v \wedge (\delta_v v) = 0. \tag{4.7}$$

Accordingly, we interpret a geodesic connecting two points as the *straightest* curve connecting those points.

This property (4.7) is independent of the parametrization given to the curve. Any monotonic scalar function $\tau = \tau(\tau')$ defines a change of parameter. The 'velocity' of the curve with respect to the parameter τ' is

$$v' = \frac{\mathrm{d}x}{\mathrm{d}\tau'} = \frac{\mathrm{d}\tau}{\mathrm{d}\tau'}\frac{\mathrm{d}x}{\mathrm{d}\tau} = \frac{\mathrm{d}\tau}{\mathrm{d}\tau'}\, v. \tag{4.8a}$$

Differentiating again, we get

$$\frac{\delta v'}{\delta \tau'} = \frac{\delta^2 x}{\delta \tau'^2} = \frac{\mathrm{d}^2 \tau}{\mathrm{d}\tau'^2}\, v + \left(\frac{\mathrm{d}\tau}{\mathrm{d}\tau'}\right)^2 \frac{\delta v}{\delta \tau}\,. \tag{4.8b}$$

Hence, $v' \wedge \delta v'/\delta\tau' = 0$ as claimed, and there exists a scalar function $\psi = \psi(\tau')$ such that

$$\frac{\delta v'}{\delta \tau'} = \psi v'. \tag{4.9}$$

Given the curve $\mathscr{C}$ parametrized by τ', we can introduce a new parameter τ defined by the differential equation

$$\frac{\mathrm{d}\tau}{\mathrm{d}\tau'} = \kappa\, \mathrm{e}^{\int \psi\, \mathrm{d}\tau'}, \tag{4.10a}$$

where κ is a constant scalar. This integrates to

$$\tau = \tau(\tau') = \kappa \int \mathrm{e}^{\int \psi\, \mathrm{d}\tau'}\, \mathrm{d}\tau'. \tag{4.10b}$$

Differentiating (4.10a), we get

$$\frac{\mathrm{d}^2\tau}{\mathrm{d}\tau'^2} = \psi\, \frac{\mathrm{d}\tau}{\mathrm{d}\tau'}. \tag{4.10c}$$

Substituting (4.10c) into (4.8b) and equating the result to (4.9) we determine that

$$\frac{\delta v}{\delta \tau} = v \cdot \nabla v = \frac{\delta^2 x}{\delta \tau^2} = 0. \tag{4.11}$$

Thus every geodesic possesses a *preferred parameter* for which the function $x = x(\tau)$ describing the curve satisfies Eqn. (4.11). Equation (4.11) is called the *geodesic equation*. Dotting (4.11) by v we determine that

$$v \cdot \frac{\delta v}{\delta \tau} = \frac{1}{2} \frac{\delta v^2}{\delta \tau} = \frac{1}{2} \frac{\mathrm{d}|v|^2}{\mathrm{d}\tau} = 0. \tag{4.12}$$

Thus, a preferred parameter assigns a constant speed $|v|$ to the geodesic.

Every vector field $v = v(x)$ satisfies the identity

$$v \cdot (\nabla \wedge v) = v \cdot \nabla v - \tfrac{1}{2} \nabla v^2. \tag{4.13}$$

Hence, a vector field $u = u(x)$ is everywhere tangent to a 'bundle of geodesics' if and only if there exists a scalar field $\psi = \psi(x)$ such that $v = \psi u$ satisfies

$$v \cdot \nabla v = v \cdot (\nabla \wedge v) + \tfrac{1}{2} \nabla v^2 = 0. \tag{4.14}$$

If condition (4.6) is satisfied for *every* curve in the region $\mathscr{R}$, then the vector field $a = a(x)$ satisfies the equation

$$\delta_v a = v \cdot \nabla a = \alpha a = v \cdot A a, \tag{4.15a}$$

where $v = v(x)$ is any vector field on $\mathscr{R}$ and we have used the fact that the 'scale factor' α must be a linear function of v and so can be written in the form $\alpha = v \cdot A$ where $A = A(x)$ is some vector field. Computing the commutator of second co-differentials from (4.15a) and recalling (4.4) we get

$$[\delta_u, \delta_v] a = (v \wedge u) \cdot (\nabla \wedge A) a = R(v \wedge u) \cdot a. \tag{4.15b}$$

The last term vanishes when dotted with a, so, since $v \wedge u$ is arbitrary, (4.15b) holds only if

$$\nabla \wedge A = 0, \tag{4.15c}$$

and thus

$$R(u \wedge v) \cdot a = 0. \tag{4.16a}$$

Since $[R(u \wedge v) \cdot a] \cdot b = R(u \wedge v) \cdot (a \wedge b) = (u \wedge v) \cdot R(a \wedge b)$, we can replace (4.16a) by the condition

$$R(a \wedge b) = 0, \tag{4.16b}$$

where b is any vector field.

Since $\nabla \wedge A = 0$, there exists a scalar field $\lambda = \lambda(x)$ such that

$$A = \nabla \log \lambda. \tag{4.17}$$

So Eqn. (4.15a) becomes

$$\delta_v a = v \cdot \nabla a = av \cdot \nabla \log \lambda. \tag{4.18a}$$

This equation can be simplified by introducing a field $b = \lambda a$ which satisfies

$$\delta_v b = v \cdot \nabla b = 0. \tag{4.18b}$$

A vector field $a = a(x)$ satisfying (4.18a) at every point x of a region $\mathscr{R}$ is said to be a *parallel field* on $\mathscr{R}$. From (4.15b) it can be shown that parallel displacement of the tangent vector $a(x)$ at a point x in $\mathscr{R}$ to a point y is independent of the curve along which it is displaced.

From (4.16) we can conclude that the curvature tensor of an m-dimensional manifold vanishes if and only if it admits $m - 1$ linearly independent parallel vector fields.

Manifolds $\mathscr{M}$ and $\mathscr{M}'$ are said to be *projectively related* if there exists a transformation f of $\mathscr{M}$ to $\mathscr{M}'$ which 'preserves' geodesics, that is, if

$$x'(\tau) = f(x(\tau)) \tag{4.19a}$$

is a geodesic on $\mathscr{M}'$ when $x(\tau)$ is a geodesic on $\mathscr{M}$. The velocities of corresponding curves on $\mathscr{M}$ and $\mathscr{M}'$ are related by the differential transformation $\underline{f}$, for, by differentiating (4.19a) we get

$$v' = \frac{\mathrm{d}x'}{\mathrm{d}\tau} = \frac{\mathrm{d}x}{\mathrm{d}\tau} \cdot \partial f = v \cdot \nabla f = \underline{f}(v). \tag{4.19b}$$

Differentiating again we have

$$\frac{\mathrm{d}v'}{\mathrm{d}\tau} = v' \cdot \partial' v' = \underline{f}_{v'}(v) + \underline{f}(v \cdot \partial v). \tag{4.19c}$$

Projecting this equation into the tangent algebra of $\mathscr{M}'$, we get

$$\frac{\delta v'}{\delta \tau} = \delta_{v'} v' = \delta_{v'} \underline{f}(v) + \underline{f}(\delta_v v), \tag{4.19d}$$

where $\delta_{v'} \underline{f}$ is the generalized codifferential defined by (3.7).

Equations (4.19) apply to the transformation of any curve. From (4.19b) and (4.19d) we get

$$v' \wedge \delta_{v'} v' = v' \wedge \delta_{v'} \underline{f}(v) + \underline{f}(v \wedge \delta_v v). \tag{4.20}$$

Invoking now the geodesic condition (4.7) on both manifolds, we conclude from (4.20) that the manifolds are projectively related by f if and only if

$$v' \wedge \delta_{v'} \underline{f}(v) = 0 \qquad \text{where } v' = \underline{f}(v). \tag{4.21}$$

Condition (4.21) holds for the transformation of any vector field $a = a(x)$ on $\mathcal{M}$ to a field $a' = \underline{f}(a)$ on $\mathcal{M}'$, because, at any given point x, $a(x)$ can be chosen tangent to a geodesic. This fact enables us to derive an expression for the codifferential $\delta_{a'}\underline{f}$. With the help of the integrability condition in the form

$$\delta_{a'}\underline{f}(b) = \delta_{b'}\underline{f}(a), \tag{4.22}$$

we determine that

$$\partial_{a'} \cdot [a' \wedge \delta_{a'}\underline{f}(a)] = \partial_{a'} \cdot [a' \wedge \delta_{a'}\underline{f}\underline{f}^{-1}(a')]$$
$$= (m+1)\,\delta_{a'}\underline{f}(a) - 2a'\,\partial_{a'} \cdot \underline{f}_{a'}(a) = 0,$$

where m is the dimension of the manifolds $\mathcal{M}$ and $\mathcal{M}'$. This result can be put in the form

$$\delta_{a'}\underline{f}(a) = 2\phi(a)a' = 2\underline{f}(\phi(a)a), \tag{4.23}$$

where, because of (4-5.35),

$$\phi(a) \equiv \frac{1}{m+1}\,\partial_{a'} \cdot \underline{f}_{a'}(a) = \frac{1}{m+1}\,a \cdot \nabla \log J_f = \frac{1}{m+1}\,a' \cdot \nabla' \log J_f, \tag{4.24}$$

and J_f is the Jacobian of the transformation f. Operating on (4.23) with $c' \cdot \partial_{a'}$ and using (4.22) we get the desired result

$$\delta_{a'}\underline{f}(c) = \underline{f}(\phi(c)a + \phi(a)c). \tag{4.25}$$

This enables us to evaluate the codifferential $\delta_{a'}\underline{f}$ algebraically from $\underline{f}$ and $\phi(a)$.

It is now a simple matter to compute the second codifferential of $\underline{f}$ and relate the curvature tensors on $\mathcal{M}$ and $\mathcal{M}'$. Codifferentiating (4.25), we get

$$\delta_{b'}\delta_{a'}\underline{f}(c) = \delta_{b'}\underline{f}(\phi(c)a + \phi(a)c) + \underline{f}(\phi_b(c)a + \phi_b(a)c)$$
$$= \underline{f}([\phi_b(c) + \phi(b)\phi(c)]a + 2\phi(a)\phi(c)b + [\phi_b(a) + \phi(b)\phi(a)]c). \tag{4.26}$$

From (4.24) we see that

$$\phi_b(a) = b \cdot \dot{\nabla}\dot{\phi}(a) = \phi_a(b). \tag{4.27}$$

Hence, from (4.26), we get for the commutator of codifferentials

$$[\delta_{a'}, \delta_{b'}]\underline{f}(c) = \underline{f}(\psi(a,c)b - \psi(b,c)a)$$
$$= \underline{f}((a \wedge b) \cdot \partial_v \psi(v,c)) = [\delta_a, \delta_b]\underline{f}(c), \tag{4.28}$$

where

$$\psi(a,c) \equiv \phi_a(c) - \phi(a)\phi(c). \tag{4.29}$$

Comparing (4.28) with (3.12), we get the fundamental relation

$$R'(a' \wedge b') \cdot c' - \underline{f}(R(a \wedge b) \cdot c) = \underline{f}((a \wedge b) \cdot \partial_v \psi(v, c)). \tag{4.30}$$

Since

$$\partial_a \cdot [R(a \wedge b) \cdot c] = [\partial_a \cdot R(a \wedge b)] \cdot c = R(b) \cdot c = b \cdot R(a),$$

and

$$\partial_a \cdot [(a \wedge b) \cdot \partial_v] = [\partial_a \cdot (a \wedge b)] \cdot \partial_v = (m-1) b \cdot \partial_v,$$

the contraction of (4.30) with $\partial_{a'}$ yields

$$b' \cdot R'(c') - b \cdot R(c) = (m-1)\psi(b, c). \tag{4.31}$$

This implies that

$$(a' \wedge b') \cdot R'(c') - \underline{f}((a \wedge b) \cdot R(c)) = (m-1)\underline{f}((a \wedge b) \cdot \partial_v \psi(v, c)). \tag{4.32}$$

Combining (4.32) with (4.30) we find

$$\begin{aligned} R'(a' \wedge b') \cdot c' &- \frac{1}{m-1}(a' \wedge b') \cdot R(c') \\ &= \underline{f}(R(a \wedge b) \cdot c - \frac{1}{m-1}(a \wedge b) \cdot R(c)). \end{aligned} \tag{4.33}$$

The tensor $R(a \wedge b) \cdot c - 1/(m-1)(a \wedge b) \cdot R(c)$ is commonly called the *Projective Weyl Tensor*. According to (4.33), its form is preserved by the differential of a projective transformation.

We can characterize projectively related curvatures in a different way. We write (4.30) in the form

$$R'(A') \cdot c' - \underline{f}(R(A) \cdot c) = \underline{f}(A \cdot \partial_v)\psi(v, c), \tag{4.34}$$

where $A' = \underline{f}(A)$ is a bivector field. Since $\underline{f}(A) \wedge \underline{f}(A \cdot \partial_v) = \underline{f}(A \wedge (A \cdot \partial_v)) = 0$, from (4.34), we get

$$A' \wedge (R'(A') \cdot c') = \underline{f}(A \wedge (R(A) \cdot c)). \tag{4.35}$$

Equation (4.35) is equivalent to Eqn. (4.33). And the trivector $A \wedge (R(A) \cdot c)$ is fully equivalent to the Projective Weyl Tensor, though, of course, it is not a tensor because it is a quadratic function of A; let us call it the Weyl trivector. Equivalence of the Weyl trivector to the Weyl tensor can be proved by carrying out the differentiation

$$\begin{aligned} (\partial_b \wedge \partial_a) &\cdot [a \wedge b \wedge (R(a \wedge b) \cdot c)] \\ &= m(m-1)R(a \wedge b) \cdot c - m(a \wedge b) \cdot R(c). \end{aligned} \tag{4.36}$$

The Weyl trivector has some advantages. For example, when $m = 2$, it vanishes identically; whence the projective Weyl tensor also vanishes.

As a simple but important application of the general results we have established, we determine the functional form of the curvature tensor for projectively flat manifolds. A manifold is said to be *flat* if its curvature tensor vanishes everywhere. A manifold is said to be *projectively flat* if it is related to a flat manifold by a projective transformation. From (4.33) it follows immediately that a manifold $\mathscr{M}$ is projectively flat if and only if its Weyl tensor vanishes identically. Thus, the curvature tensor satisfies the identity

$$(m-1)R(a \wedge b) \cdot c = (a \wedge b) \cdot R(c). \tag{4.37}$$

From (4.37) we easily derive

$$R(a \wedge b) = \frac{R}{m(m-1)}\, a \wedge b, \tag{4.38}$$

where R is the scalar curvature. The contraction of (4.38) is the Ricci tensor

$$R(a) = \frac{R}{m}\, a. \tag{4.39}$$

Using (1.23) we find $2\nabla R = m\nabla R$, which implies that the scalar curvature must be constant if $m > 2$. Hence we have proved (for $m > 2$) that a manifold is projectively flat if and only if it is a space of constant curvature. It can be proved that this result obtains for the case $m = 2$ as well.

5-5. Conformally Related Geometries

This section develops the general theory of changes in geometric quantities induced by conformal transformations. The method of Section 5-3 is used, and the special advantages of the spinor representation of rotations are fully exploited. This enables us to derive simple equations for the transformed curvature and related quantities.

A transformation f of a vector manifold $\mathscr{M}$ into a manifold $\mathscr{M}'$ is said to be *conformal* if its differential satisfies the relation

$$\underline{f}(a) \cdot \underline{f}(b) = \mathrm{e}^{2\phi} a \cdot b, \tag{5.1}$$

where $\phi = \phi(x)$ is a definite scalar function and $a = a(x)$ and $b = b(x)$ are any vector fields on $\mathscr{M}$. The manifold $\mathscr{M}$ is said to be *conformally related* to $\mathscr{M}'$ by f. If $\mathrm{e}^{\phi} \equiv 1$ in (5.1), then f is said to be an *isometric* transformation and $\mathscr{M}$ is *isometrically related* to $\mathscr{M}'$.

We will ultimately arrive at a more general conception of conformally related geometries by considering an extensor function h, as defined in Section 5-3, which satisfies

$$h(a) \cdot h(b) = \mathrm{e}^{2\phi} a \cdot b. \tag{5.2}$$

After examining the consequences of (5.2), we will consider the implications of assuming that h is also a differential $\underline{f}$.

Our analysis of isometries in Chapter 3 enables us to infer at once from (5.2) that h can be put in the canonical form

$$h(a) = \psi a \psi^\dagger = \mathrm{e}^{\phi} U a U^\dagger , \tag{5.3a}$$

where

$$\psi = \mathrm{e}^{\phi/2} U, \tag{5.3b}$$

and

$$UU^\dagger = 1. \tag{5.3c}$$

If $U = \langle U \rangle_-$, that is, if U is an odd multivector, then $UaU^\dagger$ is a reflection and we say that $h(a)$ is an *anticonformal* tensor. But we will confine our attention to the more interesting case when U is even, that is, when

$$U = \langle U \rangle_+ . \tag{5.3d}$$

Then, in the parlance of Section 3-8, $\psi = \psi(x)$ is a *spinor* function and we say that $h(a)$ is a *conformal* tensor.

Strictly speaking, instead of (5.3a) we should have written

$$h(a) = \psi P(a) \psi^\dagger \tag{5.4}$$

to insure that $h^{-1}h = P$, but as long as we understand that $a = P(a)$, (5.3a) is sufficient.

We extend the tensor $h(a)$ to an extensor by assuming the outermorphism property $h(a \wedge b) = h(a) \wedge h(b)$. Thus from (5.3a) we get

$$h(\langle A \rangle_r) = \mathrm{e}^{r\phi} U \langle A \rangle_r U^\dagger = \mathrm{e}^{r\phi} \langle UAU^\dagger \rangle_r \tag{5.5}$$

for any multivector field $A = A(x) = P(A)$. In particular, for the unit pseudoscalar $I = I(x)$ of the m-dimensional manifold $\mathscr{M}$, we have

$$h(I) = \mathrm{e}^{m\phi} I' , \tag{5.6a}$$

where

$$I' = UIU^\dagger \tag{5.6b}$$

is the pseudoscalar of $\mathscr{M}'$ if h is assumed to have its values in the tangent algebra of $\mathscr{M}'$. This is weaker than the assumption that $h = \underline{f}$.

The canonical form (5.3a) enables us to compute the codifferentials of h from differentials of the spinor ψ, and this makes the special properties of conformal transformations explicit. We exploit the fact that the differential $a \cdot \partial U$ of the spinor U in (5.3) can be written in the general form

$$a \cdot \partial U = \tfrac{1}{2} U B_a = \tfrac{1}{2} B'_a U , \tag{5.7}$$

where B_a and $B_a' = UB_aU^\dagger$ *must be bivectors* to be consistent with the constraint (5.3c). We work with B_a, but occasionally mention results in terms of B_a'. According to (5.3b) and (5.7), the differential $\psi_a = a \cdot \partial\psi$ of the spinor can be written

$$\psi_a = \tfrac{1}{2}\psi(\phi_a + B_a) = \tfrac{1}{2}(\phi_a + B_a')\psi, \tag{5.8a}$$

where $\phi_a = \delta_a\phi = a \cdot \partial\phi$. Since $B_a^\dagger = -B_a$, the reverse of (5.8a) is

$$\psi_a^\dagger = \tfrac{1}{2}(\phi_a - B_a)\psi^\dagger = \tfrac{1}{2}\psi^\dagger(\phi_a - B_a'). \tag{5.8b}$$

For the differential of $h(c) = \psi c \psi^\dagger$, we obtain

$$h_a(c) = \psi_a c\psi^\dagger + \psi c\psi_a^\dagger = \psi(B_a \cdot c + c\phi_a)\psi^\dagger. \tag{5.9}$$

So for the codifferential of h, as defined by (3.7), we have

$$\begin{aligned}\delta_a h(c) &= P'h_aP(c) = P'(\psi(B_a \cdot c + c\phi_a)\psi^\dagger)\\ &= \psi P(B_a \cdot c + c\phi_a)\psi^\dagger.\end{aligned} \tag{5.10}$$

This assumes the simple form

$$\delta_a h(c) = h(\Omega_a \cdot c + c\phi_a) = \Omega_a' \cdot h(c) + h(c)\phi_a \tag{5.11}$$

if we introduce

$$\Omega_a \equiv P(B_a) = U^\dagger \Omega_a' U. \tag{5.12}$$

From (5.11), we readily compute higher codifferentials of h; thus

$$\begin{aligned}\delta_b\,\delta_a h(c) &= h((\delta_b\Omega_a) \cdot c + c\phi_{ab}) + \delta_b h(\Omega_a \cdot c + c\phi_a)\\ &= h((\delta_b\Omega_a) \cdot c + \Omega_b \cdot (\Omega_a \cdot c) + \Omega_a \cdot c\phi_b +\\ &\quad + \Omega_b \cdot c\phi_a + c(\rho_{ab} + \rho_a\rho_b)).\end{aligned} \tag{5.13}$$

Using the Jacobi identity

$$\Omega_a \cdot (\Omega_b \cdot c) - \Omega_b \cdot (\Omega_a \cdot c) = (\Omega_a \times \Omega_b) \cdot c,$$

we find for the commutator of codifferentials

$$[\delta_a, \delta_b]h(c) = h(C(a \wedge b) \cdot c) = C'(a \wedge b) \cdot h(c), \tag{5.14a}$$

where

$$C(a \wedge b) \equiv \delta_a\Omega_b - \delta_b\Omega_a + \Omega_a \times \Omega_b, \tag{5.14b}$$

$$C'(a \wedge b) \equiv \delta_a\Omega_b' + \delta_b\Omega_a' - \Omega_a' \times \Omega_b'. \tag{5.14c}$$

Recalling (3.11), we get from (5.14a)

$$R'(a' \wedge b') \cdot h(c) - h(R(a \wedge b) \cdot c) = h(C(a \wedge b) \cdot c) = C'(a' \wedge b') \cdot h(c), \quad (5.15)$$

where $a' = \underline{f}(a)$, $b' = \underline{f}(b)$ and the primes on a and b in the last term are justified by our discussion in Section 5-3. Observing that, for any bivector field $B = P(B)$ and vector field $c = P(c)$, we have

$$h(B \cdot c) = \mathrm{e}^{\phi} U(B \cdot c) U^{\dagger} = \mathrm{e}^{\phi}(UBU^{\dagger}) \cdot (UcU^{\dagger})$$
$$= (UBU^{\dagger}) \cdot h(c) = \mathrm{e}^{-2\phi} h(B) \cdot h(c),$$

we are able to replace (5.15) by the simpler relation

$$R'(a' \wedge b') = U(R(a \wedge b) + C(a \wedge b))U^{\dagger} = \mathrm{e}^{-2\phi} h(R(a \wedge b) + C(a \wedge b)), \quad (5.16a)$$

or

$$R'(a' \wedge b') - C'(a' \wedge b') = \mathrm{e}^{-2\phi} h(R(a \wedge b)). \quad (5.16b)$$

This is the general equation relating the curvature tensors of conformally related manifolds.

It may be of interest to note that if we define a codifferential of U by

$$\delta_a U \equiv \tfrac{1}{2} U \Omega_a = \tfrac{1}{2} \Omega'_a U, \quad (5.17a)$$

so that

$$\delta_a \psi = \tfrac{1}{2} \psi(\phi_a + \Omega_a) = \tfrac{1}{2}(\phi_a + \Omega'_a)\psi, \quad (5.17b)$$

then

$$[\delta_a, \delta_b]\psi = \psi \Omega_{ab} = \Omega'_{ab}\psi, \quad (5.18a)$$

and this is related to (5.14a) by

$$[\delta_a, \delta_b] h(c) = [\delta_a, \delta_b]\psi c \psi^{\dagger} + \psi c [\delta_a, \delta_b]\psi^{\dagger}. \quad (5.18b)$$

We now consider the consequences of assuming that $h = f$. In the first place, the integrability condition $h_a(b) = \underline{f}_a(b) = \underline{f}_b(a)$ implies, by (5.11), that

$$\Omega_a \cdot b + b\phi_a = \Omega_b \cdot a + a\phi_b, \quad (5.19a)$$

or

$$\Omega_a \cdot b - \Omega_b \cdot a = (b \wedge a) \cdot \partial\phi. \quad (5.19b)$$

On the other hand, from (5.3) the adjoint transformation is seen to have the canonical form

$$\bar{f}(a') = \psi^{\dagger} a' \psi, \quad (5.20)$$

and its differential is

$$\delta_b \bar{f}(a') = \bar{f}(a') \cdot \Omega_b + \bar{f}(a')\phi_b. \tag{5.21}$$

But, according to (4-5.22),

$$\partial_b \wedge \delta_b \bar{f}(a') = 0.$$

Hence, from (5.11) we get

$$\partial_b \wedge (a \cdot \Omega_b) = a \wedge \partial\phi. \tag{5.22}$$

From (5.19a) we get, using $\partial_b \wedge b = 0$ and $\partial_b \wedge (\Omega_a \cdot b) = -2\Omega_a$,

$$2\Omega_a = \partial_b \wedge (a \cdot \Omega_b) + a \wedge \partial\phi. \tag{5.23}$$

Finally, substituting (5.22) into (5.23), we get the remarkably simple result

$$\Omega_a = a \wedge \partial\phi. \tag{5.24}$$

If ϕ is constant, that is, if $\partial\phi = 0$, then $\Omega_a = 0$ by (5.24), so $C(a \wedge b) = 0$ by (5.14b), and, by (5.16),

$$R'(\underline{f}(a) \wedge \underline{f}(b)) = \underline{f}(R(a \wedge b)). \tag{5.25}$$

This is the equation relating the curvature tensors of isometrically related manifolds.
We also note that $\partial\phi = 0$ in (5.24) and (5.11) implies that

$$\delta_a \underline{f}(b) = 0. \tag{5.26}$$

Hence the condition (4.21) for a projective transformation is trivially satisfied, and we may conclude that every isometric transformation is projective. The metric on $\mathscr{M}'$ defines a metric tensor $g(a, b)$ on $\mathscr{M}$ according to

$$g(a, b) \equiv \underline{f}(a) \cdot \underline{f}(b) = a' \cdot b'. \tag{5.27}$$

Since g is scalar valued its coderivative is equal to its differential; thus,

$$\begin{aligned} \delta_c g(a, b) &= c \cdot \dot{\partial}\dot{g}(a, b) = \underline{f}_c(a) \cdot \underline{f}(b) + \underline{f}(a) \cdot \underline{f}_c(b) \\ &= \delta_c \underline{f}(a) \cdot \underline{f}(b) + \underline{f}(a) \cdot \delta_c \underline{f}(b), \end{aligned} \tag{5.28}$$

the last equality following from the definition (3.7) for the codifferential. Hence (5.26) is equivalent to $\delta_c g(a, b) = 0$; this is another way to express the condition that a transformation f be isometric.

The simplest significant application of (5.26) is to determine the isometric transformations of a flat manifold into itself. In this case, the codifferentials of f are equal to its differentials, so (5.26) becomes

$$\underline{f}_a(b) = 0. \tag{5.29a}$$

Since $\underline{f}(b) = UbU^\dagger$, (5.29a) can be reduced to the equation

$$a \cdot \partial U = 0, \tag{5.29b}$$

hence U is constant. But

$$\underline{f}(b) = b \cdot \partial f = UbU^\dagger \tag{5.30}$$

is a differential equation for f with the general solution

$$x' = f(x) = UxU^\dagger + c, \tag{5.31}$$

where $c = P(c)$ is a constant vector. Thus, every isometric transformation of a flat manifold into itself can be expressed as a rotation about the origin followed by a translation, as exhibited explicitly by (5.31).

Now we evaluate the projective change in curvature $C(a \wedge b)$ and study its properties. For this purpose, it is convenient to write (5.24) in the form

$$\Omega_a = a \wedge w, \tag{5.32a}$$

where the vector field w satisifes

$$\nabla \wedge w = 0, \tag{5.32b}$$

which follows from $w = \partial\phi = \nabla\phi$. The differential of Ω_a is

$$\delta_b \Omega_a = a \wedge (\delta_b w), \tag{5.33}$$

where, of course, $\delta_b w = a \cdot \nabla w$. We find that the tensor $\delta_b \Omega_a$ has the contractions

$$\partial_b\, \delta_b \Omega_a = \dot{\nabla}\dot{\Omega}_a = \dot{\nabla} a \wedge \dot{w} = \delta_a w - a\nabla \cdot w, \tag{5.34a}$$

$$\partial_a\, \delta_b \Omega_a = \partial_a a \wedge (\delta_b w) = (m-1)\delta_b w, \tag{5.34b}$$

$$\partial_a\, \partial_b\, \delta_b \Omega_a = -(m-1)\nabla \cdot w. \tag{5.34c}$$

With the help of (1-1.69) we see that

$$\begin{aligned}\Omega_a \times \Omega_b &= (a \wedge w) \times (b \wedge w) = w \cdot (a \wedge w \wedge b) \\ &= w \cdot aw \wedge b - w^2 a \wedge b + w \cdot ba \wedge w.\end{aligned} \tag{5.35}$$

This tensor has the contractions

$$\partial_a \Omega_a \times \Omega_b = (w \wedge \partial_a) \cdot (a \wedge b \wedge w) = (m-2) w \cdot (b \wedge w), \tag{5.36a}$$

$$\partial_b\, \partial_a \Omega_a \times \Omega_b = -(m-w)(w \wedge \partial_b) \cdot (b \wedge w) = -(m-2)(m-1)w^2. \tag{5.36b}$$

Substituting (5.33) and (5.35) into (5.14b), we get

$$C(a \wedge b) = b \wedge \delta_a w - a \wedge \delta_b w + w \cdot (a \wedge w \wedge b). \tag{5.37}$$

With the help of (5.34) and (5.36) we determine the contractions

$$C(b) \equiv \partial_a C(a \wedge b) = -(m-2)\,(\delta_b w + (b \wedge w) \cdot w) - b\nabla \cdot w, \tag{5.38a}$$

$$C \equiv \partial_b C(b) = -(m-1)\,(2\nabla \cdot w + (m-2)w^2). \tag{5.38b}$$

Comparing (5.38a, b) with (5.37), we find that

$$C(a \wedge b) = \frac{a \wedge C(b) - b \wedge C(a)}{m-2} - \frac{a \wedge bC}{(m-1)\,(m-2)}. \tag{5.39}$$

Thus, $C(a \wedge b)$ is completely determined by its contraction $C(a)$.

We now have a general apparatus sufficient to determine all conformal transformations of flat manifolds. The group of all conformal transformations of a flat manifold into itself is called the *special conformal group* of the manifold. We are concerned here with manifolds of dimension $m \geqslant 3$, and our results hold for any signature. The case $m = 2$ is discussed in Section 4-7.

A flat manifold is characterized by a vanishing curvature tensor. So, from Eqn. (5.16) we find immediately that $C(a \wedge b) = 0$ is the condition for a special conformal transformation. From Eqn. (5.38a) we get the simpler condition $C(b) = 0$ and the differential equation

$$\delta_b w + (b \wedge w) \cdot w + \frac{b\nabla \cdot w}{(m-2)} = 0.$$

From Eqn. (5.38b) we get $C = 0$ and

$$2\nabla \cdot w + (m-2)w^2 = 0.$$

Eliminating $\nabla \cdot w$ from these equations, we get

$$\delta_b w = b \cdot ww - \tfrac{1}{2} w^2 b,$$

or

$$\delta_b w = b \cdot \nabla w = \tfrac{1}{2} wbw. \tag{5.40}$$

Our problem now is to solve this differential equation for $w = \nabla\phi$. We have already considered the case $\phi = 0$ and found the general form (5.31) for the corresponding isometric transformations. It is also trivial to show that the dilatations $x \to e^{\alpha}x$ correspond to constant $\phi = \alpha$. By direct differentiation we can verify that the nontrivial solution to (5.40) is

$$w = \nabla\phi = 2c(1 - xc)^{-1} = 2(1 - cx)^{-1}c = \frac{2(1 - xc)}{\sigma(x)}, \tag{5.41}$$

where c is a constant vector and

$$\sigma(x) \equiv (1 - xc)\,(1 - cx) = 1 - 2x \cdot c + c^2x^2 = e^{-\phi}. \tag{5.42}$$

The verification is easy when we have the following important formula at our disposal:

$$b \cdot \nabla (1 - xc)^{-1} = (1 - xc)^{-1} b (1 - xc)^{-1}. \tag{5.43}$$

Note that the factor bc does not commute with the other factors except in the case $m = 2$, when all vectors are in the same plane. The formula (5.43) is most easily derived by using (5.41) and (5.42) as follows

$$b \cdot \nabla (1 - xc)^{-1} = b \cdot \dot{\nabla} \left(\frac{1 - \dot{c}x}{\sigma} \right)$$

$$= -\frac{cb}{\sigma} - (1 - cx) \frac{[-b(1 - cx) - (1 - xc)cb]}{\sigma^2}$$

$$= \frac{(1 - cx)bc(1 - cx)}{\sigma^2}.$$

The remaining steps in establishing (5.41) are left to the reader.

Our next problem is to determine the differential of the special conformal transformation from the equation

$$\underline{f}(a) = \psi a \psi^\dagger .$$

According to (5.8), (5.12) and (5.24), the spinor ψ satisfies the differential equation

$$a \cdot \nabla \psi = \tfrac{1}{2} \psi a \nabla \phi. \tag{5.44}$$

For $\nabla\phi$ given by (5.41) this becomes the specific equation

$$a \cdot \nabla \psi = \psi ac (1 - xc)^{-1}.$$

Comparing this with (5.43), we see immediately that it has the particular solution

$$\psi = (1 - xc)^{-1} = \frac{1 - xc}{\sigma}. \tag{5.45}$$

The general solution is obtained by multiplying (5.45) on the left by a constant spinor, but this has already been accounted for in our discussion of transformations with constant ϕ. From (5.45) we get the differential

$$\underline{f}(a) = a \cdot \nabla f(x) = (1 - xc)^{-1} a (1 - xc)^{-1}. \tag{5.46}$$

Equation (5.46) is a differential equation for $f(x)$ with solution

$$f(x) = x(1 - cx)^{-1} = (1 - xc)^{-1} x, \tag{5.47}$$

where an additive constant describing a translation has been omitted. To verify (5.47) we differentiate as follows

$$a\cdot\nabla[x(1-cx)^{-1}] = a(1-cx)^{-1} + x(1-cx)^{-1}ca(1-cx)^{-1}$$
$$= (1-xc)^{-1}\,[(1-xc)a + xca]\,(1-cx)^{-1}$$
$$= (1-xc)^{-1}a(1-cs)^{-1}.$$

The special conformal transformation (5.47) can be further analyzed into the composite of an *inversion*

$$x \to -x^{-1} = -\frac{1}{x} \tag{5.48}$$

followed by a translation and another inversion. Thus,

$$f(x) = \frac{-1}{-x^{-1}+c} = \frac{1}{(1-cx)x^{-1}} = x\,\frac{1}{1-cx}. \tag{5.49}$$

We can summarize our results with the conclusion that any transformation in the special conformal group can be expressed in the form

$$x \to f(x) = \psi x(1-cx)^{-1}\psi^{\dagger} + b, \tag{5.50}$$

where $\psi = \mathrm{e}^{\alpha/2}U$ is a constant spinor. This is a composite of a *special conformal transformation*

$$x \to x(1-cx)^{-1}, \tag{5.51a}$$

a *rotation* about the origin

$$x \to UxU^{\dagger}, \tag{5.51b}$$

a *dilatation*

$$x \to \mathrm{e}^{\alpha}x, \tag{5.51c}$$

and a *translation*

$$x \to x + b. \tag{5.51d}$$

Next in generality to the problem we have just solved is the problem of determining the conformal transformations from a flat space to a curved space. The most important transformation of this kind is the stereographic projection, which we have already found to be given by the explicit Eqn. (4-6.31). We will not study the extent to which this exhausts the possibilities. But we will use it in Section 8-3 to prove that the special conformal group is isomorphic to a rotation group in higher dimensions.

For general applications to Riemann manifolds it is convenient to have conformally invariant relations among the curvature tensors. For $h = \underline{f}$, Eqn. (5.16a) can be written

$$\begin{aligned} e^{2\phi}R'(a' \wedge b'') &= \underline{f}(R(a \wedge b) + C(a \wedge b)) \\ &= e^{2\phi}U(R(a \wedge b) + C(a \wedge b))U^{\dagger}. \end{aligned} \tag{5.52}$$

To help us derive the corresponding equation relating the Ricci curvature tensors $R'(b') = \partial_{a'}R(a' \wedge b')$ and $R(b) = \partial_b(R(a \wedge b)$, we use (5.3) to invert the general equation $\partial_a = \underline{f}(\partial_{a'})$, obtaining

$$\partial_{a'} = e^{-2\phi}\underline{f}(\partial_a) = e^{-\phi}U\,\partial_a U^{\dagger}. \tag{5.53}$$

Using (5.30) to contract (5.52), we get

$$\begin{aligned} e^{2\phi}R'(b') &= \underline{f}(R(b) + C(b)) \\ &= e^{\phi}U(R(b) + C(b))\,U^{\dagger}. \end{aligned} \tag{5.54}$$

And, contracting (5.54) in the same way, we find that the scalar curvature $R = \partial_b R(b)$ on $\mathscr{M}$ is related to the scalar curvature R' on $\mathscr{M}'$ by

$$e^{2\phi}R' = R + C. \tag{5.55}$$

Now, considering (5.39), we are led to introduce a new tensor

$$W(a \wedge b) = R(a \wedge b) + \frac{b \wedge R(a) - a \wedge R(b)}{m-2} + \frac{a \wedge bR}{(m-2)(m-1)}. \tag{5.56}$$

By combining (5.52), (5.54), (5.55) and using (5.39) we easily prove that

$$e^{2\phi}W'(a' \wedge b') = \underline{f}(W(a \wedge b)) = e^{2\phi}UW(a \wedge b)U^{\dagger}. \tag{5.57}$$

By dotting (5.57) with $c' = \underline{f}(c) = e^{\phi}UcU^{\dagger}$ we get

$$W'(a' \wedge b') \cdot c' = \underline{f}(W(a \wedge b) \cdot c) = e^{\phi}UW(a \wedge b) \cdot cU^{\dagger}. \tag{5.58}$$

The tensor $W(a \wedge b) \cdot c$ is commonly called the *Conformal Weyl Tensor*; we use the same name for the equivalent tensor $W(a \wedge b)$.

Recalling our analysis of linear multivector functions in Section 3-9, we recognize (5.56) as the equation for the tractionless part $W(a \wedge b)$ of the tensor $R(a \wedge b)$; thus,

$$\partial_a W(a \wedge b) = 0. \tag{5.59}$$

Examining (5.57), we note another important property of the Weyl tensor, namely that its magnitude is preserved by conformal transformations, that is,

$$[W'(a' \wedge b')]^2 = [W(a \wedge b)]^2. \tag{5.60}$$

The Weyl tensor has important applications in the theory of gravitation.

5-6. Induced Geometries

The intrinsic geometry of vector manifolds developed in preceding sections is just the conventional Riemannian geometry expressed without coordinates. This section shows how our method can be used to formulate *affine geometry*.

Affine geometry generalizes Riemannian geometry. However, we prefer not to regard affine manifolds as essentially different from Riemannian manifolds. Instead, we develop affine geometry by enlarging the class of differential operators defined on a given vector manifold. This approach helps us compare the Riemannian geometries of related manifolds by showing us how to express them as different affine geometries on a single manifold. Conversely, our approach clearly exhibits the necessary conditions for an affine geometry on one manifold to be equivalent to a Riemannian geometry on another manifold.

In preceding sections we expressed the intrinsic (Riemannian) geometry of a vector manifold in terms of properties of the directional coderivative $a \cdot \nabla$. We can define $a \cdot \nabla$ by the operator equation

$$a \cdot \nabla = Pa \cdot \partial P, \tag{6.1}$$

where P and $a \cdot \partial$ are defined as before. Equation (6.1) suggests that we generalize $a \cdot \nabla$ by replacing the projection operator P in (6.1) by a more general extensor function.

Our approach here is a variation of our approach in Section 5-3. We begin with an *invertible extensor function* $h(a) = h(x, a(x))$ defined on the tangent algebra of the manifold $\mathcal{M}$. We assume that h is an *outermorphism*, though it will suffice for us to consider the action of h on vector fields. Since h is invertible, we have

$$h^{-1}h = P. \tag{6.2a}$$

We can define a new projection operator P' by the equation

$$P' \equiv hh^{-1}. \tag{6.2b}$$

We do not assume here, as we did in Section 5-3, that P' is necessarily associated with some manifold $\mathcal{M}'$. We can express (6.2a, b) in the alternative forms

$$h = hP = P'h = P'hP, \tag{6.3a}$$

$$Ph^{-1} = h^{-1}P' = Ph^{-1}P'. \tag{6.3b}$$

We define the codifferential of h as in Section 5-3:

$$\delta_a h = P'h_a P, \tag{6.4}$$

where $h_a = a \cdot \dot{\partial}\dot{h}$ is the differential of h. By differentiating (6.2a) we can get $\delta_a h^{-1}$ from $\delta_a h$; since $\delta_a P = PP_a P = 0$, we have

$$(\delta_a h^{-1})h + h^{-1}\,\delta_a h = 0$$

or

$$\delta_a h^{-1} = -h^{-1}(\delta_a h)h^{-1} = -h^{-1}h_a h^{-1}. \tag{6.5}$$

Now we are prepared to generalize (6.1) by introducing an operator

$$a \cdot D \equiv h^{-1} a \cdot \partial h \tag{6.6}$$

to be applied to multivector fields on $\mathscr{M}$. Using (6.1) and (6.2a) we can write (6.6) in the form

$$a \cdot D = a \cdot \nabla + L_a, \tag{6.7}$$

where L_a is a linear operator defined by

$$L_a \equiv h^{-1} h_a P = h^{-1}\, \delta_a h. \tag{6.8}$$

The operator P appears in (6.8) because $a \cdot D$ is understood to act on fields only. In standard terminology, the operator $a \cdot D$ is called an *affine connexion* or simply a *connexion* on $\mathscr{M}$.*

We have seen how Riemannian geometry is determined by properties of the operator $a \cdot \nabla$. Similarly, $a \cdot D$ determines an *affine geometry* for the manifold $\mathscr{M}$. Since $a \cdot D$ is determined by h according to (6.6), we say that the connexion $a \cdot D$ and its corresponding affine geometry are *induced* by h. Furthermore, if $h = \underline{f}$ is the differential outermorphism of a transformation f of the manifold $\mathscr{M}$ into a manifold $\mathscr{M}'$, we say that $a \cdot D$ and its geometry are *induced* by f. It will be clear that the affine geometry *induced* on $\mathscr{M}$ by f is equivalent to the Riemannian geometry on $\mathscr{M}'$ *related* to the Riemannian geometry on $\mathscr{M}$ by f in the sense of Section 5-3.

We call the operator L_a an *affine extensor*. By studying its properties, we learn important things about induced geometries. According to (6.7), L_a can be expressed as the difference of the two connexions $a \cdot D$ and $a \cdot \nabla$, and this is the way it generally arises in the literature, but we determine the properties of L_a from (6.8). Since h is an outermorphism, $h(\phi) = \phi$ if ϕ is a scalar field; hence

$$L_a(\phi) = 0, \tag{6.9a}$$

and

$$a \cdot D\phi = a \cdot \nabla\phi = a \cdot \partial\phi. \tag{6.9b}$$

It also follows from the outermorphism property of h that

$$L_a(b \wedge c) = L_a(b) \wedge c + b \wedge L_a(c), \tag{6.10}$$

where b and c are vector fields. So the properties of L_a are determined by its action on vector fields.

* The terminology and approach in this section may be compared with Chapter 5 in Ref. [Hi].

When L_a operates on vector fields, we call it an *affine tensor*. The affine tensor $L_a(b)$ can be decomposed into the form

$$L_a(b) = \Gamma_a(b) + T_a(b), \tag{6.11a}$$

where

$$\Gamma_a(b) = \tfrac{1}{2}(L_a(b) + L_b(a)), \tag{6.11b}$$

$$T_a(b) = \tfrac{1}{2}(L_a(b) - L_b(a)). \tag{6.11c}$$

The skewsymmetric tensor $T_a(b) = -T_b(a)$ is called the *torsion tensor*. Using (6.7), we can express it in terms of the affine connexion; thus

$$2T_a(b) = a \cdot Db - b \cdot Da - [a, b]. \tag{6.12}$$

Recall that the Lie bracket can be written $[a, b] = a \cdot \partial b - b \cdot \partial a$, so (6.12) does not actually involve coderivatives. On the other hand, with (6.8) we can express torsion in the form

$$T_a(b) = \tfrac{1}{2}h^{-1}(h_a(b) - h_b(a)). \tag{6.13}$$

If $h = \underline{f}$ is the differential of a transformation f, then the integrability condition implies the symmetry $h_a(b) = \underline{f}_a(b) = \underline{f}_b(a) = h_b(a)$, and $T_a(b)$ vanishes by (6.13). Thus, every affine geometry induced by a transformation has vanishing torsion.

The extensor h determines a symmetric extensor *field* g on $\mathscr{M}$ defined by

$$g = \bar{h}h, \tag{6.14}$$

where $\bar{h}$ is the adjoint of h. The symmetry of g is evident from

$$a \cdot g(b) = a \cdot \bar{h}h(b) = h(a) \cdot h(b) = g(a) \cdot b. \tag{6.15}$$

Thus $a \cdot g(b)$ can be regarded as a metric tensor on $\mathscr{M}$. We say that g is the metric *induced* by h, or by f if $h = \underline{f}$.

By differentiating (6.14) and using (6.8) with (6.2a), we easily derive

$$\delta_a g = \bar{L}_a g + g L_a, \tag{6.16}$$

where $\bar{L}_a$ is the adjoint of L_a. From (6.16) we can solve for the symmetric affine tensor $\Gamma_a(b)$ in terms of the metric tensor g and its codifferential. From (6.16) we get

$$\delta_a g(b) = \bar{L}_a g(b) + g L_a(b), \tag{6.17a}$$

$$\delta_b g(a) = \bar{L}_b g(a) + g L_b(a), \tag{6.17b}$$

$$\dot{\partial} a \cdot \dot{g}(b) = \partial_c a \cdot g_c(b) = \partial_c a \cdot \delta_c g(b) = \bar{L}_a g(b) + \bar{L}_b g(a). \tag{6.17c}$$

Adding (6.17a) to (6.17b) and subtracting (6.17c), we get

$$g(L_a(b) + L_b(a)) = \delta_a g(b) + \delta_b g(a) - \dot{\partial} a \cdot \dot{g}(c).$$

So from (6.11b) we get

$$\Gamma_a(b) = \tfrac{1}{2} g^{-1}(\delta_a g(b) + \delta_b g(a) - \dot{\partial} a \cdot \dot{g}(b))$$

$$= \tfrac{1}{2} g^{-1}(g_a(b) + g_b(a) - \partial_c a \cdot g_c(b)). \tag{6.18}$$

This should be compared with the expression for $\Gamma_a(b)$ in terms of h,

$$\Gamma_a(b) = \tfrac{1}{2} h^{-1}(h_a(b) + h_b(a)). \tag{6.19}$$

In the special case that $h = \underline{f}$ and f is a transformation of a *flat* manifold $\mathscr{M}$ to a manifold $\mathscr{M}'$, Eqn. (6.18) reduces to an expression for the 'Christoffel symbols' of classical Riemannian geometry (a manifold is said to be flat if its Riemann curvature vanishes). This relation can be regarded as a specification of a coordinate system for $\mathscr{M}'$; it represents the intrinsic geometry of $\mathscr{M}'$ as an induced geometry of the flat manifold $\mathscr{M}$. Of course, there are many ways to supply $\mathscr{M}'$ with a coordinate system, that is, there are many ways to transform a flat manifold into $\mathscr{M}'$, so there are many different representations of the intrinsic geometry of $\mathscr{M}'$ by a metric and Christoffel symbols on flat space.

Recall that the Riemann curvature was introduced in Section 5-1 as a measure of the degree of commutivity of codifferentials; specifically,

$$[\delta_a, \delta_b]c = [a \cdot \nabla, b \cdot \nabla]c - [a, b] \cdot \nabla c = R(a \wedge b) \cdot c. \tag{6.20}$$

The Lie bracket $[a, b]$ appears in (6.20) so the commutator of derivatives will result in a tensor. Similarly, we define the *affine curvature tensor* $L(a \wedge b, c)$ by

$$L(a \wedge b, c) \equiv [a \cdot D, b \cdot D]c - [a, b] \cdot Dc. \tag{6.21}$$

Using (6.7) and (6.20), we find that the affine curvature is related to the Riemann curvature by

$$L(a \wedge b, c) = R(a \wedge b) \cdot c + (\delta_a L_b - \delta_b L_a + [L_a, L_b])(c). \tag{6.22}$$

This shows that $L(a \wedge b, c)$ is a tensor field, since the right-hand side is obviously a linear function of vector fields a, b, c.

If we define the *affine differential* d_a by the operator equation

$$\mathrm{d}_a = \delta_a + L_a, \tag{6.23}$$

then the significant feature of (6.22) can be expressed by the operator equation

$$[\mathrm{d}_a, \mathrm{d}_b] = [\delta_a, \delta_b] + \delta_a L_b - \delta_b L_a + [L_a, L_b]. \tag{6.24}$$

At any rate, the change in curvature induced by the extensor h is given by the operator $\delta_a L_b - \delta_b L_a + [L_a, L_b]$. Let us express this in terms of h itself. Differentiating (6.8) we get

$$\delta_b L_a = h^{-1}\,\delta_b\,\delta_a h + (\delta_b h^{-1})\,\delta_a h,$$

and using (6.5) we find

$$L_b L_a = h^{-1}(\delta_b h)h^{-1}\,\delta_a h = -(\delta_b h^{-1})\,\delta_a h.$$

Hence,

$$\delta_b L_a + L_b L_a = h^{-1}\delta_b\,\delta_a h. \tag{6.25}$$

So the desired result is

$$\delta_a L_b - \delta_b L_a + [L_a, L_b] = h^{-1}[\delta_a, \delta_b]h. \tag{6.26}$$

This makes it easy to compare the results of the present section with those of Section 5-3. In particular, for $h = \underline{f}$ Eqn. (6.24) is seen to be equivalent to (3.11), the commutator of affine differentials $[\mathrm{d}_a, \mathrm{d}_b]$ on $\mathscr{M}$ corresponding to the commutator of codifferentials $[\delta_{a'}, \delta_{b'}]$ on $\mathscr{M}'$, while the operator (6.26) describing the induced change in curvature is obviously equivalent to the operator (3.10).

The results of Sections 5-4 and 5-5 concerning the geometries of projectively and conformally related manifolds can now be easily re-expressed as features of induced affine geometries on a single manifold. For example, comparison of (6.8) with (5.11) shows that an induced conformal geometry is characterized by

$$L_a(c) = h^{-1}\,\delta_a h(c) = \Omega_a \cdot c + c\phi_a. \tag{6.27}$$

And comparison of (6.26) with (5.14a) shows that

$$\delta_a L_b(c) - \delta_b L_a(c) + [L_a, L_b]c = C(a \wedge b)\cdot c, \tag{6.28}$$

where $C(a \wedge b) \equiv \delta_a \Omega_b - \delta_b \Omega_a + \Omega_a \times \Omega_b$.

Chapter 6

The Method of Mobiles

This chapter develops an efficient method for expressing the intrinsic geometry of a manifold in terms of local properties of vector fields. The method is actually a special case of the theory in Section 5-6, but we develop it *ab initio* here to make its relation to the classical method of tensor analysis as direct and clear as possible. This chapter does not depend on results of Chapters 4 and 5, though it does presume the basic properties of a vector manifold and the definitions and notations for differentials and codifferentials established in Sections 4-1 and 4-3, and a couple of results from Chapter 4 are used without taking the trouble to rederive them by the method of this chapter.

A *mobile* is a frame of orthonormal vector fields smoothly attached to a manifold. The method of mobiles has significant advantages in physical applications primarily because a frame can be directly associated with a rigid body. The formalism in this chapter is designed to be immediately applicable to Einstein's geometrical theory of gravitation, but specific physical applications must be considered elsewhere.

Nothwithstanding its direct applications to physics, the method of mobiles suffers the drawbacks of requiring reference to a coordinate system and of completely overlooking relations between intrinsic and extrinsic geometry. These drawbacks are necessary concomitants of a theory that deals exclusively with quantities in the tangent algebra. They were avoided in the general theory of the last two chapters by working outside the tangent algebra. Section 6-2 shows how the two approaches are related.

Our method should be compared with Cartan's formulation of differential geometry in terms of differential forms, which has enjoyed increasing popularity in recent years. Section 6-4 shows that the entire calculus of differential forms is included in Geometric Calculus. This perspective reveals that the calculus of forms is subject to unnecessary limitations which are removed by adopting the more general Geometric Calculus.

6-1. Frames and Coordinates

Let $\mathscr{M}$ be an m-dimensional vector manifold as defined in Section 4-1. The signature of a vector space was defined in Section 1-5. We say that the manifold $\mathscr{M}$ has *signature* $(m-q, q)$ if the tangent space at each point of $\mathscr{M}$ has signature $(m-q, q)$. To make the results of this chapter immediately applicable to the 'spacetime manifold' of physics, which has signature $(1,3)$, it is desirable to take account of signature from the beginning. However, it will be seen that signature has little influence on the form of the general theory.

This section expresses local properties of $\mathscr{M}$ in terms of frames, coordinates and their derivatives. After such preparation, the fundamental notion of a *fiducial system* is introduced.

A set of m vector fields $\{e_k = e_k(x); k = 1, 2, \ldots, m\}$ is said to be a *frame* for a region $\mathscr{R}$ of $\mathscr{M}$ if the pseudoscalar field $e = e(x)$, defined by

$$e \equiv e_1 \wedge e_2 \wedge \ldots \wedge e_m, \tag{1.1}$$

does not vanish at any point of $\mathscr{R}$. We say that e is *the pseudoscalar of the frame* $\{e_k\}$. As we will be concerned only with local properties of $\mathscr{M}$, it will be unnecessary to repeat that we require the existence of frames on subregions of $\mathscr{M}$.

The m vectors $\{e_k(x)\}$ constitute a *frame* or basis for the tangent space at x in $\mathscr{R}$. Thus, a frame for the region $\mathscr{R}$ is a set of frames for the tangent spaces at each point of $\mathscr{R}$, and is sometimes referred to as a *frame field* to distinguish it from a frame at a point.

A frame $\{e^k\}$ *reciprocal* to the frame $\{e_k\}$ is determined by the set of m^2 equations

$$e^k \cdot e_j = \delta_j^k, \tag{1.2}$$

where $i, j = 1, \ldots, m$. As shown in Section 1-3, Eqn. (1.2) can be explicitly solved for the reciprocal vectors e^k, with the result

$$e^k = (-1)^{k-1} e_1 \wedge \ldots \wedge \check{e}_k \wedge \ldots \wedge e_m e^{-1}. \tag{1.3}$$

Moreover,

$$e^{-1} = \frac{e}{e^2} = e^m \wedge \ldots \wedge e^2 \wedge e^1 \tag{1.4}$$

is the pseudoscalar for the reciprocal frame $\{e^k\}$.

In accordance with standard usage we refer to the quantities

$$g_{ij} \equiv e_i \cdot e_j \tag{1.5}$$

as the metric tensor of the frame $\{e_k\}$. Different senses in which g_{ij} is to be regarded as a tensor will be explained at the end of this section and the next. Following the approach to determinants in Section 1-4, we find that the determinant of the metric tensor is related to the pseudoscalar of a frame by

$$g \equiv \det \{g_{ij}\} = (e_m \wedge \ldots \wedge e_1) \cdot (e_1 \wedge \ldots \wedge e_m)$$
$$= e^\dagger e = (-1)^q |e|^2, \tag{1.6a}$$

where q is the number of vectors in the frame with negative square. The quantity

$$|e| = |e_1 \wedge \ldots \wedge e_m| = |g|^{1/2} \tag{1.6b}$$

is a measure of volume associated with the frame $\{e^k\}$.

Having established the basic algebraic properties of a frame, we now determine general properties of its derivatives. Because of (1.2), from the differential identity (4-3.19) we get the relation

$$(e_i \wedge e_j) \cdot (\partial \wedge e^k) = (e_i \wedge e_j) \cdot (\nabla \wedge e^k) = [e_i, e_j] \cdot e^k, \tag{1.7}$$

where we have used the notations of Section 4-2 for the cocurl $\nabla \wedge e^k = P(\partial \wedge e^k)$ and the Lie bracket

$$[e_i, e_j] = e_i \cdot \partial e_j - e_j \cdot \partial e_i = e_i \cdot \nabla e_j - e_j \cdot \nabla e_i. \tag{1.8}$$

According to (4-3.18), $P([e_i, e_j]) = [e_i, e_j]$, hence (1.7) can be solved for the Lie bracket, giving

$$[e_i, e_j] = e_k (e_i \wedge e_j) \cdot (\partial \wedge e^k) = e_k (e_i \wedge e_j) \cdot (\nabla \wedge e^k), \tag{1.9}$$

where sum over repeated upper and lower indices is understood. Thus the Lie brackets of a frame are determined by the curls or cocurls of its reciprocal frame. Conversely, the cocurls of a frame are determined by the Lie brackets of the reciprocal frame, according to

$$\nabla \wedge e^k = \tfrac{1}{2} e^j \wedge e^i [e_i, e_j] \cdot e^k. \tag{1.10}$$

The general duality relation between codivergence and cocurl implies that the divergences of a frame are determined by the cocurls of its reciprocal frame. Precise mathematical relations can be determined in the following way. Noting that $e = \pm|e|I$ where I is the unit pseudoscalar of $\mathscr{M}$, we use (4-3.9) to obtain

$$|e| \nabla \cdot (|e|^{-1} e_k) = [\nabla \wedge (e_k e^{-1})] e. \tag{1.11}$$

On the other hand, from (1.3) we have $e_k e^{-1} = e_k \cdot e^{-1} = (-1)^{m-k} e^m \wedge \ldots \wedge \check{e}^k \wedge \ldots \wedge e^1$, whence

$$\nabla \wedge (e_k e^{-1}) = (\nabla \wedge e^k) \wedge (e_j \wedge e_k e^{-1}) = (\nabla \wedge e^k) \cdot (e_j \wedge e_k) e^{-1}. \tag{1.12}$$

Combining (1.11) and (1.12), we obtain the desired 'duality relations'

$$\nabla \cdot e_k - e_k \cdot \nabla \log |e| = (\nabla \wedge e^j) \cdot (e_j \wedge e_k). \tag{1.13a}$$

Because of (4-3.6), we can write this instead as

$$\partial \cdot e_k - e_k \cdot \partial \log |e| = (\partial \wedge e^j) \cdot (e_j \wedge e_k). \tag{1.13b}$$

If we use (1.7) to express the right side of (1.13) in terms of the brackets, and note from the definition (1.8) of the bracket that

$$e^j \cdot [e_j, e_k] = \partial \cdot e_k - e^j \cdot (e_k \cdot \partial e_j),$$

then we find

$$e_k \cdot \partial \log |e| = e^j \cdot (e_k \cdot \nabla e_j) = -e_j \cdot (e_k \cdot \nabla e^j), \tag{1.14}$$

the last equality being obtained by differentiating $e^j \cdot e_j = m$. Equation (1.14) is equivalent to a well-known result of tensor analysis, as can be seen by using (1.6b) to express the left side in terms of the metric tensor.

Let the points $\{x\}$ in some region of $\mathscr{M}$ be parametrized by scalar *coordinates* $x^1, x^2, \ldots, x^m$; thus,

$$x = x(x^1, x^2, \ldots, x^m). \tag{1.15a}$$

A frame $\{e_k\}$ is said to be a *coordinate frame* if the e_k are tangents to coordinate curves, that is, if

$$e_k = e_k(x) = \partial_k x, \tag{1.15b}$$

where, by an abuse of our notation, ∂_k is the (partial) derivative with respect to the scalar x^k, that is,

$$\partial_k \equiv \partial_{x^k} = \frac{\partial}{\partial x^k}. \tag{1.16a}$$

By virtue of the chain rule, directional derivatives by vectors in a coordinate frame are equivalent to partial derivatives by the coordinates, that is,

$$e_k \cdot \partial = (\partial_k x) \cdot \partial_x = \partial_k. \tag{1.16b}$$

The inverse of the function (1.15a) is a set of m scalar functions on

$$x^k = x^k(x) \tag{1.17a}$$

defined on $\mathscr{M}$ and called *coordinate functions*. The gradients of a set of coordinate functions comprise a frame

$$e^k = \partial x^k = \nabla x^k. \tag{1.17b}$$

Indeed, the frame $\{e^k = \nabla x^k\}$ is reciprocal to the frame $\{e_k = \partial_k x\}$, as is easily proved by using (1.16b) to show that (1.2) is a consequence of the chain rule.

A frame $\{e_k\}$ for a region of $\mathscr{M}$ is a coordinate frame if and only if its Lie brackets vanish;

$$[e_i, e_j] = 0. \tag{1.18a}$$

For the coordinate frame (1.15b), we prove (1.18a) by applying (1.16b); thus,

$$e_i \cdot \partial e_j = \partial_i\, \partial_j x = \partial_j\, \partial_i x = e_j \cdot \partial e_i.$$

We shall not attempt the more difficult converse proof that (1.18a) implies (1.15), because we will not make use of it.

By virtue of (1.10) and (1.9), Eqn. (1.18a) is equivalent to

$$\nabla \wedge e^k = \nabla \wedge \nabla x^k = P(\partial \wedge \partial x^k) = 0. \tag{1.18b}$$

Hence, (1.18a, b) are equivalent expressions of the integrability conditions for a coordinate frame.

To every frame $\{e_k\}$ there corresponds a *unique* orthonormal frame $\{\gamma_k\}$ such that

$$e_k = h(\gamma_k) = h^j_{\ k}\gamma_j, \tag{1.19a}$$

where h is the *unique* tensor field determining, at each point x, a *positive symmetric* linear transformation of the tangent space which takes $\{\gamma_k(x)\}$ into $\{e_k(x)\}$. Being symmetric, h is equal to its adoint $\bar{h}$, and we have

$$\gamma^k = \bar{h}(e^k) = h(e^k) = h^k_{\ j} e^j. \tag{1.19b}$$

The matrix elements of this transformation are

$$h^j_{\ k} = \gamma^j \cdot e_k = \gamma^j \cdot h(\gamma_k) = h(e^j) \cdot e_k. \tag{1.20}$$

The metric tensor $g_{ij} = e_i \cdot e_j$ can be regarded as a symmetric tensor field $a \cdot g(b)$ with components

$$g_{ij} = \gamma_i \cdot g(\gamma_j) = h(\gamma_i) \cdot h(\gamma_j) = \gamma_i \cdot h^2(\gamma_j). \tag{1.21}$$

Thus, the tensor h is the positive square root of the metric tensor.

The metric tensor g of a frame $\{e_k\}$ can be regarded as a linear transformation of its reciprocal frame $\{e^k\}$ into $\{e_k\}$. To express this explicitly, we first take account of the metric signature by introducing the indicators $\eta_k = \gamma_k^2$, so the orthonormality of the γ_k is expressed by

$$\gamma_j \cdot \gamma_k = \eta_k\, \delta_{jk}. \tag{1.22}$$

Then from (1.19b) we have

$$\gamma_k = \eta_k \gamma^k = \eta_k h(e^k), \tag{1.23}$$

and by substituting this in (1.19a), we get

$$e_k = \eta_k h^2(e^k) = \eta_k g(e^k) = g_{kj} e^j. \tag{1.24}$$

The matrix elements of this linear transformation are

$$g_{jk} = e_j \cdot e_k = \eta_k e_j \cdot g(e^k). \tag{1.25a}$$

The inverse of the transformation g is expressed by

$$e^k = \eta_k g^{-1}(e_k) = g^{kj} e_j, \tag{1.25b}$$

with matrix elements

$$g^{jk} = e^j \cdot e^k = \eta_k e^j \cdot g^{-1}(e_k). \tag{1.26}$$

When $\{x^k\}$ is a system of coordinates with coordinate frame $\{e_k = \partial_k x\}$, we call the corresponding tensor h and the frame $\{\gamma_k\}$ just defined, respectively, the *fiducial tensor* and the *fiducial frame of the coordinate system* $\{x^k\}$. We will refer to this whole set of quantities as a *fiducial system*. We shall see that there are significant advantages to working with the fiducial tensor instead of the metric tensor as has been traditional in classical differential geometry.

6-2. Mobiles and Curvature

This section develops the method of mobiles as a general tool for the study of intrinsic differential geometry. The main result is a systematic procedure for computing the curvature tensor from a frame of vector fields. The efficiency of this procedure will be obvious to anyone who applies it to explicit calculations in gravitation theory, but such claculations will not be carried out here. The method of mobiles will be related to the more general method of Chapter 5, so results obtained by either method can be applied with impunity in connexion with the other.

The method of mobiles can be formulated in other languages, such as tensor analysis or differential forms, but only Geometric Calculus relates the method of mobiles directly to spinors. This feature greatly simplifies applications to physics where spinors are necessary. At the end of this section, the relation of our method to the standard method of tensor analysis is established. Relation to the method of differential forms is established in Section 6-4.

Given the codifferentials of a frame of vector fields on $\mathscr{M}$, the coderivatives of any tensor field are determined, because the frame provides a basis for the tangent algebra of multivector fields on $\mathscr{M}$. The curvature tensor is then determined as usual by the commutator of codifferentials. As we shall see, it is especially convenient to begin with the codifferentials of a fiducial frame.

We continue to use here the definitions of coderivative and codifferential introduced in Section 4-3. We introduce the term *mobile* to refer to an orthonormal frame of vector fields. Since the vectors of a mobile $\{\gamma_k\}$ are orthonormal, their codifferentials can be expressed in the form

$$\delta_a \gamma_k = a \cdot \nabla \gamma_k = P(a \cdot \partial \gamma_k) = \omega_a \cdot \gamma_k, \tag{2.1}$$

where ω_a is a bivector-valued linear function of the vector a. We can interpret ω_a at any point x as the 'angular velocity' of the frame $\{\gamma_k(x)\}$ as it is displaced in the direction $a(x)$. We do not present a proof of (2.1) just now, as it will be supplied later on when we solve Eqn. (2.1) explictly for ω_a.

It is convenient to define the *mobile differential* $\mathrm{d}_a B$ of a multivector field B by the equation

$$\delta_a B = \mathrm{d}_a B + \omega_a \times B, \tag{2.2}$$

where, as before, the cross denotes commutator product. Consider, for example, the codifferential of a vector field b. Writing $b = b^k \gamma_k$, where $b^k = b \cdot \gamma^k$ are the components of b relative to the mobile $\{\gamma_k\}$, we have, using (2.1),

$$\begin{aligned} \delta_a b &= \delta_a (b^k \gamma_k) = (\delta_a b^k) \gamma_k + b^k \, \delta_a \gamma_k \\ &= (a \cdot \partial b^k) \gamma_k + \omega_a \cdot b. \end{aligned}$$

Comparison with (2.2) shows that, since $\omega_a \cdot b = \omega_a \times b$,

$$\mathrm{d}_a b = (a \cdot \partial b^k) \gamma_k. \tag{2.3}$$

Thus, the mobile differential of a vector field is just the derivative of its *mobile components*. The same is true of the mobile differential of any multivector field. Thus, if B is a bivector field, with *mobile components* $B^{ij} = \gamma^i \cdot B \cdot \gamma^j = B \cdot (\gamma^j \wedge \gamma^i)$, we have

$$B = \tfrac{1}{2} B^{ij} \gamma_i \wedge \gamma_j. \tag{2.4a}$$

Note that, with the help of the algebraic identity (1-1.67), from (2.1) we get

$$\delta_a (\gamma_i \wedge \gamma_j) = \omega_a \times (\gamma_i \wedge \gamma_j). \tag{2.4b}$$

Hence,

$$\delta_a B = \tfrac{1}{2} (a \cdot \partial B^{ij}) \gamma_i \wedge \gamma_j + \omega_a \times B \tag{2.4c}$$

and

$$\mathrm{d}_a B = \tfrac{1}{2} (a \cdot \partial B^{ij}) \gamma_i \wedge \gamma_j \tag{2.4d}$$

as advertised.

For a scalar field ϕ, Eqn. (2.2) reduces to

$$\delta_a \phi = \mathrm{d}_a \phi = a \cdot \nabla \phi, \tag{2.5a}$$

since $\omega_a \times \phi = 0$. The integrability condition (4-1.12) then gives us

$$[\delta_a, \delta_b] \phi = [\mathrm{d}_a, \mathrm{d}_b] \phi = 0. \tag{2.5b}$$

Since the mobile differential of any multivector field B reduces to the mobile differential of scalars (the mobile components of B), we have the general result

$$[\mathrm{d}_a, \mathrm{d}_b] B = 0. \tag{2.6}$$

Thus, mobile differentials always commute.

One other basic property of mobile differentials should be mentioned; like codifferentials they conserve linearity when they operate on tensors, as follows by requiring (2.2) to apply if B is a tensor. We have tacitly used this property in obtaining (2.6).

Now we use (2.2) to obtain an expression for the curvature tensor. With the help of (2.6) and the Jacobi identity

$$\omega_b \times (\omega_a \times B) - \omega_a \times (\omega_b \times B) = (\omega_b \times \omega_a) \times B,$$

we easily obtain

$$[\delta_a, \delta_b]B = R(a \wedge b) \times B, \tag{2.7}$$

where the curvature tensor $R(a \wedge b)$ is given by

$$R(a \wedge b) = \mathrm{d}_a \omega_b - \mathrm{d}_b \omega_a + \omega_a \times \omega_b \,. \tag{2.8}$$

Since $\delta_a \omega_b = \mathrm{d}_a \omega_b + \omega_a \times \omega_b$, we can replace mobile differentials in (2.8) by codifferentials to get

$$R(a \wedge b) = \delta_a \omega_b - \delta_b \omega_a + \omega_a \times \omega_b \,. \tag{2.9}$$

The expression (2.8) is most useful for computations, while (2.9) is valuable for theoretical considerations.

When (2.8) is used for computations, it is essential to remember that d_a is required to preserve the linearity of tensors; since ω_b is bivector valued, we can write explicitly

$$\mathrm{d}_a \omega_b - \mathrm{d}_b \omega_a = \tfrac{1}{2}(a \cdot \partial \omega_b^{ij} - b \cdot \partial \omega_a^{ij}) \gamma_i \wedge \gamma_j - \omega_{[a,\, b]} \,. \tag{2.10}$$

This should be compared with (2.4d). The last term in (2.10) vanishes if the commutator $[a, b]$ vanishes. For this reason, evaluation of curvature from (2.8) may be simplified by using coordinate frames.

To prove that the representation of curvature given by (2.9) is equivalent to the one derived in Chapter 5, we use (2.1) and (4-3.16) to get

$$a \cdot \partial \gamma_k = a \cdot \nabla \gamma_k - S_a \cdot \gamma_k = (\omega_a - S_a) \cdot \gamma_k .$$

Hence, we can write

$$a \cdot \partial \gamma_k = \Omega_a \cdot \gamma_k , \tag{2.11a}$$

where

$$\Omega_a = \omega_a - S_a \tag{2.11b}$$

gives the decomposition of the total mobile angular velocity Ω_a into a tangential component $\omega_a = P(\Omega_a)$, which describes the rotation of the mobile within the manifold, and a normal component S_a, which describes the rotation of the mobile with the manifold as it is displaced. The integrability condition for differentials of the mobile can be written

$$[a \cdot \partial, b \cdot \partial]\gamma_k - [a, b] \cdot \partial \gamma_k = 0,$$

which we use to deduce from (2.11a) that

$$\delta_a \Omega_b - \delta_b \Omega_a + P(\Omega_a \times \Omega_b) = 0. \tag{2.12}$$

Finally, by substituting (2.11b) in (2.12) and using (4-4.17) as well as (4-2.28), we obtain

$$\delta_a \omega_b - \delta_b \omega_a + \omega_a \times \omega_b = P(S_a \times S_b). \tag{2.13}$$

This completes the proof that (2.9) is equivalent to the previously derived expression (5-1.5a) for the curvature in terms of the curl tensor S_a. So all the general properties of the curvature tensor established in Section 5-1 are properties of the tensor (2.9) constructed from the mobile angular velocity ω_a. Of course, Eqn. (2.13) also shows that the angular velocity of any mobile can be used to compute the curvature tensor, since the curl of the manifold S_a is independent of a choice of mobile.

Now let us study how derivatives of a mobile are related to its angular velocity. Setting $a = \gamma_j$ in (2.1) we have

$$\gamma_j \cdot \nabla \gamma_k = \omega_{\hat{j}} \cdot \gamma_k, \tag{2.14a}$$

where we have introduced the abbreviation

$$\omega_{\hat{j}} \equiv \omega_{\gamma_j}; \tag{2.14b}$$

the 'hat' on the subscript reminds us that it indicates a tensor argument evaluated at the *unit* vectors of the mobile. Algebraic manipulations of (2.14) are expedited by the formula

$$\gamma^k \langle A \rangle_r \gamma_k = (-1)^r (m - 2r) \langle A \rangle_r, \tag{2.15}$$

which holds for any multivector field A by virtue of (2-1.40). Thus, since $\omega_{\hat{j}}$ is a bivector, we have

$$\gamma^k \omega_{\hat{j}} \cdot \gamma_k = \tfrac{1}{2}(\gamma^k \omega_{\hat{j}} \gamma_k - \gamma^k \gamma_k \omega_{\hat{j}}) = -2\omega_{\hat{j}}, \tag{2.16}$$

so if we multiply (2.14a) and γ^k and sum over k we get

$$\omega_{\hat{j}} = -\tfrac{1}{2} \gamma^k \gamma_j \cdot \nabla \gamma_k = -\tfrac{1}{2} \gamma^k \wedge (\gamma_j \cdot \nabla \gamma_k) = \tfrac{1}{2} (\gamma_j \cdot \nabla \gamma_k) \gamma^k. \tag{2.17}$$

Note that the wedge in (2.17) can be retained or dropped as desired, because $\gamma^k \cdot (\gamma_j \cdot \nabla \gamma_k) = 0$ holds with or without a sum on k as a consequence of orthonormality. The quantity

$$\omega \equiv \gamma^j \omega_{\hat{j}} = \tfrac{1}{2} \gamma_k \nabla \gamma^k - \gamma_k \cdot \nabla \gamma^k \tag{2.18}$$

describes properties of the angular veclocity which are independent of direction. The right side of (2.18) is obtained from (2.17) by observing that

$$\gamma^j\gamma_k\gamma_j \cdot \nabla\gamma^k = (2\gamma^j \cdot \gamma_k - \gamma_k\gamma^j)\gamma_j \cdot \nabla\gamma^k = 2\gamma_k \cdot \nabla\gamma^k - \gamma_k\nabla\gamma^k.$$

If we multiply (2.14a) on the left by γ^j and sum, we see that

$$\begin{aligned}\nabla\gamma_k &= \gamma^j \tfrac{1}{2}(\omega_{\hat{j}}\gamma_k - \gamma_k\omega_{\hat{j}})\\ &= \tfrac{1}{2}(\gamma^j\omega_{\hat{j}}\gamma_k - (2\gamma^j \cdot \gamma_k - \gamma_k\gamma^j)\omega_{\hat{j}})\\ &= \tfrac{1}{2}(\omega\gamma_k + \gamma_k\omega) - \omega_{\hat{k}}.\end{aligned}$$

Hence,

$$\nabla\gamma_k = \omega \cdot \gamma_k - \omega_{\hat{k}}. \tag{2.19}$$

The scalar part of (2.19) is

$$\nabla \cdot \gamma_k = \langle\omega\rangle_1 \cdot \gamma_k = \langle\omega\gamma_k\rangle. \tag{2.20}$$

Thus the divergence of the mobile vectors is completely specified by the vector part of ω.

We can solve (2.19) for $\langle\omega\rangle_3$. With the help of (2.15) we find

$$\begin{aligned}\gamma^k\,\nabla\gamma_k &= \tfrac{1}{2}(\gamma^k\omega\gamma_k + \gamma^k\gamma_k\omega) - \omega\\ &= \tfrac{1}{2}(2\langle\omega\rangle_1 + 6\langle\omega\rangle_3) - (\langle\omega\rangle_1 + \langle\omega\rangle_3).\end{aligned}$$

Hence

$$\langle\omega\rangle_3 = -\gamma^k \wedge \omega_{\hat{k}} = \tfrac{1}{2}\gamma^k\nabla\gamma_k = \tfrac{1}{2}\gamma^k \wedge \nabla \wedge \gamma_k. \tag{2.21}$$

Comparison of (2.21) with (2.18) and (2.20) shows that

$$\langle\omega\rangle_1 = \gamma^k \cdot \omega_{\hat{k}} = \gamma^k\nabla \cdot \gamma_k = -\gamma_k \cdot \nabla\gamma^k = -\gamma^k \cdot (\nabla \wedge \gamma_k). \tag{2.22}$$

Finally, substituting (2.21) into the bivector part of (2.19), we obtain

$$\omega_{\hat{k}} = \tfrac{1}{2}(\gamma^j \wedge \nabla \wedge \gamma_j) \cdot \gamma_k - \nabla \wedge \gamma_k. \tag{2.23}$$

This is a most important result, for it tells us how to obtain the angular velocity of a mobile from the cocurl of the mobile vectors. Note that the free index in (2.23) can be raised or lowered at will, because γ^k differs from γ_k by at most a sign depending on signature.

A fiducial frame $\{\gamma_k\}$ is a mobile 'tied' to a coordinate frame $\{e_k\}$ by a fiducial tensor h, as described in the preceding section. Since $[e_j, e_k] = 0$ for a coordinate frame, it is most convenient to deal with coordinate differentials of a fiducial frame (this point was explained previously in connexion with Eqn. (2.10)). Accordingly, we set $a = e_j$ in (2.1) and write

$$\delta_j\gamma_k = e_j \cdot \nabla\gamma_k = \omega_j \cdot \gamma_k, \tag{2.24a}$$

with the abbreviations

$$\delta_j \equiv \delta_{e_j}, \qquad \omega_j \equiv \omega_{e_j}. \tag{2.24b}$$

Using (1.19a), we get an explicit expression for ω_j from (2.23), namely,

$$\omega_j = h_j^k \omega_{\hat{k}} = \tfrac{1}{2}(\gamma^k \wedge \nabla \wedge \gamma_k) \cdot e_j - h_j^k \nabla \wedge \gamma_k. \tag{2.25}$$

Thus, the ω_j are determined by the $\nabla \wedge \gamma_k$.

We can calculate $\nabla \wedge \gamma_k$ from derivatives of the fiducial tensor. Taking the cocurl of (1.23) and using (1.18b) we find

$$\nabla \wedge \gamma_k = \eta_k(\nabla h_j^k) \wedge e^j = \eta_k \dot{\nabla} \wedge \dot{h}(e^k) = \dot{\nabla} \wedge \dot{h}h^{-1}(\gamma_k). \tag{2.26}$$

This equation becomes particularly simple for an orthogonal coordinate system, for then

$$e_k = h(\gamma_k) = h_k\gamma_k, \tag{2.27a}$$

and

$$\gamma^k = h(e^k) = h_k e^k, \tag{2.27b}$$

that is, each γ_k and each e^k is an eigenvector of the fiducial tensor with eigenvalue h_k. In this case, (2.25) and (2.26) reduce to

$$\omega_k = -\tfrac{1}{2} h_k \nabla \wedge \gamma_k = \tfrac{1}{2} \gamma_k \wedge \nabla h_k. \tag{2.28}$$

These results are most useful for explicit calculations of curvature in gravitation theory.

The present method should be compared with the conventional approach of covariant tensor analysis. Tensor analysis proceeds by specifying the codifferentials of a coordinate frame $\{e_k\}$ with the equation

$$e_j \cdot \nabla e_k = L^i_{jk} e_i. \tag{2.29}$$

This equation amounts to no more than the requirement that codifferentials preserve tangency of vector fields. Many expositions of tensor analysis do not write an equation like (2.29) explicitly, but such an equation must nevertheless be tacitly assumed. The 'coefficients of connection' L^i_{jk} can be related to the metric tensor by using (2.29) to differentiate $g_{jk} = e_j \cdot e_k$, with the result

$$\partial_i g_{jk} = L^n_{ij} g_{nk} + L^n_{ik} g_{nj} \tag{2.30}$$

where, in accordance with (1.16), $\partial_i = \partial/\partial x^i$. Under the assumption $L^i_{jk} = L^i_{kj}$, Eqn.

(2.30) can be solved for the coefficients yielding the famous 'Christoffel symbols' $\{{}^{\,i}_{jk}\}$;

$$L^i_{jk} = \begin{Bmatrix} i \\ jk \end{Bmatrix} \equiv \tfrac{1}{2} g^{in}(\partial_j g_{nk} + \partial_k g_{nj} - \partial_n g_{kj}), \tag{2.31}$$

where $g^{in} = e^i \cdot e^n$. This symmetry of the coefficients of connexion entails the vanishing of the torsion tensor described in the last chapter.

Expression of the 'coordinate connexion' in terms the metric tensor by (2.31) corresponds to specification of the 'fiducial connexion' in terms of the fiducial tensor by (2.25) and (2.26). The coordinate and fiducial connexions can be directly related by using (2.25) and (2.14) as well as (1.19) and (1.16) in (2.29); we find

$$(\partial_j h^i_k)\gamma_i + h^i_j \omega_j \cdot e_k = L^i_{jk} e_i. \tag{2.32}$$

These equations are easily solved for either L^i_{jk} or ω_i.

Covariant derivatives of tensors are usually defined by requiring 'covariance' under coordinate transformations. Our codifferential, defined without reference to coordinates or transformations by (4-3.3), is completely equivalent to the usual covariant derivative. It is worthwhile to see this exhibited by some examples. Let v and u be vector fields. The product $v \cdot u$ can alternatively be regarded as a scalar field

$$\phi \equiv v \cdot u, \tag{2.33a}$$

as a tensor field of rank 1,

$$v(u) \equiv v \cdot u \tag{2.33b}$$

or as a tensor field of rank 2,

$$g(v, u) \equiv v \cdot u. \tag{2.33c}$$

By (4-3.3) and (4-3.1), the codifferentials of these tensors are respectively,

$$\delta_a \phi = a \cdot \nabla\phi = a \cdot \partial\phi = a \cdot \partial(v \cdot u), \tag{2.34a}$$

$$\delta_a v(u) = a \cdot \partial(v \cdot u) - v \cdot (a \cdot \nabla u), \tag{2.34b}$$

$$\delta_a g(u, v) = a \cdot \partial(v \cdot u) - (\delta_a v) \cdot u - v \cdot (\delta_a u). \tag{2.34c}$$

Now, the quantities $v_k \equiv v(e_k) = v \cdot e_k$ are coordinate components of v. Writing $a = a^j e_j$ and using (2.29) in (2.34b) we obtain

$$\delta_a v_k = a^j(\partial_j v_k - L^i_{jk} v_i). \tag{2.35}$$

The quantity in parenthesis here will be recognized as the classical expression for the covariant derivative of a vector field.

Similarly, by applying (2.34c) to $g_{ij} = g(e_i, e_j) = e_i \cdot e_j$, one sees that (2.34c) is equivalent to (2.30) if

$$\delta_a g(v, u) = 0. \tag{2.36}$$

This is the classical condition that the covariant derivative of the metric tensor vanishes on a Riemannian manifold.

6-3. Curves and Comoving Frames

This section adapts the method of mobiles to curves imbedded in an m-dimensional vector manifold $\mathscr{M}$. It provides us with an efficient apparatus for comparing directions at different points on a curve as well as for describing geometrical properties of the curve itself. Among other things, it enables us to cast the classical Frenet equations for a curve in a particularly useful form. The method has important applications to physics, for example, to describe the precession of a satellite in a gravitational field.

Let $x = x(\tau)$ describe a smooth curve $\mathscr{C}$ in $\mathscr{M}$ parametrized by the scalar τ. The *velocity*

$$v = \frac{\mathrm{d}x}{\mathrm{d}\tau} \tag{3.1}$$

is a nonvanishing vector field on $\mathscr{C}$. Indeed, v is a pseudoscalar for the one-dimensional manifold $\mathscr{C}$. If $v^2 = 0$ on $\mathscr{C}$, we say that the curve is *null*. If $\mathscr{C}$ is not null, we can always parametrize the curve so that $|v|^2 = 1$, in which case $\tau_2 - \tau_1$ is the *arc length* between points $x(\tau_2)$ and $x(\tau_1)$ on $\mathscr{C}$. We automatically do this whenever convenient.

As we have noted before, for any function $f = f(x(\tau))$ defined on $\mathscr{C}$, the *coderivative* by τ is equal to the *codifferential* by v; that is, by virtue of the chain rule,

$$\nabla_\tau = \left(\frac{\mathrm{d}x}{\mathrm{d}\tau}\right) \cdot \nabla_x = v \cdot \nabla = \delta_v. \tag{3.2}$$

An orthonormal frame $\{e_k\}$ of vector fields $e_k = e_k(x(\tau)) = e_k(\tau)$ on the curve $\mathscr{C}$ is called a *mobile* or *comoving frame* on $\mathscr{C}$ if

$$e_1 = v = \frac{\mathrm{d}x}{\mathrm{d}\tau} \tag{3.3}$$

when $\mathscr{C}$ is not null, or

$$e_1 + e_m = v = \frac{\mathrm{d}x}{\mathrm{d}\tau} \tag{3.4}$$

when $\mathscr{C}$ is null. Many properties of mobiles determined in the preceding section obtain here as well. Thus, by the same argument that established (2.1), it can be proved that there exists a unique bivector Ω such that

$$\dot{e}_k \equiv \nabla_\tau e_k = \delta_v e_k = \Omega \cdot e_k. \tag{3.5}$$

It is convenient to indicate differentiation by an overdot here, because we will be concerned with only a single parameter. We call Ω the ***angular velocity*** of the mobile. The argument that gave us (2.17), also yields from (3.5)

$$\Omega = \tfrac{1}{2}\dot{e}_k e^k = \tfrac{1}{2}\dot{e}_k \wedge e^k. \tag{3.6}$$

For the rest of our discussion we assume that $v^2 = 1$, since this is the case of most interest in physics, and the argument is easily adjusted to take care of the other cases. From $v^2 = 1$, we obtain $v \cdot \dot{v} = 0$ and note the identity

$$\dot{v} = (\dot{v} \wedge v) \cdot v. \tag{3.7}$$

Since $v = e_1$, we see by comparing (3.7) with (3.5) that Ω can be written uniquely in the form

$$\Omega = \dot{v} \wedge v + B, \tag{3.8a}$$

where B is a bivector satisfying

$$B \cdot v = 0. \tag{3.8b}$$

This shows the high degree of freedom available in the selection of a comoving frame, for any value of B is allowed provided only that (3.8b) is satisfied.

We interpret the mobile $\{e_k = e_k(\tau)\}$ as a frame whose 'rigid motion' along the curve $\mathscr{C}$ is described by the equations of motion (3.5). The term $\dot{v} \wedge v$ in (3.8a) is the angular velocity needed for the frame to 'follow' the curve while preserving frame orthogonality and the tangency of $v = e_1$ to the curve. The term B in (3.8a) is an additional angular velocity of vectors orthogonal to v. Hence, a mobile with angular velocity

$$\Omega = \dot{v} \wedge v \tag{3.9}$$

can be interpreted as a frame moving without rotation relative to the curve along which it is transported. Such a frame is called *Fermi–Walker frame* in the physics literature, and its displacement along a curve is called *Fermi–Walker transport*. A geodesic is defined by $\dot{v} = 0$, whence $\dot{e}_k = 0$ by (3.9) and (3.5), so Fermi–Walker transport reduces to parallel transport when the curve is a geodesic. Fermi–Walker transport provides a good description of the motion of a gyroscope. See [MTW] for further discussion and references.

Instead of beginning with a curve to which we wish to attach a mobile, we are often supplied with a mobile determined by specifying the angular velocity Ω. The problem then is to integrate Eqn. (3.5) to get $e_k = e_k(\tau)$ and then to integrate $e_1 = v(\tau) = dx/d\tau$ to get the 'orbit' $x = x(\tau)$ of the mobile. The general method described here has wide applicability to geometry as well as physics.

Suppose our manifold $\mathcal{M}$ is described by specifying a fiducial frame $\{\gamma_k\}$ as described in the preceding section. According to the results of Section 3-8, the frame $\{e_k(x)\}$ at any point x of $\mathcal{M}$ is related to the fiducial frame $\{\gamma_k(x)\}$ at the same point by the equation

$$e_k = U\gamma_k U^\dagger, \tag{3.10}$$

where U is a rotor satisfying

$$U^\dagger U = 1. \tag{3.11}$$

Thus the mobile $e_k = e_k(\tau)$ is completely determined by the rotor-valued function $U = U(\tau)$ which relates it to a given fiducial frame at each point along the orbit.

According to (2.2), we can write Eqn. (3.5) in the form

$$\delta_v e_k = \mathrm{d}_v e_k + \omega_v \cdot e_k = \Omega \cdot e_k. \tag{3.12}$$

One easily verifies by differentiating (3.10) that the set of m equations (3.12) is equivalent to the single rotor equation

$$\mathrm{d}_v U = \tfrac{1}{2}(\Omega - \omega_v)U. \tag{3.13}$$

Since $\mathrm{d}_v \gamma_k = 0$, integration of (3.13) can be carried out as if it were an equation on an m-dimensional flat manifold for which $\{\gamma_k\}$ is a 'cartesian frame'. It is only necessary to remember that fiducial components of the integrated rotor $U = U(x(\tau)) = U(\tau)$ are relative to the fiducial frame $\{\gamma_k(x(\tau))\}$ at the point of interest on the curve.

Expressed in terms of the codifferential, Eqn. (3.13) takes the form

$$\delta_v U = \mathrm{d}_v U + \omega_v \times U = \tfrac{1}{2}\Omega U - \tfrac{1}{2}U\omega_v. \tag{3.14}$$

However, Eqn. (3.13) is more useful than Eqn. (3.14) because it can be integrated by conventional methods.

Classical differential geometry defines the *Frenet Frame* $\{e_k\}$ of a non-null curve by the so-called *Frenet equations*

$$\begin{aligned}
\dot{x} &= e_1 = v,\\
\ddot{x} &= \dot{e}_1 = \kappa_1 e^2,\\
\dot{e}_2 &= -\kappa_1 e^1 + \kappa_2 e^3,\\
\dot{e}_3 &= -\kappa_2 e^2 + \kappa_3 e^4,\\
&\ldots\\
\dot{e}_m &= -\kappa_{m-1} e^{m-1},
\end{aligned} \tag{3.15}$$

where the κ_k are scalars chosen so that the e_k are orthonormal. Substituting Eqns. (3.15) into (3.16), we get

$$\Omega = \kappa_1 e^2 e^1 + \kappa_2 e^3 e^2 + \kappa_3 e^4 e^3 + \ldots + \kappa_{m-1} e^m e^{m-1}. \tag{3.16}$$

We call the angular velocity Ω defined by (3.16) the ***Darboux bivector***, because it directly generalizes the classical Darboux vector.

The Darboux bivector summarizes the geometrical properties of a curve in a single convenient quantity. From (3.16) we see that the *kth curvature* κ_k is the projection of Ω on the $e_k \wedge e_{k+1}$ plane, thus,

$$\kappa_k = e_{k+1} \cdot \Omega \cdot e_k = \Omega \cdot (e_k \wedge e_{k+1}) = \dot{e}_k \cdot e_{k+1}. \tag{3.17}$$

Curves can be classified according to properties of their curvatures $\{\kappa_k\}$, but an equivalent classification of Darboux bivectors would evidently be more systematic and compact.

The greatest benefit of the Darboux bivector comes from the replacement of the coupled system of Eqns. (3.15) by the single rotor Eqn. (3.13) relating the Frenet frame to a fiducial frame. Applications of Frenet frames in the literature are easily reformulated in the present language, so further discussion is unnecessary. It should be remarked, however, that in spite of the geometrical significance of Frenet frames, other choices of comoving frames are often more significant in physical applications.

6-4. The Calculus of Differential Forms

In the last decade or so, formulations of integral and differential calculus and geometry in terms of Cartan's *calculus of differential forms* have steadily worked their way into the mathematics curriculum. Recently, an influential physics text-book [MTW] has broadened the base of this movement with the claim that Cartan's calculus is superior in many physical applications to conventional vector and tensor analysis. Since the calculus of forms calls for a substantial revision of the mathematical language used by physicists, any commitment to it should be fully justified.

The section compares the calculus of forms to Geometric Calculus in sufficient detail so that any expression can be easily translated from one language to the other. It will be seen that the calculus of forms need not be regarded so much a separate language as a special set of notations for a limited part of Geometric Calculus. From the more general perspective of Geometric Calculus the limitations of the calculus of forms and the drawbacks of its special notations become apparent. A critique of the calculus of forms is given at the end of this section. Chapter 7 discusses the role of forms in geometric integration theory.

For the sake of comparison, we introduce some common notations used in the calculus of forms. We use such notations only in this section, because they are superfluous and inappropriate in the general Geometric Calculus.

In Section 4-1 we encountered differential forms as scalar valued multiform fields. Let us repeat and elaborate on our discussion of forms with a notation which is closer to conventional practice.

We use the definition and properties of r-forms developed in Section 1-4. A differential r-form α_r is a scalar valued skewsymmetric linear function $\alpha_r = \alpha_r(\mathrm{d}x_1, \mathrm{d}x_2, \ldots, \mathrm{d}x_r)$, where the 'differentials' $\mathrm{d}x_1, \ldots, \mathrm{d}x_r$ are arbitrary vector *fields.* * According to Section 1-4, α_r can always be written,

$$\alpha_r = \alpha_r(\mathrm{d}X_r) = A_r^\dagger(x) \cdot \mathrm{d}X_r(x) = \mathrm{d}X_r^\dagger \cdot A_r, \tag{4.1a}$$

where $\mathrm{d}X_r = \mathrm{d}x_1 \wedge \mathrm{d}x_2 \wedge \ldots \wedge \mathrm{d}x_r$, and $A_r = A_r(x)$ is a definite r-vector field. It should be noted that A_r need not be a tangent field; however, since $\mathrm{d}X_r$ is tangent only the tangential component of A_r contributes to α_r, that is,

$$\alpha_r = \mathrm{d}X_r^\dagger \cdot A_r = \mathrm{d}X_r^\dagger \cdot P(A_r), \tag{4.1b}$$

where, as before, P is the projection into the tangent algebra. Hence, we suppose A_r is a tangent field in the following, unless stated otherwise.

According to Section 1-4, the exterior product of forms $\alpha_r = \mathrm{d}X_r^\dagger \cdot A_r$ and $\beta_r = \mathrm{d}X_s^\dagger \cdot B_s$ is the $(r+s)$-form

$$\alpha_r \wedge \beta_s = \mathrm{d}X_{r+s}^\dagger \cdot (A_r \wedge B_s). \tag{4.2}$$

The exterior product is not sufficient for routine applications of the algebra of forms, so a duality operation is commonly introduced which can be expressed in terms of Geometric Algebra as follows. An $(n-r)$-form 'dual' to α_r is defined by

$$*\alpha_r \equiv \mathrm{d}X_{n-r}^\dagger \cdot (A_r^\dagger I) = (A_r \wedge \mathrm{d}X_{n-r})^\dagger \cdot I, \tag{4.3}$$

where I is the unit pseudoscalar and $\mathrm{d}X_{n-r} = \mathrm{d}x_{r+1} \wedge \mathrm{d}x_{r+2} \wedge \ldots \wedge \mathrm{d}x_n$. Since by (1-1.23a), $A_r I = A_r \cdot I = (-1)^{r(n-1)} I A_r$, the 'double dual' is

$$\begin{aligned} **\alpha_r &= \mathrm{d}X_r^\dagger \cdot [(A_r^\dagger I)^\dagger I] \\ &= \mathrm{d}X_r^\dagger \cdot [I^\dagger A_r I] = (-1)^{r(n-1)} \alpha_r. \end{aligned} \tag{4.4}$$

Combining (4.2) and (4.3), we have

$$\begin{aligned} \alpha_r \wedge *\beta_r &= (\mathrm{d}X_r \wedge \mathrm{d}X_{n-r})^\dagger \cdot (A_r \wedge (B_r^\dagger I)) \\ &= A_r \cdot B_r^\dagger I^\dagger \cdot \mathrm{d}X_n, \end{aligned} \tag{4.5}$$

where $\mathrm{d}X_n = \mathrm{d}X_r \wedge \mathrm{d}X_{n-r} = \mathrm{d}x_1 \wedge \ldots \wedge \mathrm{d}x_n$. And with $\mathrm{d}X_0 \equiv 1$, we get

$$*(\alpha_r \wedge *\beta_b) = A_r \cdot B_r^\dagger. \tag{4.6}$$

* It might be better to write $\mathrm{d}_1 x, \ldots, \mathrm{d}_r x$ instead of $\mathrm{d}x_1, \ldots, \mathrm{d}x_r$ to indicate more clearly that we have here r distinct tangent vectors at each point x rather than r components of a single vector $\mathrm{d}x$ or 'differentials' at different points x_k. As a rule, it is best to use the notation $\mathrm{d}x_r$ for a tangent vector only when it stands 'under' an integral sign where it represents a directed measure (see Chapter 7). We use it here hoping to clarify its relation to similar usage in the literature.

Thus the *-operator on differential forms can be used to do the work of the inner product on tangent fields. However, the inner product is more convenient than the *-operator for many reasons; for instance, the *-operator can be applied only to tangent fields unless the manifold is imbedded in some manifold of higher dimension. But, in the preceding two chapters we have seen important applications of the inner product to non-tangent fields. More important, use of the *-operator helps to disguise the existence of the elementary combination of inner and outer products $ab = a \cdot b + a \wedge b$, which cannot even be expressed with differential forms.

As we pointed out in Section 4-1, Cartan's *exterior derivative* $\mathrm{d}\alpha_r$ of a form α_r can be expressed in Geometric Algebra by the formula

$$\begin{aligned} \mathrm{d}\alpha_r &= \mathrm{d}X^\dagger_{r+1} \cdot (\partial \wedge A_r) = (\mathrm{d}X^\dagger_{r+1} \cdot \partial) \cdot A_r \\ &= \mathrm{d}X^\dagger_{r+1} \cdot (\nabla \wedge A_r) = (\mathrm{d}X^\dagger_{r+1} \cdot \nabla) \cdot A_r. \end{aligned} \tag{4.7}$$

Note that, because of (4.1b), either the derivative ∂ or the coderivative $\nabla = P(\partial)$ can be used in (4.7). Rather than show that (4.7) reduces to the usual expression for $\mathrm{d}\alpha_r$ when coordinates are introduced, we merely point out that (4.7) implies the fundamental properties of the exterior derivative:

$$\mathrm{d}(\alpha_r \wedge \beta_r) = \mathrm{d}\alpha_r + \mathrm{d}\beta_r, \tag{4.8}$$

$$\mathrm{d}(\alpha_r \wedge \beta_s) = \mathrm{d}\alpha_r \wedge \beta_s + (-1)^r \alpha_r \wedge \mathrm{d}\beta_s, \tag{4.9}$$

$$\mathrm{d}(\mathrm{d}\alpha_r) = 0. \tag{4.10}$$

Equation (4.8) is an obvious consequence of the linearity of the curl and the scalar product; (4.9) follows from (2-1.50); and (4.10) follows from (4-3.11).

Equation (4.7) shows that $\mathrm{d}\alpha_r$ is equivalent to the projected curl $\nabla \wedge A_r = P(\partial \wedge A_r)$; this reveals the chief limitation of the exterior derivative. In Chapters 4 and 5, we saw that even if A_r is a tangent field, $\partial \wedge A_r$ has a nontangential component which may contain significant information about intrinsic as well as extrinsic properties of the manifold. Such information is suppressed from the beginning when calculus on manifolds is formulated in terms of differential forms. Thus, the calculus of forms is not sufficiently general to give a complete account of manifold theory. This limitation is overcome in the literature by extending the formalism in various ways, for example, with the theory of fibre bundles. But such complications are unnecessary if Geometric Calculus is used from the beginning. The transformation of a differential r-form induced by a point transformation $f : x \to x' = f(x)$ is determined by assuming that the numerical value of the r-form is unchanged. According to Chapter 4, the induced transformation of an r-vector field $\mathrm{d}X_r$ is

$$\mathrm{d}X'_r = \underline{f}(\mathrm{d}X_r), \tag{4.11}$$

so, by virtue of (3-1.15) relating differential and adjoint transformations, we have

$$\begin{aligned} \alpha'_r(x') = f^*(\alpha_r) &\equiv A'_r \cdot \mathrm{d}X'_r = A'_r \cdot \underline{f}(\mathrm{d}X_r) \\ &= \overline{f}(A'_r) \cdot \mathrm{d}X_r = A_r \cdot \mathrm{d}X_r = \alpha'_a(f(x)) = \alpha_r(x), \end{aligned} \tag{4.12}$$

where $A_r = \bar{f}(A'_r)$. In the next chapter we show that Eqn. (4.12) is just a special case of the rule for transforming integrals by substitution, and we avoid the common notation f^* for the induced transformation of forms. But just now, let us verify that, when defined by (4.12), f^* has the usual properties attribted to it. Obviously, Eqn. (4.12) entails the linearity

$$f^*(\alpha_r + \beta_r) = f^*(\alpha_r) + f^*(\beta_r), \tag{4.13}$$

and because of (3-1.13),

$$f^*(\alpha_r \wedge \beta_s) = f^*(\alpha_r) \wedge f^*(\beta_s). \tag{4.14}$$

The fact that exterior differentiation commutes with f^* follows directly from (4-5.30b) and (4.7); thus,

$$\mathrm{d}\alpha'_r = \mathrm{d}f^*(\alpha_r) = f^*(\mathrm{d}\alpha_r). \tag{4.15}$$

Ordinarily $f^*(\mathrm{d}_r)$ is defined by using coordinates and coordinate substitutions. We have just shown that this is unnecessary if differential forms are obtained from Geometric Calculus.

The 'adjoint' δ of the exterior derivative is sometimes defined by the operator equation

$$\delta \equiv (-1)^{n(r+1)+1} * \mathrm{d} *. \tag{4.16}$$

With the help of (4.3), (4.7) and (4-3.9), we find

$$\begin{aligned} *\mathrm{d}*\alpha_r &= \mathrm{d}X^\dagger_{r-1} \cdot ([\nabla \wedge (\nabla^\dagger I)]^\dagger I \\ &= (-1)^{n-r}(-1)^{r(n-1)}\mathrm{d}X^\dagger_{r-1}([\nabla \wedge (A_r I^\dagger)]I) \\ &= (-1)^{n(r+1)}\mathrm{d}X^\dagger_{r-1} \cdot (\nabla \cdot A_r). \end{aligned} \tag{4.17a}$$

Hence,

$$\delta\,\alpha_r = -\mathrm{d}X^\dagger_{r-1} \cdot (\nabla \cdot A_r). \tag{4.17b}$$

Thus, δ is equivalent to the codivergence ∇ defined in Section 4-3, and, comparison of (4.17b) with (4-1.28) shows that δ can be regarded more generally as the 'interior derivative' mentioned in Section 4-1.

Obviously, 'differential identities' in the language of differential forms can be obtained by multiplying the identities in Sections 2-1 and 4-2 by arbitrary multivector fields to obtain scalar equations. Given the identifications made above, it is an easy matter to derive and relate any of the identities in the literature on differential forms to identities in Geometric Algebra. We give only one example. Especially with the help (1-1.38), (4-3.23) and (4-3.27), we find

$$\begin{aligned} \mathrm{d}\alpha_r &= (\mathrm{d}X_{r+1} \cdot \nabla) \cdot A_r^\dagger \\ &= (\mathrm{d}\dot{X}_{r-1} \cdot \dot{\nabla}) \cdot A_r^\dagger + (\mathrm{d}\dot{X}_{r-1} \cdot \dot{\nabla}) \cdot A_r^\dagger \\ &= \sum_{k=1}^{r+1} \mathrm{d}x_k \cdot \nabla\alpha_r(\mathrm{d}x_1 \wedge \ldots \mathrm{d}\check{x}_k \ldots \wedge \mathrm{d}x_{r+1}) + \\ &\quad + \sum_{j<k} (-1)^{j+k}\alpha_r([\mathrm{d}x_j, \mathrm{d}x_k] \wedge \mathrm{d}x_1 \wedge \ldots \mathrm{d}\check{x}_j \ldots \mathrm{d}\check{x}_k \ldots \wedge \mathrm{d}x_{r+1}). \end{aligned} \tag{4.18}$$

The right side of (4.18) is sometimes given as a definition of the exterior derivative in terms of the Lie bracket.

What we have said so far about the calculus of forms could as well have been said at the end of Chapter 4. But Cartan's formulation of differential geometry in terms of forms (see, for example, p. 61 of ref. [Hi]) should be compared with our formulation in Section 6.2.

A frame field $\{e_k\}$ with reciprocal frame $\{e^k\}$ determines the so-called *dual frame of one-forms*:

$$\alpha^k = \alpha^k(a) \equiv a \cdot e^k, \tag{4.19}$$

which would be referred to as the components of a vector field $a = a(x)$ in our language. According to (4.7), the exterior differential of α^k is a two-form equivalent to the cocurl of e^k,

$$\mathrm{d}\alpha^k = \mathrm{d}\alpha^k(a \wedge b) = (b \wedge a) \cdot (\nabla \wedge e^k). \tag{4.20}$$

The so-called *connexion one-forms* W_j^i are defined by

$$W_j^i = W_j^i(a) \equiv a \cdot L_j^i, \tag{4.21a}$$

with

$$L_j^i \equiv L_{jk}^i e^k, \tag{4.21b}$$

where the L_{jk}^i are the coefficients of connexion introduced by Eqn. (2.29). When we wrote Eqn. (2.29), we were concerned with a coordinate frame. However, an equation of that kind obviously holds for the arbitrary frame $\{e_k\}$ which concerns us here.

By differentiating $e^k \cdot e_j = \delta_j^k$ and using (2.29), we see that

$$(e_i \cdot \nabla e^k) \cdot e_j = -e^k \cdot (e_i \cdot \nabla e_j) = -L_{ij}^k = -L_i^k \cdot e_j.$$

Thus,

$$e_i \cdot \nabla e^k = -L_i^k, \tag{4.22}$$

whence,

$$\nabla \wedge e^k = L_i^k \wedge e^i. \tag{4.23}$$

'Dotting' (4.23) with $b \wedge a$ and using (4.20) and (4.21), we get a corresponding equation in terms of differential forms:

$$d\alpha^k = W_i^k \wedge \alpha^i, \tag{4.24}$$

where we have used the outer product of forms defined by (4.2), that is,

$$W_i^k \wedge \alpha^i = (b \wedge a) \cdot (L_i^k \wedge e^i) = W_i^k(a)\alpha^i(b) - W_i^k(b)\alpha^i(b). \tag{4.25}$$

Equation (4.25) is called *Cartan's first structural equation.** It is, of course, equivalent to Eqn. (4.23). We have already noted that

$$\nabla \wedge \nabla \phi = 0 \tag{4.26}$$

for scalar ϕ implies that $\nabla \wedge e^k = \nabla \wedge \nabla x^k = 0$ for a coordinate frame; whence, by (4.23),

$$L_i^k \wedge e^i = 0, \tag{4.27}$$

and this is equivalent to $L_{ij}^k = L_{ji}^k$, or, in words, 'the coefficients of connexion for a coordinate frame are symmetric'. Thus, Eqn. (4.25) merely expresses the elementary property of coderivatives (4.26) as a property of the coefficients of connexion for an arbitrary frame.

Cartan's second structural equation is

$$dW_i^k + W_j^k \wedge W_i^j = \Omega_j^k, \tag{4.28}$$

where, in accordance with (4.21), (4.7) and (4.2), we have

$$dW_i^k = (b \wedge a) \cdot (\nabla \wedge L_i^k), \tag{4.29}$$

and

$$W_j^k \wedge W_i^j = (b \wedge a) \cdot (L_j^k \wedge L_i^j), \tag{4.30}$$

* Cartan's first structural equation is often written with a torsion term (see p. 62 of [Hi]). This is not so much a generalization of Eqn. (4.25) as it is a different definition of a connexion on the manifold. Torsion can be introduced or eliminated at will by our method in Section 5-6.

while

$$\Omega_j^k \equiv (b \wedge a) \cdot R(e^k \wedge e_j) = (e_j \wedge e^k) \cdot R(a \wedge b), \tag{4.31}$$

and $R(a \wedge b)$ is the curvature tensor specified by Eqn. (2.9). Obviously Eqn. (4.28) is equivalent to the equation

$$\nabla \wedge L_j^k + L_j^k \wedge L_i^j = R(e^k \wedge e_i). \tag{4.32}$$

For a coordinate frame $\{e_k\}$, Eqns. (4.29), (4.30), and (4.31) give

$$(e_n \wedge e_m) \cdot (\nabla \wedge L_i^k) = \partial_m L_{in}^k - \partial_n L_{im}^k,$$

$$(e_n \wedge e_m) \cdot (L_j^k \wedge L_i^j) = L_{jm}^k L_{in}^j - L_{jn}^k L_{im}^j,$$

$$R_{imn}^k \equiv (e_i \wedge e^k) \cdot R(e_m \wedge e_n),$$

whence (4.28) yields the classical covariant form of the curvature tensor:

$$R_{imn}^k = \partial_m L_{in}^k - \partial_n L_{im}^k + L_{jm}^k L_{in}^j - L_{jn}^k L_{in}^j. \tag{4.33}$$

Having established this relation to tensor analysis, we will not bother with a more formal proof of Eqn. (4.28).

Though we have seen that Cartan's structural equations are readily formulated with Geometric Calculus, we shall not make use of them, because alternative expressions for the curvature tensor developed in Sections 6-2 and 5-1 exploit the Geometric Calculus more efficiently.

The calculus of forms is insufficient for most applications. It must be supplemented by other algebraic systems such as matrices or tensors. Thus, though [MTW] strongly advocates the use of forms in physics, it frequently resorts to tensor methods. The algebra of forms is not competent to handle the theory of linear transformations in Chapter 3, though it is intimately related to the subject. Only in Geometric Calculus are spinors, tensors and linear transformations as well as forms developed in a unified mathematical system.

The exterior product is the fundamental algebraic operation in the calculus of forms. It is generally defined first for forms; consequently, an independent definition will be required when the similar product for vectors is needed. Such redundancy in the algebraic axioms is unnecessary. Geometric Algebra relates the two kinds of product by Eqn. (4.2), which shows that the exterior product of forms is readily obtained from the outer product of vectors but not vice versa. Thus, if we regard the product of vectors as fundamental, the product of forms is quite superfluous.

Save for integration theory, which is handled in the next chapter, we have shown that the basic relations and equations in the calculus of differential forms can be formulated with equal or greater efficiency in terms of Geometric Calculus. At the same time we have noted that, by itself, the calculus of forms has several serious drawbacks and limitations which are avoided by Geometric Calculus. These points should have decisive bearing on the selection of a mathematical system, so let us review and discuss them further.

The virtues of Cartan's exterior differential are undoubtedly responsible for the popularity of his calculus. But, according to (4.7) and (4.17), the exterior differential dα and its 'adjoint' $\delta\alpha$ are respectively equivalent to the cocurl $\nabla \wedge A$ and the codivergence $\nabla \cdot A$. This brings to light two major deficiencies of the exterior differential. First, since $\nabla \wedge A = P(\partial \wedge A)$ is the projection of the curl $\partial \wedge A$ into the tangent algebra, it follows that the exterior differential projects away all the information about extrinsic geometry and its relation to intrinsic geometry discovered in Chapters 4 and 5. The lost information can be restored only at the cost of unnecessary complications in the theory.

The second major deficiency of the exterior differential and its calculus is the fact that it does not discover any relation corresponding to

$$\nabla A = \nabla \cdot A + \nabla \wedge A, \tag{4.34}$$

which makes it possible to regard $\nabla \cdot A$ and $\nabla \wedge A$ as two distinct parts of a single more fundamental quantity, the coderivative ∇A. The importance of (4.34) is apparent in many places in this book and in applications to physics. For example, ref. [Hl] shows that (4.34) makes it possible to reduce Maxwell's equations for the electromagnetic field to a single equation, whereas tensor analysis and the calculus of forms must deal with two separate equations corresponding to the two parts of (4.34). The combination of $\nabla \cdot A$ and $\nabla \wedge A$ in (4.34) is more than a mere abbreviation or triviality, because, as shown in Section 7-4, when $\nabla A = \partial A$, it can be directly solved for A, whereas $\nabla \cdot A$ and $\nabla \wedge A$ cannot, because each is only part of the derivative of A.

The calculus of forms was developed expressly to unify and simplify integration theory. But it fails to provide a straightforward account of one of the most remarkable and useful results in mathematics, namely, Cauchy's integral formula. The reason is that the calculus of forms works with $\delta\alpha$ and dα separately, and, for two-dimensional manifolds, this is equivalent to working separately with the real and imaginary parts of complex functions. On the other hand, by exploiting (4.34), Geometric Calculus provides an efficient formulation not only of the celebrated 'Generalized Stokes' Theorem', but also of Cauchy's integral formula and its generalization to manifolds of any dimension. The details are given in Chapter 7.

Besides the major deficiencies just mentioned, the calculus of forms suffers a number of lesser defects. Compare, for example, the exterior derivative dϕ of a scalar-valued function $\phi = \phi(x)$ with the vector derivative $\nabla\phi = \nabla \wedge \phi$. According to (4.7), they are related by the equation

$$\mathrm{d}\phi = \mathrm{d}x \cdot \nabla\phi = \mathrm{d}x \cdot (\nabla\phi). \tag{4.35}$$

The utility of the vector $\nabla\phi$ (the gradient) as indicating the direction of change of ϕ is evident to everyone. Yet the calculus of forms can represent the gradient only indirectly in terms of its projection dϕ onto an arbitrary direction indicated by the vector dx in (4.35). The notation dϕ hides the dependence of the exterior derivative on the vector dx, which for many purposes is superfluous. More generally, Eqn.

(4.7) shows that the exterior derivative implicitly introduces the superfluous variable dX_r in its representation of the curl of a vector or a multivector. Of course, the dX_r is needed under integrals, but even there it must be inserted in a more general way than dictated by (4.7) if it is to yield results like Cauchy's integral (see Chapter 7). However, as ref. [Hl] shows, the dX_r is quite superfluous in the basic differential equations of physics. Moreover, for three-dimensional manifolds, Geometric Calculus yields the well-established vector calculus due to Gibbs without alternation of definitions, notations or point of view (see, for example, p. 72 of Ref. [Hl]). In contrast, the calculus of forms does not mesh well with vector analysis; it calls for a translation of all the standard vector equations of physics, a translation which is as unnecessary as it is troublesome.

Cartan's calculus is often said to provide a coordinate-free formulation of differential geometry. But actually, it is only nominally coordinate-free. By this we mean that, though some equations and manipulations do not explicitly refer to coordinates, nevertheless, for some basic definitions, proofs and manipulations, reference to coordinates cannot be avoided. There are two distinct reasons for the implicit coordinate dependence of Cartan's calculus. First, manifolds are usually defined as sets to which coordinates can be assigned. Having been included in the very definition of a manifold, coordinates must then necessarily be employed in the development of calculus. However, we have seen in Chapter 4 that manifolds can be defined without reference to coordinates. Hence Eqn. (4.7) can be regarded as a coordinate-free definition of the exterior derivative. But, of course, we used Geometric Calculus to achieve this.

A second reason that Cartan's calculus must resort to coordinates is the fact that it works only with scalar quantities defined on the tangent algebra of a manifold. It was only by going outside the tangent algebra that we were able to achieve a completely coordinate-free formulation of differential geometry in Chapters 4 and 5. Coordinates are as essential to our treatment of intrinsic geometry in Section 6-2 as they are to the formulation and the application of Cartan's structural equations described in this section.

To sum up, we have shown that Geometric Calculus retains the advantages of the conventional calculus of forms while removing its deficiencies and exhibiting a greater range and flexibility.

Chapter 7

Directed Integration Theory

This chapter describes some basic contributions of Geometric Calculus to the theory of integration. The directed integral enables us to formulate and prove a few comprehensive theorems from which the main results of both real and complex variable theory are easily obtained.

The main feature of this chapter is the new form given to the fundamental theorem of calculus in Section 7-3. This leads to the remarkable generalization of Cauchy's integral formula in Section 7-4. The power of this result in general theoretical arguments is demonstrated by the new constructive proofs of the inverse and implicit function theorems in Section 7-6.

Modern integration theory on manifolds is usually couched in the language of differential forms. To facilitate contact with the literature, this chapter conforms as closely to that language as is consistent with the integrity of the present system. Aside from some important notational deviations discussed in Section 4-1, the geometric integration theory in this chapter differs from conventional approaches chiefly in its explicit use of directed measure, the vector derivative and multivector-valued forms.

7-1. Directed Integrals

This section introduces the notion of a directed integral. To facilitate comparison with more conventional theories and explain the main ideas with a minimum of argument, the directed integral is defined by relating it to the standard Riemann integral. Although the Riemann integral is commonly regarded as less satisfactory than the Lebesgue integral, it has the advantage of a straightforward geometric interpretation. A completely satisfactory theory of integration should combine the geometric features of the Riemann theory with the generality of the Lebesgue theory. A unified theory of Riemann and Lebesgue integration has been proposed by McShane [Mc]. We submit that integration theory could be improved substantially by exploiting the unique features of Geometric Algebra from the outset.

For a more detailed treatment of integration theory, the reader is referred to the book by Whitney [Wh], which is especially related to the spirit of Geometric Calculus.

We will be concerned with integrals over an m-dimensional manifold $\mathcal{M}$ with boundary $\partial\mathcal{M}$. Our discussion presumes the definition of a *smooth oriented* vector manifold given in Section 4-1.

Let $f = f(x)$ be a multivector-valued function defined on the manifold $\mathcal{M}$. The *directed* integral of f over $\mathcal{M}$ is defined by

$$\int_{\mathcal{M}} \mathrm{d}Xf = \int_{\mathcal{M}} \mathrm{d}X(x)f(x) \equiv \lim_{n\to\infty} \sum_{i=1}^{n} \Delta X(x_i)f(x_i). \tag{1.1}$$

The limit on the right side of (1.1) is to be understood in the usual sense of Riemann integration theory. It is well defined, because multivectors have a unique norm defined by (1-1.49) or, more generally, by (1-5.1).* The 'magnitudes' $|\mathrm{d}X|$ and $|\Delta X|$ are each to be understood as the usual Riemann measure of 'volume' for $\mathcal{M}$. The direction of a volume element at x is characterized by the unit pseudoscalar $I(x)$; thus,**

$$\Delta X(x_i) = |\Delta X(x_i)|I(x_i), \tag{1.2a}$$

$$\mathrm{d}X(x) = |\mathrm{d}X(x)|I(x). \tag{1.2b}$$

Obviously, the *directed integral of f* is equivalent to the *Riemann integral of If*;

$$\lim_{n\to\infty} \sum_{i=1}^{n} |\Delta X(x_i)|I(x_i)f(x_i) = \int_{\mathcal{M}} |\mathrm{d}X|If = \int_{\mathcal{M}} \mathrm{d}Xf. \tag{1.3}$$

Note that the directed integral is an *oriented integral*. The orientation is determined by the unit pseudoscalar field $I = I(x)$, which, as we have pointed out before, assigns an orientation to $\mathcal{M}$.

The directed integral (1.1) is a nontrivial generalization of the Riemann integral, in spite of the fact that the two are related by (1.3). The reason is that a complete description of the local direction and orientation of the manifold has been built in to the directed integral, whereas it must be supplied *ad hoc* to the usual Riemann integral. The directed integral would best be founded on a generalization of measure theory characterizing 'directed measure' instead of 'scalar measure'. A 'directed measure' associates a direction and a dimension as well as a magnitude to a set. Accordingly, the directed integral (1.1) may be said to use 'directed Riemann measure' instead of the usual 'scalar Riemann measure'. Of course, it is the geometric product that enables us to represent direction algebraically. Geometric Algebra is essential to the theory of directed integration.

* Later on we will suppose that integrals are defined in the slightly more general sense of distribution theory, but the reader is referred to the literature [GS] for details.

** Sometimes it is of interest to integrate over *null* manifolds, such as the light cone in spacetime. For such manifolds all pseudoscalars have vanishing square, so a measure cannot be specified by using (1.2) to relate it to Riemann measure. The integral (1.1) may be well defined nevertheless, but we need not go into such details here.

The 'volume' or '*constant*' $|\mathcal{M}|$ of the manifold $\mathcal{M}$ is given by

$$|\mathcal{M}| \equiv \int_{\mathcal{M}} \mathrm{d}XI^{-1} = \int_{\mathcal{M}} |\mathrm{d}X|. \tag{1.4}$$

The integral

$$\vec{\mathcal{M}} \equiv \int_{\mathcal{M}} \mathrm{d}X \tag{1.5}$$

may be called the 'directed volume' or the 'directed content' of $\mathcal{M}$. It defines an 'average direction' for $\mathcal{M}$. For a flat manifold $I = I(x)$ is constant, hence

$$\vec{\mathcal{M}} = I|\mathcal{M}|. \tag{1.6}$$

On the other hand, for any *closed* manifold, the definition (1.1) implies that

$$\vec{\mathcal{M}} = \oint_{\mathcal{M}} \mathrm{d}X = 0, \tag{1.7}$$

where $\oint$ indicates that the integral is over a closed manifold. Intuitively, it is clear that (1.7) obtains because on a closed manifold 'directed volume elements' occur in pairs with opposite orientations which cancel when added.

Since the geometric product is noncommutative, the position of the volume element 'under the integral' is important. Accordingly, we generalize (1.1) to

$$\int_{\mathcal{M}} g\, \mathrm{d}Xf \equiv \lim_{n\to\infty} \sum_{i=1}^{n} g(x_i)\, \Delta X(x_i) f(x_i), \tag{1.8}$$

where, of course, f and g are multivector-valued functions defined on $\mathcal{M}$. Actually, we can reduce (1.8) to (1.1) by writing $g\, \mathrm{d}Xf = \mathrm{d}Xg'f$, where, for nonnull manifolds, $g' = I^{-1}gI$. But this device is usually awkward as well as unnecessary.

The most general form for a directed integral is

$$\int_{\mathcal{M}} L(\mathrm{d}X) = \int_{\mathcal{M}} L(x, \mathrm{d}X(x)) \equiv \lim_{n\to\infty} \sum_{i=1}^{n} L(x_i, \Delta X(x_i)), \tag{1.9}$$

where $L = L(x, \mathrm{d}X(x))$ is a multiform of degree m on $\mathcal{M}$, that is, L is a multivector valued linear function of the pseudoscalar *field* $\mathrm{d}X = \mathrm{d}X(x)$. From our study of multiforms in Section 3-9, we know that there exist functions $g_k(x)$ and $f_k(x)$ such that L can be expressed in the form

$$L(\mathrm{d}X) = \sum_k g_k\, \mathrm{d}Xf_k. \tag{1.10}$$

Hence, the integral (1.9) can be reduced to a sum of integrals of type (1.8). However, such a reduction is not always necessary or desirable.

Defining f by $f = \Sigma_k f_k g_k$, it follows from (1.10) that the scalar part of L can be written in the form

$$\langle L\rangle = \langle \mathrm{d}Xf\rangle = \mathrm{d}X \cdot \langle f\rangle_m . \tag{1.11}$$

As observed in Section 4-1, this is the general expression for a differential form. Thus, the scalar parts of the integrals (1.9) and (1.1) are equivalent expressions for the integral of a differential form, that is

$$\left\langle \int_{\mathscr{M}} L \right\rangle = \int_{\mathscr{M}} \langle L\rangle = \int_{\mathscr{M}} \langle \mathrm{d}Xf\rangle. \tag{1.12}$$

From this view, conventional integration theory of differential forms is seen as a special case of 'directed integration theory'.

Of course, one could adopt the alternative view that a directed integral is no more general than the integral of a differential form, because a multivector-valued integral can be reduced to a set of scalar-valued integrals.

It may be noticed that many notions of homology theory are more readily expressed by directed integrals than by the scalar-valued integrals of cohomology theory. For example, Eqn. (1.7) could be regarded as a condition that a connected manifold be closed. It remains to be seen to what degree homology theory can be regarded as a calculus of directed integrals.

This chapter will show that the directed integral extends the well-known virtues of complex integrals to the whole of integration theory. The directed integral literally brings new dimensions to the theory of integration.

7-2. Derivatives from Integrals

Considering the performance of the vector derivative in preceding chapters, there can be no doubt that it should be regarded as the fundamental differential operator on vector manifolds. Yet in Chapter 4 we obtained the vector derivative from the directional derivative. The directed integral provides us with a more fundamental approach. It enables us to define the vector derivative directly in terms of a limit process. Then, as we have seen, the directional derivative obtains by elementary algebra.

The derivative $\partial f = \partial_x f(x)$ of a function $f = f(x)$ on $\mathscr{M}$ can be defined by any of the equivalent limits

$$\partial f(x) \equiv \lim_{\boldsymbol{\mathscr{R}} \to 0} \frac{1}{\boldsymbol{\mathscr{R}}} \oint \mathrm{d}Sf = \lim_{|\mathscr{R}| \to 0} \frac{I^{-1}(x)}{|\mathscr{R}(x)|} \oint \mathrm{d}Sf, \tag{2.1}$$

where: (1) The set $\mathscr{R} = \mathscr{R}(x)$ is a *neighborhood* of the point x in $\mathscr{M}$, that is, $\mathscr{R}$ is an open smooth m-dimensional submanifold of $\mathscr{M}$ with x as an interior point. In the limit, the directed content $\boldsymbol{\mathscr{R}}$ (x) can be replaced by $|\mathscr{R}(x)| I(x)$, where $I(x)$ is the unit pseudoscalar at x.

(2) The directed integral of f is taken over the boundary of $\mathscr{R}$. The $(m-1)$-vector $\mathrm{d}S$, representing a directed volume element of $\partial\mathscr{R}$, is oriented so that

$$\mathrm{d}S(x')n(x') = I(x')\,|\mathrm{d}S(x')| \tag{2.2a}$$

or, equivalently,

$$\mathrm{d}S = In^{-1}|\mathrm{d}S| = (-1)^{m-1}n^{-1}I|\mathrm{d}S|, \tag{2.2b}$$

where $n = n(x')$ is the *outward unit* normal vector at a point x' of $\partial\mathscr{M}$.* According to (2.2b) the direction of $\mathrm{d}S$ is In^{-1}, which is obtained by 'dividing out' the normal direction from the unit pseudoscalar.

(3) The limit is taken by shrinking $\mathscr{R}$ and hence its volume $|\mathscr{R}|$ to zero at the point x. We allow the limit to be proportional to a delta function or its derivatives, so the limit is well defined in the sense of distribution theory, ref. [GS]. A precise discussion of the limiting process is too involved to be given here. It requires, however, only standard arguments of analysis.

The definition of the derivative $\partial = \partial_x$ in terms of an integral by (2.1) is equivalent to the definition of ∂ in Section 4-1, but the proof will be left to the reader. The definition of the derivative as the limit of an integral is not only independent of coordinates, but, as we shall see, it clarifies and simplifies the relation of integration to differentiation. These are compelling reasons to introduce integrals before derivatives in a systematic development of calculus.

Without taking the trouble to re-establish all the properties to the derivative found in preceding chapters, let us consider some examples to show how the basic definition (2.1) can be applied.

Consider the evaluation of the limit (2.1) in the simplest case, when $\mathscr{M}$ is a one-dimensional manifold. In this case, $\mathscr{R}$ is an oriented curve passing through the point x with unit tangent vector $I(x)$ and arc length $s \equiv |\mathscr{R}|$. The boundary of the curve consists of the end points, say x_1 at the beginning and x_2 at the end of the curve. Since the boundary of $\mathscr{R}$ consists of only two points, the integral over the boundary is evaluated by trivial appeal to the definition (1.1). The only problem is to ascertain the volume element ΔS on the boundary. To do this, note that at the boundary points the outward normals are given by $n(x_2) = I(x_2)$ and $n(x_1) = -I(x_1)$, so, from (2.2), we get*

$$\Delta S(x_2) = I^2 = 1, \qquad \Delta S(x_2) = -I^2 = -1. \tag{2.3}$$

(Note that to get (2.3) we have assumed the Euclidean norm $I^2 = 1$. We could as well have assumed $I^2 = -1$. Indeed, the latter convention obtains for space-like curves imbedded in a space-time manifold. But the overall sign is not a matter of concern here.)

* It should be noted that n^{-1} in (2.2b) depends on the signature of $\mathscr{M}$. For Euclidean signature $n^2 = 1$, so $n^{-1} = n$. For other signatures, one may have $n^2 = -1$, whence $n^{-1} = -n$. This is important in applications to space-time physics.

Thus,

$$\oint dSf = \Delta S(x_2)f(x_2) + \Delta S(x_1)f(x_1) = f(x_2) - f(x_1). \tag{2.4}$$

Hence, (2.1) reduces to

$$\partial f(x) = I(x) \lim_{s\to 0} \frac{1}{s}(f(x_2) - f(x_1)). \tag{2.5}$$

The limit on the right side of (2.5) will be recognized as the average of left and right derivatives with respect to arc length.

As another important example, let us apply (2.1) to evaluate the derivative of the unit pseudoscalar $I = I(x)$. Recall that ∂I was previously expressed in terms of the *spur* N by Eqn. (4-4.7), which says

$$\partial I = -NI = (-1)^{m+1} IN. \tag{2.6}$$

Comparing this with what we obtain by using (2.1), we see that

$$N = \lim_{|\mathscr{R}|\to 0} (-1)^{m+1} \oint \frac{dSI^{-1}}{|\mathscr{R}|} = \lim_{|\mathscr{R}|\to 0} \oint \frac{|dS|n^{-1}}{|\mathscr{R}|}. \tag{2.7}$$

This result gives us new insight into the significance of the spur $N = N(x)$, for the right side of (2.7) can be interpreted as specifying the average direction of the normal on the boundary of an infinitestimal neighborhood of the point x. A simple figure showing the normals on the boundary of a neighborhood to a point in a one- or two-dimensional manifold makes it easy to see why N is always orthogonal to the manifold (i.e. $N \cdot I = 0$). Indeed, for the one-dimensional case, we can use (2.5) to evaluate (2.7), and we find that

$$N = \frac{dI}{ds}. \tag{2.8}$$

Thus, the spur of a curve is just the 'acceleration' of the curve (with respect to arc length). The concept of spur generalizes the concept of acceleration for curves to manifolds of arbitrary dimension. This links the geometrical theory of curves to the theory of higher dimensional manifolds. According to (6-3.15) and (6-3.17), the magnitude of its acceleration is the *first curvature* of a curve. Just as the inflection points of a curve are determined by the vanishing of its acceleration, so a general class of inflection points on any manifold are determined by a vanishing spur.

To allow for differentiation to the left as well as to the right, we generalize (2.1) to the definition

$$\dot{g}\,\partial\dot{f} \equiv \lim_{|\mathscr{R}|\to 0} \oint g' \frac{I^{-1}\, dS'f'}{|\mathscr{R}|} = \lim_{|\mathscr{R}|\to 0} (-1)^{m-1} \oint g' \frac{dS'I^{-1}f'}{|\mathscr{R}|}. \tag{2.9}$$

In (2.9) the primes indicate quantities evaluated at a point x' on the boundary of $\mathscr{R}$; thus, $f' = f(x')$. It is essential to note that $I^{-1} = I^{-1}(x)$ in (2.9) is evaluated at the limit point x and not on the boundary, even though it appears 'under the integral'. We cannot 'pull' I^{-1} out to the left in (2.9) like we did in (2.1), because I^{-1} may not commute with g'. Also, we cannot use (2.2b) to replace $I^{-1}\,\mathrm{d}S' = I^{-1}(x)\,\mathrm{d}S(x')$ in (2.9) by $I^{-1}(x')\,\mathrm{d}S(x') = n^{-1}(x')|\mathrm{d}S(x')|$, although, as indicated by (2.9), $I^{-1}\,\mathrm{d}S'$ is equivalent in the limit to $(-1)^{m-1}\,\mathrm{d}S'I^{-1}$. It is most important to distinguish the directed measure $\mathrm{d}S$ from the measure $I^{-1}\,\mathrm{d}S = n^{-1}|\mathrm{d}S|$. The two measures are equivalent only when I is independent of x, that is, when the manifold is flat. Thus, although $\oint \mathrm{d}S = 0$ for any bounding manifold, Eqn. (2.7) shows that, in general,

$$\oint |\mathrm{d}S| n^{-1} = \oint I^{-1}\,\mathrm{d}S \neq I^{-1} \oint \mathrm{d}S = 0. \tag{2.10}$$

The derivative operator $\partial = \partial_x$ defined by (2.9) is fully equivalent to the derivative as it was defined in Section 4-1. Instead of presenting a proof of this fact, we point out that the basic properties of the derivative can be proved directly from (2.9). Consider for instance, the 'Leibnitz Rule' for differentiating a product in the form

$$\dot{g}\,\dot{\partial} \dot{f} = \dot{g}\,\dot{\partial} f + g\,\dot{\partial} \dot{f}. \tag{2.11}$$

This is easily proved from the identity

$$\oint g' \frac{I^{-1}\,\mathrm{d}S' f'}{|\mathscr{R}|} = \left\{\oint g' \frac{I^{-1}\,\mathrm{d}S'}{|\mathscr{R}|}\right\} f + g \left\{\oint \frac{I^{-1}\,\mathrm{d}S' f'}{|\mathscr{R}|}\right\} +$$

$$+ \oint (g' - g)\frac{I^{-1}\,\mathrm{d}S'}{|\mathscr{R}|}(f - f') - g\,\frac{I^{-1}}{|\mathscr{R}|}\left\{\oint \mathrm{d}S'\right\} f. \tag{2.12}$$

The last term in (2.12) is identically zero while the next to last term vanishes in the limit.

Our definition of the derivative deserves a final generalization. Let $T = T(n) = T(x, n(x))$ be a tensor function on $\mathscr{M}$. In accord with our definition in Section 4-1, the *divergence* of T is given by

$$\dot{T}(\dot{\partial}) \equiv \lim_{|\mathscr{R}|\to 0} |\mathscr{R}|^{-1} \oint T(I^{-1}\,\mathrm{d}S). \tag{2.13}$$

Obviously this includes (2.1) and (2.6) as special cases.

Our definition of exterior differential in Section 4-1 can be founded on (2.13). If L is a multiform of degree k, and A is a multivector field of grade $k + 1$, then $L(A \cdot n)$ is a tensor function of a vector n. According, the *exterior differential* $\mathrm{d}L$ is given by

$$\mathrm{d}L(A) = \dot{L}(A \cdot \dot{\partial}) = \dot{L}(\dot{A} \cdot \dot{\partial}) - L(\dot{A} \cdot \dot{\partial}). \tag{2.14}$$

The last two terms in (2.14) can be evaluated directly from (2.9) and (2.1).

7-3. The Fundamental Theorem of Calculus

The Fundamental Theorem: Let $\mathcal{M}$ be an m-dimensional smooth oriented vector manifold with a piecewise smooth boundary $\partial\mathcal{M}$. Let $L = L(A)$ be a differentiable multiform of degree $(m - 1)$ on $\mathcal{M}$ and $\partial\mathcal{M}$. Then

$$\int_{\mathcal{M}} \dot{L}(\mathrm{d}X\, \dot{\partial}) = \int_{\partial\mathcal{M}} L(\mathrm{d}S), \tag{3.1}$$

where $\mathrm{d}X$ and $\mathrm{d}S$ are directed volume elements on $\mathcal{M}$ and $\partial\mathcal{M}$.

The derivative ∂ in (3.1) is, of course, the derivative with respect to a point on $\mathcal{M}$, as defined in Section 7-2. The term 'differentiable' means that the derivative of L exists in the sense of distribution theory. Since $\mathrm{d}X$ is a pseudoscalar, we have, in accord with (2-1.4),

$$\mathrm{d}X \wedge \dot{\partial} = 0, \tag{3.2a}$$

$$\mathrm{d}X\, \dot{\partial} = \mathrm{d}X \cdot \dot{\partial} = (-1)^{m-1} \dot{\partial}\, \mathrm{d}X. \tag{3.2b}$$

The property (3.2) entails that $\mathrm{d}X\, \dot{\partial}$, like $\mathrm{d}S$, is pseudovector-valued.

The great virtue of the formulation (3.1) of the fundamental theorem is that it holds for manifolds of any dimension. As a consequence, the famous special cases of the fundamental theorem, such as the theorems of Gauss and Stokes, can all be written in the same form and established by a single proof. The fundamental theorem formulated for arbitrary manifolds is frequently referred to as the 'generalized Stokes' Theorem', or simply *Stokes' Theorem*.

If we adopt the notation (1.11) for exterior differential, we can write (3.1) in the abbreviated form

$$\int_{\mathcal{M}} \mathrm{d}L = \int_{\partial\mathcal{M}} L. \tag{3.3}$$

Many modern mathematics texts formulate Stokes' theorem with the notation (3.3), but they limit their considerations to the case where L is scalar-valued. If the differential argument were not suppressed in (3.3) we would have the peculiar expression $\mathrm{d}L = \mathrm{d}L(\mathrm{d}X)$. Of course, the symbol 'd' is used here in two quite different senses, which must not be confused. Following long standing tradition, the symbol 'd' in 'dX' serves to indicate an *increment* of directed measure, and it is not well-defined except in connexion with integration. On the other hand, the 'd' indicating the exterior differential $\mathrm{d}L$ is well defined whether or not it appears 'under' an integral.

The employment of directed integrals is essential to the formulation of the fundamental theorem, although this fact can be disguised by various devices. For example, define a tensor function $T = T(n)$ by

$$T(n) = L(In), \tag{3.4}$$

where I is the unit pseuoscalar and L is the extensor function of (3.1). Then

$$\dot{T}(\dot{\partial}) = \dot{L}(I\,\dot{\partial}) + L(\dot{I}\,\dot{\partial}) = \dot{L}(I\,\dot{\partial}) - T(N), \tag{3.5}$$

where we have used

$$\dot{I}\dot{\partial} = -IN = (-1)^{m-1}NI, \tag{3.6}$$

obtained from Eqn. (4-4.7), to introduce the *spur* N of $\mathscr{M}$. Also

$$\mathrm{d}X = |\mathrm{d}X|I, \tag{3.7a}$$

and, assuming Euclidean signature, from (2.2b) we have

$$\mathrm{d}S = In|\mathrm{d}S|, \tag{3.7b}$$

where n is the outward normal. Consequently, the fundamental theorem (3.1) can be put in the form

$$\int_{\mathscr{M}} |\mathrm{d}X|[\dot{T}(\dot{\partial}) + T(N)] = \int_{\partial\mathscr{M}} |\mathrm{d}S|T(n). \tag{3.8}$$

The spur N appears in (3.8) as a consequence of the fact that $\mathrm{d}X$ is not differentiated in (3.1).

The fundamental theorem will frequently be applied in the special cases

$$\int \mathrm{d}X\,\partial f = \oint \mathrm{d}Sf \tag{3.9}$$

and

$$\int g\,\mathrm{d}X\,\partial f + (-1)^{m-1}\int \dot{g}\,\dot{\partial}\,\mathrm{d}Xf = \oint g\,\mathrm{d}Sf, \tag{3.10}$$

where we have suppressed the reference to $\mathscr{M}$, and, of course, f and g are differentiable functions on $\mathscr{M}$. To get (3.10), we used (2.7) and (3.2b).

Taking $g = I^{-1}$ in (3.10) and using (3.6) and (3.7), we get

$$\int |\mathrm{d}X|[\partial f + Nf] = \oint |\mathrm{d}S|nf. \tag{3.11}$$

This is just what we would obtain from (3.8) with $T(n) = nf$. Note how (3.11) simplifies for flat manifolds, because $N = 0$ when I is constant.

Since $\mathrm{d}S$ had grade $m - 1$, only the $(m - 1)$-vector part of f contributes to the scalar part of 3.9). Thus, writing $A = \langle f \rangle_{m-1}$, we have

$$\langle \mathrm{d}Sf \rangle = \mathrm{d}S \cdot A,$$

$$\langle \mathrm{d}X\,\partial f \rangle = (\mathrm{d}X \cdot \dot{\partial}) \cdot \dot{A} = \mathrm{d}X \cdot (\partial \wedge A).$$

Hence the scalar part of (3.9) can be written in the form

$$\int dX \cdot (\partial \wedge A) = \oint dS \cdot A. \qquad (3.12)$$

Following (6-4.7), we can identify $d\alpha = dX \cdot (\partial \wedge A)$ as the exterior derivative of the differential form $\alpha = dS \cdot A$. Thus (3.12) is fully equivalent to the conventional formulation of Stokes' theorem in terms of differential forms.

Alternatively, if the vector part of f is denoted by $a = \langle f \rangle_1$, then the scalar part of (3.11) can be written

$$\int |dX|[\partial \cdot a + N \cdot a] = \oint |dS| n \cdot a. \qquad (3.13)$$

This could fairly be called the generalized Gauss theorem. If a is a vector field then $N \cdot a = 0$ in (3.13), because $N \cdot I = 0$ according to (4-4.8). The same result could be obtained by substituting $A = Ia = I \cdot a$ into (3.12).

We have adopted the name 'fundamental theorem', because it has long been used in the conventional differential and integral calculus of scalars for what amounts to the one-dimensional case of (3.1). Thus, for an oriented curve $\mathscr{e}$ with end points x_1 and x_2, by virtue of (2.4) and (3.2), the fundamental theorem (3.1) can be put in the familiar form

$$\int_{\mathscr{e}} dX \cdot \partial f = f(x_2) - f(x_1). \qquad (3.14)$$

Our stipulation in the fundamental theorem that the boundary must be 'piecewise continuous' is already 'degenerate' in this case.

Having examined alternative formulations of the fundamental theorem, let us now discuss its proof. Suppose that $\mathscr{M}$ is partitioned into n regions and let ΔX_i denote the directed content of the ith region $\mathscr{M}_i$. A proof of the fundamental theorem in the form (3.9) is achieved by establishing the following sequence of equations:

$$\int_{\mathscr{M}} dX\, \partial f = \lim_{n \to \infty} \sum_{i=1}^{n} \Delta X_i \left\{ \frac{1}{\Delta X_i} \int_{\partial \mathscr{M}_i} dSf \right\}$$

$$= \lim_{n \to \infty} \sum_{i=1}^{n} \int_{\partial \mathscr{M}_i} dSf = \int_{\partial \mathscr{M}} dSf. \qquad (3.15)$$

Obviously, the proof is greatly simplified by using the expression (2.1) for the derivative as the limit of an integral instead of the original definition of derivative adopted in Chapter 4. Equally important is the fact that details of the proof do not depend on the dimension of $\mathscr{M}$. The limits in (3.15) can be established by standard techniques, so we need not consider them further. However, it is worth mentioning that most proofs of Stokes' theorem to be found even in advanced mathematics texts fail to establish the theorem in its utmost generality. For a careful discussion of this issue see [He] and [Ca].

The fundamental theorem gives us a number of useful results immediately. For example, using $f = 1$ in (3.9) we recover (1.7), and using $f = 1$ in (3.11), we get the generalization of (2.7):

$$\oint |\mathrm{d}S|n = \int |\mathrm{d}X|N. \tag{3.16}$$

Again, since we know from Chapter 5 that $\partial x = \partial \cdot x = m$, we get from (2.1) a formula for the directed content of $\mathscr{M}$:

$$\vec{\mathscr{M}} \equiv \int \mathrm{d}X = \frac{1}{m} \oint \mathrm{d}Sx = \frac{1}{m} \oint \mathrm{d}S \wedge x. \tag{3.17}$$

The $(m - 2)$-vector part of (3.17) is simply

$$\oint \mathrm{d}S \cdot x = 0. \tag{3.18}$$

Similarly, from (3.11) we get a general formula for the content of $\mathscr{M}$:

$$\begin{aligned} |\mathscr{M}| \equiv \int |\mathrm{d}X| &= \frac{1}{m} \oint |\mathrm{d}S| nx - \frac{1}{m} \int |\mathrm{d}X| Nx \\ &= \frac{1}{m} \oint |\mathrm{d}S| n \cdot x - \frac{1}{m} \int |\mathrm{d}X| N \cdot x. \end{aligned} \tag{3.19}$$

The bivector part of (3.19) is just

$$\int |\mathrm{d}X| N \wedge x = \oint |\mathrm{d}S| n \wedge x. \tag{3.20}$$

It is easy to prove that the last few formulas are independent of a choice of origin in spite of the explicit appearance of x. For flat manifolds $N = 0$, so (3.19) or (3.17) can be used to compute volume from a 'surface integral'. For example, for the m-ball $\mathscr{B}_m$ of radius R, (3.19) gives

$$|\mathscr{B}_m| = \frac{R}{m} |\partial \mathscr{B}_m|. \tag{3.21}$$

7-4. Antiderivatives, Analytic Functions and Complex Variables

The fundamental problem of calculus on manifolds is to invert the 'differential equation'

$$\partial f = s \tag{4.1}$$

to get the solution

$$f = \partial^{-1} s. \tag{4.2}$$

In this section we show how to use the fundamental theorem to get an explicit expression for the *antiderivative* operator ∂^{-1}. Actually, it is something of a misnomer to speak of (4.2) as a 'solution' to (4.1) unless certain 'boundary conditions' on f are specified; otherwise, (4.2) will be an integral equation equivalent to the differential equation (4.1).

It should be pointed out that a method for solving (4.1) enables us to solve the more general differential equation

$$h(\dot{\partial})\dot{f} = s, \tag{4.3}$$

when h can be expressed as the differential of a transformation, for then, because of (4-5.13), Eqn. (4.3) can be put in the form (4.1) by a change of variable. It may not be superfluous to point out as well that any system of partial differential equations can be expressed as an equation of the form (4.3). Thus, Geometric Calculus reduces the 'theory of partial differential equations' to the theory of operator functions $h(\partial)$ of a single fundamental differential operator ∂. From our point of view, the name 'theory of partial differential equations' is unfortunate, because it emphasizes the representation of the vector derivative ∂ by the partial derivatives, which are merely components of ∂ along coordinate directions. Of course, it takes Geometric Calculus to define ∂ and deal with it directly.

We shall see that existence of an antiderivative for (4.1) requires only that $s = s(x)$ be integrable. Thus s can be singular in the sense of distribution theory. This result is important in itself, for it says that any integrable function is the derivative of some other function. Thus, to an integrable function $s = s(x)$ there corresponds a function $f = f(x)$ called a *potential* of s, such that $s = \partial f$. The potential f is, of course, not uniquely determined without further conditions. However, it will be evident from the discussion below that equivalent potentials differ only by an analytic function. A well-known special case of the existence theorem for potentials is called the *Helmholtz Theorem* in vector analysis.

Now let us see how to determine the antiderivative. Suppose there exists a multivector-valued function $g = g(x, x')$ of points x and x' on the manifolds $\mathcal{M}$ and $\partial\mathcal{M}$ which satisfies the equation

$$\partial g = -\dot{g}\,\dot{\partial}' = \delta(x - x'), \tag{4.4}$$

where ∂ and ∂_x and $\partial' = \partial_{x'}$ are derivatives with respect to points x and x' respectively, while $\delta(x - x')$ is the *δ-function*. Recall that the δ-function is a 'distribution' with the properties:

$$\delta(x - x') = 0 \quad \text{if } x \neq x', \tag{4.5}$$

$$\int_{\mathcal{R}} |\mathrm{d}X(x')|\, \delta(x - x')F(x') = F(x), \tag{4.6}$$

if $F = F(x)$ is a continuous function on $\mathcal{R}$ and $\mathcal{R}$ is a subregion of $\mathcal{M}$ containing the point x.

Suppose that $\mathscr{M}$ is a *simple* manifold. By this we mean that $\mathscr{M}$ is not self-intersecting. Then using (4.4) and (4.6) in the fundamental theorem (3.10), we get

$$(-1)^m I(x)f(x) = -\int_{\mathscr{M}} g(x, x')\, \mathrm{d}X(x')\, \partial' f(x') + \\ + \int_{\partial\mathscr{M}} g(x, x')\, \mathrm{d}S(x') f(x'), \tag{4.7}$$

where x is an interior point of $\mathscr{M}$. Equation (4.7) is the desired general expression for the antiderivative. It shows explicitly that values of f at interior points of $\mathscr{M}$ are determined by values of f on $\partial\mathscr{M}$ as well as by values of ∂f inside $\mathscr{M}$. Of course, boundary values of f are not needed on segments of $\partial\mathscr{M}$ where g vanishes.

In common parlance, the function g is said to be a *Green's function* for the differential Eqn. (4.1). But g holds a special place in the general theory of Green's functions because of its relation to the fundamental differential operator ∂. Equation (4.7) reduces the problem of solving (4.1) to that of solving the more special Eqn. (4.4). The solution of (4.4) depends on the manifold $\mathscr{M}$ and the desired boundary conditions. We shall exhibit particular solutions for flat manifolds later. Such solutions make it possible to prove existence of antiderivatives for integrable functions on any manifold, because the manifold can be locally mapped onto a flat manifold, and the mapping can be applied to (4.1) and (4.7). However, we shall not prove this formally, as our aim is simply to explain the role of Geometric Algebra.

We say that a multivector-valued function $f = f(x)$ is *analytic** on a manifold $\mathscr{M}$ if

$$\partial f(x) = 0 \tag{4.8}$$

at each point of $\mathscr{M}$. The term 'analytic' is appropriate here, because f has the fundamental properties of an analytic function in complex variable theory.* Indeed, as we have explained in Section 4-7, Eqn. (4.8) generalizes the Cauchy–Riemann equations. By substituting (4.8) into the fundamental theorem (3.9), we get

$$\oint \mathrm{d}S f = 0. \tag{4.9}$$

This generalizes Cauchy's theorem.

By substituting (4.8) into (4.7) we get

$$f(x) = \frac{(-1)^m}{I(x)} \oint g(x, x')\, \mathrm{d}S(x') f(x'). \tag{4.10}$$

* It might be better, instead, to adopt the term *monogenic* recommended by Richard Delanghe. That would avoid confusion with the common meaning of *analytic* as 'differentiable to all orders'.

This generalizes Cauchy's integral formula. It reveals the fundamental property of an analytic function, namely: If f is analytic on $\mathscr{M}$, then its value at every point of $\mathscr{M}$ is *uniquely determined* by its values on $\partial\mathscr{M}$. Equation (4.10) reduces the study of analytic functions on a manifold to the study of the function $g(x, x')$.

A function f is said to be *meromorphic* if it is analytic except at *poles* x_k in $\mathscr{M}$, that is, if

$$\partial f(x) = -\odot_m \sum_k R_k\, \delta(x - x_k), \tag{4.11}$$

where

$$\odot_m \equiv \frac{2\pi^{m/2}}{\Gamma(m/2)} \tag{4.12}$$

is the area of the $(m - 1)$-dimensional unit sphere in $\mathscr{E}_m$ expressed in terms of the gamma function. The factor $-\odot_m$ has been introduced into (4.11) for conformity with standard conventions in complex variable theory. Accordingly, the multivector R_k is called the *residue* at the pole x_k. By substituting (4.11) into (3.9), we get the *residue theorem*

$$\oint \mathrm{d}Sf = -\odot_m \sum_k I(x_k)R_k. \tag{4.13}$$

Now suppose that $\mathscr{M}$ is an m-dimensional manifold embedded in an m-dimensional Euclidean vector manifold $\mathscr{E}_m$. Then $\mathscr{M}$ is flat, and Eqn. (4.4) admits the particular solution

$$g(x, x') = \frac{1}{\odot_m} \frac{x - x'}{|x - x'|^m}, \tag{4.14}$$

where $\odot_m$ is defined by (4.12). This is easily verified by using (2-1.36) to differentiate (4.14) and show that property (4.5) is satisfied, and then using the fundamental theorem (3.9) to show that property (4.6) is satisfied for $F = 1$. Using (4.14), we get (4.7) in the more specific form

$$f(x) = \frac{(-1)^m}{\odot_m I} \left\{ \int_{\partial\mathscr{M}} \frac{(x - x')}{|x - x'|^m}\, \mathrm{d}S(x')f(x') - \int_{\mathscr{M}} \frac{(x - x')}{|x - x'|^m}\, \mathrm{d}X(x')\, \partial' f(x') \right\}. \tag{4.15}$$

Since $\mathscr{M}$ is flat, the pseudoscalar I is constant, so (1.2) and (2.2) can be used to put (4.15) in the form

$$\begin{aligned} f(x) = {} & \frac{1}{\odot_m} \int_{\mathscr{M}} |\mathrm{d}X(x')|\, \frac{(x - x')}{|x - x'|^m}\, \partial' f(x') - \\ & - \frac{1}{\odot_m} \int_{\partial\mathscr{M}} |\mathrm{d}S(x')|\, \frac{(x - x')}{|x - x'|^m}\, n(x')f(x'). \end{aligned} \tag{4.16}$$

From (4.15) and (4.16) we find that an analytic function satisfies

$$\begin{aligned} f(x) &= \frac{(-1)^m}{\odot_m I} \int_{\partial\mathscr{M}} \frac{(x - x')}{|x - x'|^m}\, \mathrm{d}S(x')f(x') \\ &= \frac{-1}{\odot_m} \int_{\partial\mathscr{M}} |\mathrm{d}S(x')|\, \frac{(x - x')}{|x - x'|^m}\, n(x')f(x'). \end{aligned} \tag{4.17}$$

With (4.17), one can easily establish many straightforward generalizations of theorems in complex variable theory: Liouville's theorem, the mean value theorem, the maximum modulus principle, power series expansions, etc. The proofs are so similar to well-known proofs based on Cauchy's integral formula that they need not be discussed here.

To see that our theory encompasses complex variable theory in a simple and natural manner, consider the case $m = 2$. For this case, Eqn. (4.15) can be written

$$f(x) = \frac{1}{2\pi I}\int_{\partial\mathcal{M}} \frac{1}{x - x'}\,\mathrm{d}x' f(x') + \frac{1}{2\pi}\int_{\mathcal{M}} \frac{|\mathrm{d}X(x')|}{x - x'}\,\partial' f(x'). \tag{4.18}$$

This can be expressed in the conventional language of complex variables. Following Section 4-7, introduce the complex variables $z = ax$, where a is a fixed unit vector in $\mathscr{E}_2$. Write $F(z) = f(x)a = f(az)a$ so from (4-7.11a)

$$\partial f = \nabla f = a(a\nabla)\,(fa)a = 2a\,\frac{\mathrm{d}F}{\mathrm{d}z\dagger}\,a.$$

Then,

$$(x - x')^{-1}\,\mathrm{d}x' = (x - x')^{-1} aa\,\mathrm{d}x' = [a(x - x')]^{-1} a\,\mathrm{d}z' = \frac{\mathrm{d}z'}{z - z'}$$

and

$$\frac{1}{x - x'}\,\partial' f(x') = \frac{2}{z - z'}\,\frac{\mathrm{d}F}{\mathrm{d}z\dagger}\,(z')a.$$

Thus, Eqn. (4.18) can be put in the form

$$F(x) = \frac{1}{2\pi I}\oint_{\partial\mathcal{M}} \frac{\mathrm{d}z' F(z')}{z - z'} + \frac{1}{\pi}\int_{\mathcal{M}} \frac{|\mathrm{d}X(z')|}{z - z'}\,\frac{\mathrm{d}F(z')}{\mathrm{d}z\dagger}. \tag{4.19a}$$

This reduces to the conventional form for Cauchy's integral formula when $\mathrm{d}F/\mathrm{d}z\dagger = 0$ and

$$I = -i. \tag{4.19b}$$

It would have been nice if we could arrange things so that $I = i$ instead. But, unfortunately, we are faced with conflicting conventions. Section 4-7 explained how to identify the bivector i as the imaginary unit of Complex Variable Theory. In accordance with Eqn. (4-7.6), the orientation of i is fixed by the convention that it is the generator of *counterclockwise* rotations. On the other hand, the orientation of I is determined by (2.2) and the convention that curves in the complex plane are oriented *counterclockwise*. According to (2.2), at a point on a closed curve in the plane, the unit tangent v is related to the outward normal n by

$$v = \frac{\mathrm{d}x}{|\mathrm{d}x|} = In = -nI. \tag{4.19c}$$

This establishes (4.19b), since v is obtained from n by a counterclockwise rotation.

Formula (4.19a), or its equivalent (4.18), is a generalization of Cauchy's integral formula which applies to any differentiable function defined on the Euclidean plane, and not to analytic functions only. In spite of its generality and power, it is seldom noted or exploited in works on complex variable theory (see [Ber] for an exception). Perhaps this is because the last term in (4.19a) involves an area integral and only line integrals are employed in conventional approaches. More probably, it is because the conventional definition (4-7.15) of the complex derivative d/dz applies only to analytic functions. The usual definition of analytic functions by the requirement that d/dz exist works in two dimensions, because it happens to be equivalent to the requirement $\partial f = 0$ in that case. However, it is not the existence of d/dz, but Cauchy's integral formula which best describes the essence of analytic functions. As we have seen, Cauchy's integral formula generalizes quite nicely to higher dimensions, nonanalytic functions and curved manifolds, while the definition of analytic functions by requiring that d/dz exist not only fails to generalize but may be accused of obscuring the essence of analytic function theory.

Our generalization of Cauchy's integral formula should be compared with the conventional one in the 'theory of analytic functions of several complex variables.' The conventional result can be expressed in terms of Geometric Calculus in the following way: Consider the Euclidean space $\mathscr{E}_{2m}$ as a 'Cartesian product' of m planes, and let i_k be the unit pseudoscalar of the kth plane. Since i_k is a bivector, we have

$$i_k^2 = -1. \tag{4.20a}$$

Moreover,

$$i_j i_k = i_k i_j, \tag{4.20b}$$

and the pseudoscalar I of $\mathscr{E}_{2m}$ can be factored into the product

$$I = i_1 i_2 \ldots i_m. \tag{4.20c}$$

A point x in $\mathscr{E}_{2m}$ can be expressed as the sum

$$x = x_1 + x_2 + \ldots + x_m, \tag{4.21a}$$

where

$$x_k = (i_k)^{-1} i_k \cdot x = -i_k i_k \cdot x \tag{4.21b}$$

is the component in the kth plane. A function $f(x)$ on $\mathscr{E}_{2m}$ can be regarded as a function $f(x_1, x_2, \ldots, x_m)$ of the m variables defined by (4.21). If f is analytic in a region $\mathscr{M}_k$ of the kth plane, then

$$\partial_k f = 0, \tag{4.22}$$

where $\partial_k \equiv \partial_{x_k}$ is the derivative with respect to a point x_k in $\mathcal{M}_k$. Let the manifold

$$\mathcal{M} = \mathcal{M}_1 \otimes \mathcal{M}_2 \otimes \ldots \otimes \mathcal{M}_m \tag{4.23}$$

be the Cartesian product of the regions $\mathcal{M}_k$. Then, for x in $\mathcal{M}$ we have, by m applications of (4.18),

$$f(x) = \frac{I}{(-2\pi)^m} \int_{\partial\mathcal{M}_1} \frac{1}{x_1 - x_1'} \, \mathrm{d}x_1' \int_{\partial\mathcal{M}_2} \frac{1}{x_2 - x_2'} \, \mathrm{d}x_2' \ldots \times$$
$$\times \int_{\partial\mathcal{M}_m} \frac{1}{x_m - x_m'} \, \mathrm{d}x_m f(x'). \tag{4.24}$$

This is equivalent to Cauchy's integral formula for an analytic function of m complex variables. The order of integrations in (4.24) is immaterial.

On the other hand, we note that (4.22) implies that

$$\partial_x f(x) = \partial_1 f(x) + \partial_2 (fx) + \cdots + \partial_m f(x) = 0. \tag{4.25}$$

Therefore, (4.17) applies and assumes the form

$$\begin{aligned} f(x) &= \frac{(m-1)!}{2\pi^m I} \int_{\partial\mathcal{M}} \frac{(x - x')}{|x - x'|^{2m}} \, \mathrm{d}S(x') f(x') \\ &= \frac{(m-1)!}{2\pi^m} \int_{\partial\mathcal{M}} \frac{|\mathrm{d}S'|(x' - x) n' f'}{|x - x'|^{2m}}. \end{aligned} \tag{4.26}$$

Evidently (4.26) is a more general formula than (4.24) and reduces to (4.24) when the special conditions (4.22) and (4.23) are satisfied.

An equation corresponding to $\partial f = s$ is never written down in conventional complex variable theory. Nevertheless, our discussion shows that complex variable theory can be regarded as a study of that equation on the Euclidean plane for the special case where s vanishes except at isolated points (poles) and lines (cuts). The equation $\partial f = s$ has also been studied, though indirectly, on higher dimensional manifolds in the branch of mathematics known as *Potential Theory*. To see the relation to Potential Theory, substitute $f = \partial\phi$ into $\partial f = s$ to get

$$\partial^2 \phi = s. \tag{4.27}$$

For Euclidean manifolds an integral form for (4.27) can be obtained from (4.16) by expressing the Green's function (4.14) in terms of a potential. Writing

$$g = \partial G = -\dot{G}\,\dot{\partial}', \tag{4.28}$$

it is readily verified that (4.14) obtains if

$$G(x, x') = \frac{|x - x'|^{2-m}}{(2-m)\Theta_m} \quad \text{when } m > 2, \tag{4.29a}$$

$$= \frac{1}{2\pi} \log |x - x'| \quad \text{when } m = 2. \tag{4.29b}$$

So, by inspection of (4.16) we can immediately write down

$$\phi(x) = \int_{\mathcal{M}} |\mathrm{d}X(x')| G(x, x')\, \partial'^2 \phi(x') - $$
$$- \int_{\partial\mathcal{M}} |\mathrm{d}S(x')| G(x, x') n(x')\, \partial' \phi(x') + \phi_0(x), \tag{4.30}$$

where

$$\partial \phi_0(x) = 0. \tag{4.31}$$

For ϕ_0 we are free to choose any function analytic on the manifold $\mathcal{M}$. With the choice

$$\phi_0(x) = \int_{\partial\mathcal{M}} |\mathrm{d}S(x')|\, \partial G(x, x') n(x') \phi(x'), \tag{4.32}$$

Eqn. (4.30) assumes the form

$$\phi = \int_{\mathcal{M}} |\mathrm{d}X| G\, \partial^2 \phi + \int_{\partial\mathcal{M}} |\mathrm{d}S|\, [(n \cdot \partial G)\phi - G n \cdot \partial\phi]\,. \tag{4.33}$$

This is the main result of Potential Theory. Of course, it obtains for other choices of G besides (4.29). Also, the potential ϕ need not be a scalar but may have multivector values of any grade. In this respect our result (4.33) is somewhat more general than the one in Potential Theory.

The Laplacian ∂^2 has been the fundamental differential operator in Potential Theory for the simple reason that its 'square root' ∂ was not available for a long time. Now that Geometric Calculus has made it possible, the advantage of dealing directly with ∂ should be evident enough.

7-5. Changing Integration Variables

The rules for 'changing the variable' in a directed integral follow almost trivially from the transformation theory developed in Section 4-5. But they are so important that we spell them out explicitly.

A nonsingular transformation f of $\mathcal{M}$ to $\mathcal{M}'$ induces a transformation $\underline{f}$ of each directed volume element $\mathrm{d}X$ on $\mathcal{M}$ to a volume element

$$\mathrm{d}X' = \pm \underline{f}(\mathrm{d}X) \tag{5.1}$$

on $\mathcal{M}'$. The appropriate sign in (5.1) is determined by a choice of orientations for $\mathcal{M}$ and $\mathcal{M}'$. We can establish (5.1) by writing

$$\mathrm{d}X = |\mathrm{d}X| I, \qquad \mathrm{d}X' = |\mathrm{d}X'| I' \tag{5.2}$$

and using Eqn. (4-5.3) for the transformation of the unit pseudoscalar:

$$\underline{f}(I) = J_f I'. \tag{5.3}$$

If J_f is positive, the transformation is said to be *orientation-preserving*. Alternatively, if J_f is assumed positive, Eqn. (5.3) determines I' and so specifies an orientation on $\mathscr{M}'$ induced by its relation to $\mathscr{M}$ by f. Using the linearity of $\underline{f}$, we get (5.1) from (5.2) and (5.3), and we obtain the famous relation

$$|\mathrm{d}X'| = |J_f|\, |\mathrm{d}X|. \tag{5.4}$$

Our argument shows that the Jacobian J_f arises from the transformation of the unit pseudoscalar I rather than from the transformation of $|\mathrm{d}X|$ as (5.4) suggests. This fact is well disguised in classical treatments.

It should be mentioned that (5.1) applies to transformations of null manifolds. Our argument omitted this case by presuming the existence of a unit pseudoscalar in writing (5.2), but this presumption is actually not essential.

Having established the rule (5.1) for the transformation of directed measure, we obtain the rule for changing the variable in the general directed integral (1.9) by simple substitution. If $L' = L'(\mathrm{d}X')$ is a differential multiform on $\mathscr{M}'$, then by substitution of (5.1) we get a differential multiform $L = L(\mathrm{d}X)$ on $\mathscr{M}$ defined by

$$L(x, \mathrm{d}X) \equiv L'(f(x), \underline{f}(\mathrm{d}X)) = L'(x', \mathrm{d}X'). \tag{5.5}$$

Thus, a one-to-one, orientation-preserving mapping f from $\mathscr{M}$ to $\mathscr{M}'$ determines the change of integration variables

$$\int_{\mathscr{M}'} L' = \int_{\mathscr{M}} L'\underline{f}, \tag{5.6a}$$

or, more explicitly,

$$\int_{\mathscr{M}'} L'(\mathrm{d}X') = \int_{\mathscr{M}} L'(\underline{f}(\mathrm{d}X)) = \int_{\mathscr{M}} L(\mathrm{d}X). \tag{5.6b}$$

If, instead, the one-to-one mapping f is orientation-reversing, we have

$$\int_{\mathscr{M}'} L' = -\int_{\mathscr{M}} L'\underline{f}. \tag{5.7}$$

We take it for granted that the integrals being considered have finite values, so we do not mention mathematical assumptions of compactness.

When the integrals are scalar valued, Eqns. (5.6) and (5.7) describe the change of variables for differential forms. Their relation to conventional formulations should be clear from the discussion in Section 6-4.

For the important special case where $L'(x', \mathrm{d}X') = |\mathrm{d}X'|F'(x')$, use of (5.4) reduces the relation (5.5) to

$$|\mathrm{d}X'|F'(x') = |\mathrm{d}X|J_f(x)F'(f(x)). \tag{5.8a}$$

For this case (5.6) yields

$$\int_{\mathscr{M}'} |\mathrm{d}X'|F'(x') = \int_{\mathscr{M}} |\mathrm{d}X|J_f(x)F(x), \tag{5.8b}$$

where $F(x) = F'(f(x))$. Thus, the standard rule for changing variables in a (non-directed) integral is a simple special case of the rule (5.6) for directed integrals.

The formulas (5.6) and (5.7) for change of integration variables admit a valuable generalization for many-to-one mappings. In this case f maps a discrete set of points $\{x_k\}$ in $\mathscr{M}$ to each point x' in $\mathscr{M}'$. A neighborhood of each x_k maps one-to-one onto a neighborhood of x'. The *degree* of f is defined by

$$\deg(f) = \sum_k \operatorname{sign} J_f(x_k), \tag{5.9}$$

where sign $J_f(x_k)$ is the sign of the Jacobian of the local mapping. For connected manifolds, the continuity of f implies that $\deg(f)$ does not depend on the choice of a point in $\mathscr{M}'$. Thus, 'degree' is a general topological property of a mapping.

A many-to-one mapping of connected manifolds determines the change of variables

$$\int_{\mathscr{M}} L'\underline{f} = \deg(f) \int_{\mathscr{M}'} L'. \tag{5.10}$$

This result is called the *degree formula*. It can be proved by decomposing f into one-to-one mappings to which (5.6) and (5.7) apply. A proof of the degree formula with due attention to mathematical detail is given in [GP]. See [Mi] for a concise but thorough development of the degree concept.

It is worth pointing out that the general change of variables formula (5.10) applies without modification to induced transformations of submanifolds. Specifically, Eqn. (5.10) applies to the case where $\mathscr{M}$ is a submanifold of some manifold $\mathscr{N}$ and f is a transformation of $\mathscr{N}$ to $\mathscr{N}'$ which maps $\mathscr{M}$ to a submanifold $\mathscr{M}'$ of $\mathscr{N}'$.

We should consider how a change of variable affects the fundamental theorem of calculus. According to (5.5), a change of variable on the right-hand side of (3.1) or (3.3) is expressed by

$$L'(\mathrm{d}S') = L'(\underline{f}(\mathrm{d}S)) = L(\mathrm{d}S), \tag{5.11}$$

while the change on the left-hand side is

$$\mathrm{d}L' = \dot{L}'(\mathrm{d}X'\,\dot{\partial}') = \dot{L}'(\underline{f}(\mathrm{d}X\,\dot{\partial})) = \dot{L}(\mathrm{d}X\,\dot{\partial}) = \mathrm{d}L. \tag{5.12}$$

In (5.12) we used

$$\underline{f}(\mathrm{d}X)\,\dot{\partial}' = \underline{f}[\mathrm{d}X\bar{f}(\dot{\partial}')] = \underline{f}(\mathrm{d}X\,\dot{\partial}), \tag{5.13}$$

which is an application of (3-1.14), and

$$\mathrm{d}\underline{f}(\mathrm{d}X) = \dot{\underline{f}}\,(\mathrm{d}X\,\dot{\partial}) = 0, \tag{5.14}$$

which is an application of (4-5.23).

Equation (5.12) may be interpreted by the statement that 'the exterior differential commutes with a change of variable'. This might be clearer if (5.12) is written in the alternative form

$$[\mathrm{d}L']\underline{f} = \mathrm{d}[L'\underline{f}] - L'\,\mathrm{d}\underline{f} = \mathrm{d}[L'\underline{f}] = \mathrm{d}L. \tag{5.15}$$

Of course the basis for this result is the integrability condition operating through (5.14).

Equations (5.11) and (5.12) together imply that the form of the fundamental theorem is not altered by a change of variables. Interpreted differently, this fact is expressed by the valuable formula

$$\int_{\mathcal{M}} \dot{L}(h(\mathrm{d}X\,\dot{\partial})) = \oint_{\partial\mathcal{M}} L(h(\mathrm{d}S)), \tag{5.16}$$

which holds if h can be identified as the extended differential of some transformation f, that is, if $h = \underline{f}$.

It should also be pointed out that by applying the fundamental theorem to (5.14) one gets

$$\oint_{\partial\mathcal{M}} \underline{f}(\mathrm{d}S) = 0. \tag{5.17}$$

Indeed, this result remains true if $\partial\mathcal{M}$ is replaced by any closed submanifold of $\mathcal{M}$. A manifold is said to be *closed* if its boundary vanishes.

7-6. Inverse and Implicit Functions

This section develops a new method for proving two venerable theorems of mathematics, the 'Inverse Function Theorem' and the 'Implicit Function Theorem'. The method is constructive, that is, it provides explicit procedures for finding the inverse of a function and solving for an implicit function. It supplies, therefore, much more than the usual existence proofs; it is itself a powerful computational tool.

We shall be concerned with the following form of *The Inverse Function Theorem: A nonsingular transformation has a unique inverse*. 'Transformations' were defined earlier as functions relating vector manifolds. We restrict our attention to transformations, because we have built up the apparatus needed to handle them efficiently. We lose little generality by this restriction, because any multivector function can be re-expressed as an equivalent vector function.

Let f be a nonsingular transformation of a vector manifold $\mathcal{M}$ to a vector manifold $\mathcal{M}'$. According to (4.7), the inverse of f, if it exists, satisfies the equation

$$f^{-1}(x') = \frac{(-1)^m}{I'}\left\{\int_{\partial\mathcal{M}'} g'(x',y')\,\mathrm{d}S'(y')f^{-1}(y') - \right.$$
$$\left. - \int_{\mathcal{M}'} g'(x',y')\,\mathrm{d}X'(y')\,\partial_{y'}f^{-1}(y')\right\}. \tag{6.1}$$

By a change of variable, we can express the right side of (6.1) as an integral on $\mathscr{M}$ determined by f. To accomplish this, recall that, according to (4-5.13), $\partial_y = \overline{f}(\partial_{y'})$ if $y' = f(y)$, hence

$$\partial_{y'} f^{-1}(y') = \overline{f}^{-1}(\partial_y) y = \overline{f}^{-1}(\dot{a})\dot{\partial}_a = [\partial_a \overline{f}^{-1}(a)]^{\dagger}, \tag{6.2}$$

where we have used the linearity of $\overline{f}$ and (2-1.19) to express the result in terms of a variable a in the tangent space at y in $\mathscr{M}$. Using (6.2) as well as (5.1) and (5.3) in (6.1), we get

$$f^{-1}(x') = \frac{(-1)^m J_f(x)}{\underline{f}(I(x))} \left\{ \int_{\partial \mathscr{M}} g'(x', f(y)) \underline{f}(\mathrm{d}S(y)) y \right.$$
$$\left. - \int_{\mathscr{M}} g'(x', f(y)) \underline{f}(\mathrm{d}X(y)) \, [\partial_a \overline{f}^{-1}(a)]^{\dagger} \right\}. \tag{6.3}$$

Of course, in (6.3) the point dependence of $\underline{f}$ and $\overline{f}$ is suppressed, as usual.

Formula (6.3) is an explicit integral representation of f^{-1} in terms of f. The conditions that f^{-1} exist are precisely the conditions that the integral in (6.3) be well-defined. The condition that f be non-singular means that $\underline{f}(I) \neq 0$, and this entails that f be continuously differentiable, as required in conventional statements of the inverse function theorem.

Chapter 4 explains how to compute the differential $\underline{f}$ and its adjoint $\overline{f}$. Evaluation of (6.2) merely requires inversion of the linear transformation $\overline{f}$, an algebraic problem solved by (3-1.21). The right side of (6.2) will be recognized as a characteristic multivector of the linear transformation $\overline{f}^{-1}$, a quantity much studied in Chapter 3. Thus, all factors on the right side of (6.3) are readily computed from f, and the only remaining problem is to find a Green's function g' for $\mathscr{M}'$. We shall see that the existence of f^{-1} can be established without finding g'.

If $\mathscr{M}' \subset \mathscr{E}_m$, that is, if $\mathscr{M}'$ is an m-dimensional submanifold of Euclidean m-space, then, using (4.14), we have

$$g'(x', f(y)) = \frac{1}{\odot_m} \frac{x' - f(y)}{|x' - f(y)|^m}, \tag{6.4}$$

where $\odot_m$ is given by (4.12). Substituting (6.4) into (6.3), we get an explicit formula for the inverse of any nonsingular transformation of a vector manifold into a Euclidean manifold with pseudoscalar $I' = J_f^{-1} \underline{f}(I)$,

$$f^{-1}(x') = \frac{(-1)^m}{\odot_m I'} \left\{ \int_{\partial \mathscr{M}} \frac{x' - f(y)}{|x' - f(y)|^m} \underline{f}(\mathrm{d}S(y)) y \; - \right.$$
$$\left. - \int_{\mathscr{M}} \frac{x' - f(y)}{|x' - f(y)|^m} \underline{f}(\mathrm{d}X(y)) \, [\partial_a \overline{f}^{-1}(a)]^{\dagger} \right\}. \tag{6.5}$$

This result has innumerable applications, for example, to inversion of series and evaluation of integrals. The reader may like to apply it to the specific transformations considered in Section 4-6.

If f^{-1} is an analytic function on $\mathcal{E}_m$, then, as (6.2) shows, the 'volume integral' in (6.5) vanishes, and we have

$$f^{-1}(x') = \frac{(-1)^m}{\odot_m I'} \oint \frac{x' - f(y)}{|x' - f(y)|^m} \underline{f}(\mathrm{d}S(y))y. \tag{6.6}$$

For the two-dimensional special case $\mathcal{M} \subset \mathcal{E}_2$ and $\mathcal{M}' \subset \mathcal{E}_2$, formula (6.6) reduces to

$$f^{-1}(x') = \frac{1}{2\pi I} \oint \frac{1}{x' - f(y)} \underline{f}(\mathrm{d}y)y. \tag{6.7}$$

By the procedure which gave us (4.19a) and with the help of (4-7.21), we can put (6.7) in the form

$$F^{-1}(z') = \frac{1}{2\pi i} \oint \frac{z\,\mathrm{d}z}{F(z) - z'} \frac{\mathrm{d}F(z)}{\mathrm{d}z}. \tag{6.8}$$

This is a well-known formula in complex variable theory. Some applications of (6.8) are discussed in Section 9.4 of ref. [Hil].

Having determined the conditions under which our general *inverse function formula* (6.3) reduces to a known result (6.8) in the literature, let us return to complete the proof of the inverse function theorem. Our proof is based on the fact that a nonsingular mapping of a manifold into a Euclidean space with the same dimension has a unique inverse. This fact is established by the explicit formula (6.5).

Let f be a nonsingular transformation of $\mathcal{M}$ on $\mathcal{M}'$. Suppose g is a nonsingular transformation of $\mathcal{M}'$ onto a region $\mathcal{R}$ in $\mathcal{E}_m$. Then the composite function $h = gf$ transforms $\mathcal{M}$ onto $\mathcal{R}$. But h is invertible, hence $f^{-1} = h^{-1}g^{-1}$ exists. It may be that g cannot be defined on all of $\mathcal{M}'$, however, our proof actually requires only that such a function can be defined on some neighborhood of any point x in the manifold. It suffices to note that we can take $g = P$ in a neighborhood of x, where P is the projection into the tangent space at x. The operator P was defined in Chapter 4 and studied in Chapter 5. This completes our proof of the inverse function theorem.

Now consider the *Implicit Function Theorem* in the following form: Suppose that

(1) $F(x, y)$ is a vector-valued function defined for x and y in vector manifolds $\mathcal{N}$ and $\mathcal{M}$ respectively.
(2) For each value of x, $F(x, y)$ is a nonsingular function of y.
(3) For a particular x_0 in $\mathcal{N}$ and y_0 in $\mathcal{M}$, $F(x_0, y_0) = 0$.

Then there is a unique function f from $\mathcal{N}$ to $\mathcal{M}$ such that

$$F(x_0, f(x_0)) = 0. \tag{6.9}$$

We plan to prove the implicit function theorem by using the inverse function formula to solve the equation

$$z = F(x, y), \tag{6.10a}$$

with x held fixed, for

$$y = F^{-1}(x, z), \tag{6.10b}$$

and then identify

$$f(x) = F^{-1}(x, 0). \tag{6.10c}$$

To make our proof as simple and explicit as possible, we assume that the values of $F(x, y)$ lie in a Euclidean manifold with the same dimension as $\mathscr{M}$. In fact most statements of the implicit function theorem in the literature make such an assumption. The assumption can be dispensed with by the same argument we used to generalize our proof of the inverse function theorem.

Applying the inverse function formula (6.5) to (6.10), we immediately get the *implicit function formula*

$$f(x) = \frac{(-1)^m}{\odot_m I'} \left\{ \int_{\mathscr{M}} \frac{F(x, y)}{|F(x, y)|^m} \underline{F}(\mathrm{d}X(y); x, y)\, [\partial_a \overline{F}^{-1}(a; x, y)]^\dagger - \right.$$
$$\left. - \int_{\partial \mathscr{M}} \frac{F(x, y)}{|F(x, y)|^m} \underline{F}(\mathrm{d}S(y); x, y) y \right\}. \tag{6.11}$$

The conditions for validity of the implicit function theorem are just the conditions that this formula be well-defined. Thus, the theorem is proved.

7-7. Winding Numbers

This section develops the concept of *winding number* for a manifold and a function, relates it to the *degree* of a mapping and uses it to describe the *index* of a vector field.

The concept of winding number arises naturally from integrals over self-intersecting manifolds. It makes its appearance in the generalization of Eqns. (4.15) and (4.16) which allows $\mathscr{M}$ and $\partial \mathscr{M}$ to be self-intersecting. That generalization has the form

$$\#_y f(y) = \frac{(-1)^m}{\odot_m I} \left\{ \int_{\partial \mathscr{M}} \frac{y - x}{|y - x|^m} \mathrm{d}S(x) f(x) - \int_{\mathscr{M}} \frac{y - x}{|y - x|^m} \mathrm{d}X(x)\, \partial f(x) \right\} \tag{7.1a}$$

$$= \frac{1}{\odot_m} \int_{\mathscr{M}} |\mathrm{d}X| \frac{y - x}{|y - x|^m} \partial f - \frac{1}{\odot_m} \int_{\partial \mathscr{M}} |\mathrm{d}S| \frac{y - x}{|y - x|^m} n f, \tag{7.1b}$$

where the *winding number* $\#_y = \#_y(\partial\mathcal{M})$ is equal to the number of times $\partial\mathcal{M}$ encloses (or, 'wraps around') the point y. The winding number, $\#_y$ can also be regarded as the *multiplicity* of the point y in $\mathcal{M}$; indeed, this interpretion is most helpful in deriving (7.1). The derivation is not different from the one in Section 4. The $f(y)$ on the left of (7.1) arises from the integral of $\delta(x - y)f(x)$; one has only to count the number of times the integral 'passes over' the point y to get $\#_y$. Evidently, this argument could be used to get from (4.7) a result more general that (7.1), but the matter will not be pursued here.

It is convenient to allow for a negative winding number by interpreting the I in (7.1a) as the positive unit pseudoscalar for the Euclidean space $\mathscr{E}_m$ in which $\mathcal{M}$ is imbedded. Then, for the directed volume elements in (7.1a), we have $\mathrm{d}X = \pm I|\mathrm{d}X|$ and $\mathrm{d}S = \pm In|\mathrm{d}S|$, with the minus sign obtaining if the orientation of $\mathcal{M}$ is opposite to that of $\mathscr{E}_m$. A positive orientation for $\mathcal{M}$ was assumed in writing the integrals in (7.1b).

Taking $f = 1$ in (7.1), we get an integral formula for the winding number of any closed hypersurface $\mathcal{N}$ in $\mathscr{E}_m$.

$$\#_y(\mathcal{N}) = \frac{(-1)^m}{\odot_m I} \oint_{\mathcal{N}} \frac{y - x}{|y - x|^m}\, \mathrm{d}S \tag{7.2a}$$

$$= \frac{1}{\odot_m} \oint_{\mathcal{N}} |\mathrm{d}S|\, \frac{x - y}{|x - y|^m}\, n. \tag{7.2b}$$

Here the sign of the winding number depends on the orientation of $\mathcal{N}$ relative to $\mathscr{E}_m$. If $\mathcal{N}$ is not self-intersecting, then $\#_y(\mathcal{N}) = +1$ or -1 if $\mathcal{N}$ encloses y with positive or negative orientation, and $\#_y(\mathcal{N}) = 0$ if y is not enclosed by $\mathcal{N}$. Formula (7.2) is known in the literature as the *Kronecker integral*. We have not allowed for the possibility that y might be located on the hypersurface $\mathcal{N}$. That can be taken care of in a natural way, yielding winding numbers with values $\pm\frac{1}{2}$. But the matter will not be pursued here.

Note that the right side of (7.2) includes the integral of a bivector function which must vanish since the left side of (7.2) is a scalar; that is,

$$\oint_{\mathcal{N}} |\mathrm{d}S|\, \frac{(x - y)\wedge n}{|x - y|^m} = \frac{(-1)^m}{I} \oint_{\mathcal{N}} \frac{(y - x)\cdot \mathrm{d}S}{|y - x|^m} = 0. \tag{7.3}$$

The winding number of a hypersurface can be interpreted or, indeed, defined as the degree of a mapping onto a hypersphere. Specifically, the winding number $\#_y(\mathcal{N})$ equals the degree of the central projection u of the hypersurface $\mathcal{N}$ onto the unit hypersphere $\mathscr{S}_y$ centered at y. The central projection takes a point x in $\mathcal{N}$ to a point

$$n = u(x) = \frac{x - y}{|x - y|} \tag{7.4}$$

in $\mathscr{S}_y$. According to (4-6.7b), the Jacobian of this transformation is

$$J_u = \frac{(x - y)\cdot n}{|x - y|^m}.$$

The increment

$$|\mathrm{d}S|\,\frac{(x-y)\cdot n}{|x-y|^m} = |\mathrm{d}S|J_u = |\mathrm{d}S'|\,\mathrm{sign}\,J_u \tag{7.5}$$

may be regarded as the element of hypersolid angle on $\mathcal{S}_y$ subtended by the surface element $\mathrm{d}S(x)$ on $\mathcal{N}$. With (7.5), the degree formula (5.10) can be applied to (7.2b), with the result

$$\#_y(\mathcal{N}) = \odot_m^{-1} \oint_{\mathcal{N}} |\mathrm{d}S(x)|\,\frac{(x-y)\cdot n}{|x-y|^m} = \odot_m^{-1} \oint_{\mathcal{N}} |\underline{u}(\mathrm{d}S(u)|\,\mathrm{sign}\,J_u$$

$$= \odot_m^{-1}\,\deg(u) \oint_{\mathcal{S}_y} |\mathrm{d}S(u)| = \deg(u), \tag{7.6}$$

as advertized.

If f is a one-to-one mapping of a manifold $\mathcal{N}'$ onto a closed hypersurface $\mathcal{N}$ in $\mathcal{E}_m$, we can define a winding number for f by identifying it with the winding number of $\mathcal{N}$. Specifically, from (7.2) and (7.6) we get the explicit expressions

$$\#_y(f) = \frac{(-1)^m}{\odot_m I} \oint_{\mathcal{N}'} \frac{y-f(x')}{|y-f(x')|^m}\,\underline{f}(\mathrm{d}S(x')) = \frac{(-1)^m}{\odot_m I} \oint_{\mathcal{N}} \frac{y-x}{|y-x|^m}\,\mathrm{d}S(x) \tag{7.7a}$$

$$= \frac{1}{\odot_m} \oint_{\mathcal{N}'} |\underline{f}(\mathrm{d}S')|\,\frac{f(x')-y}{|f(x')-y|^m}\,n(f(x')) = \frac{1}{\odot_m} \oint_{\mathcal{N}} |\mathrm{d}S|\,\frac{x-y}{|x-y|^m}\,n(x) \tag{7.7b}$$

$$= \deg\left[\frac{f(x')-y}{|f(x')-y|}\right]. \tag{7.7c}$$

Formula (7.7a) generalizes a well-known result in complex function theory. To see this, let f be a mapping of $\mathcal{E}_2$ into $\mathcal{E}_2$ which maps a closed curve $\mathcal{N}'$ onto a curve $\mathcal{N}$. Choose the origin so that $y = 0$. Then (7.7a) becomes

$$\#_0(f) = \frac{-1}{2\pi I} \oint_{\mathcal{N}'} \frac{1}{f(x')}\,\underline{f}(\mathrm{d}x') = \frac{-1}{2\pi I} \oint_{\mathcal{N}} x^{-1}\,\mathrm{d}x. \tag{7.8}$$

We can express this as a complex integral in the way we got (4.19) from (4.18). We write $z = ax$, identify $I = -i$, define $F(z) \equiv a^{-1}f(a^{-1}z) = a^{-1}f(x)$, and, following (4-7.21), observe that

$$a^{-1}\underline{f}(\mathrm{d}x) = \frac{\mathrm{d}F}{\mathrm{d}z}\,\mathrm{d}z \equiv F'(z)\,\mathrm{d}z\,.$$

With this change of variables, (7.8) assumes the form

$$2\pi i\#_0(f) = \oint_{\mathcal{N}'} \frac{F'(z)\,\mathrm{d}z}{F(z)} = \oint_{\mathcal{N}} \frac{\mathrm{d}z}{z} = \oint_{\mathcal{N}} \mathrm{d}(\log z). \tag{7.9}$$

This result is known as the *argument principle* in complex function theory.

The concept of winding number provides a valuable means to characterize *critical* points of a vector field, namely, those points at which the vector field has a zero or a singularity. Let $v = v(x)$ be a *smooth* vector field on $\mathscr{E}_m$ with an isolated critical point z. Define the z-centered hypersphere of radius ϵ by

$$\mathscr{S}_z(\epsilon) \equiv \{x : |x - z| = \epsilon\}. \tag{7.10}$$

The function $u(x) = v(x)/|v(x)|$ maps $\mathscr{S}_z(\epsilon)$ into the 0-centered unit hypersphere $\mathscr{S}_0$ in $\mathscr{E}_m$. The winding number of this mapping in the limit $\epsilon \to 0$ is said to be the *index* $\iota_z(v)$ of v at z. To get a convenient expression for the index from (7.7), recall the basic property (3-1.7) of the extended differential $\underline{u}$ of a transformation u, and note that $u^2 = 1$ implies

$$u \cdot \underline{u}(\mathrm{d}S) = 0; \tag{7.11}$$

hence

$$u\underline{u}(\mathrm{d}S) = u \wedge \underline{u}(\mathrm{d}S) = (-1)^{m-1}\underline{u}(\mathrm{d}S)u.$$

Then, from (7.7a) we get

$$\iota_z(v) = \iota_z(u) = \odot_m^{-1} \lim_{\epsilon \to 0} \oint_{\mathscr{S}_z(\epsilon)} \langle I^{-1}\underline{u}(\mathrm{d}S)u\rangle. \tag{7.12}$$

The brackets in (7.12) are not really necessary; they have been inserted to emphasize that the integrand is scalar-valued; thus, the integrand is a differential form of degree $m - 1$. If we use the degree formula to evaluate (7.12) by mapping onto the unit sphere $\mathscr{S}_0$, we have, for each value of ϵ,

$$\oint_{\mathscr{S}_z(\epsilon)} \langle I^{-1}\underline{u}(\mathrm{d}S)u\rangle = \deg(u) \oint_{\mathscr{S}_0} \langle I^{-1}\, \mathrm{d}S'(u)u\rangle. \tag{7.13}$$

By comparison with (1.4), the last integral in (7.13) can be identified as an area integral, so it has the value $\odot_m$.

We may regard (7.12) as the basic definition of the index of a vector field. The arguments leading to (7.12) show that the index is an integer and that it can be interpreted as a winding number or the degree of a mapping. The form of (7.12) has been chosen to define the index of a vector field on any vector manifold, not merely the Euclidean manifold assumed to simplify the initial argument. For an arbitrary manifold Eqn. (7.10) defines a hypersurface for ϵ sufficiently small. Though $\mathscr{S}_z(\epsilon)$ will not be a hypersphere as in the Euclidean case, it approaches a hypersphere in the tangent space at z in the limit. In the Euclidean case, Eqn. (7.13) is actually independent of ϵ for ϵ sufficiently small. In the general case, the surface $\mathscr{S}_0$ in (7.13) depends on ϵ, however, as x approaches z, $u(x)$ approaches a unit tangent vector at z, so $\mathscr{S}_0$ approaches the unit hypersphere in the tangent space at z. Therefore, in the limit the integral on the right side of (7.13) has the value found in the Euclidean case.

7-8. The Gauss–Bonnet Theorem

The most significant recent developments in geometry have been the emergence of coherent global theories such as the de Rham cohomology. We believe that a systematic reformulation of this work in the language of Geometric Calculus will be of great value, but the task is a large and difficult one, too much to be accomplished here. However, this section makes a beginning by explicitly demonstrating the efficiency of Geometric Calculus in the formulation and proof as well as the generalization of the chief global result of classical geometry, the Gauss–Bonnet theorem.

The generalized *Gauss–Bonnet Theorem*: If $\mathscr{M}$ is a closed oriented vector manifold of even dimension $m = 2p$, then

$$\oint_{\mathscr{M}} \kappa\,|\mathrm{d}X| = \tfrac{1}{2}\,\odot_{m+1}\chi, \tag{8.1}$$

where κ is the *Gaussian curvature* and χ is the *Euler characteristic* of $\mathscr{M}$. As before, $\odot_{m+1}$ is the area of the m-dimensional unit sphere.

The Gaussian curvature can be defined in terms of the Riemann curvature tensor $R(a \wedge b)$ by the equation

$$\underline{R}(I) = \kappa I, \tag{8.2a}$$

where I is the unit pseudoscalar and $\underline{R}(I)$ is an m, m-form defined by

$$\underline{R}(I) \equiv \frac{1}{m!}\langle I\,\partial_{a_m} \wedge \ldots \wedge \partial_{a_1}\rangle R(a_1 \wedge a_2) \wedge \ldots \wedge R(a_{m-1} \wedge a_1) \tag{8.2b}$$

$$= \frac{1}{m!}\langle Ie^{i_m} \wedge \ldots \wedge e^{i_1}\rangle R(e_{i_1} \wedge e_{i_2}) \wedge \ldots \wedge R(e_{i_{m-1}} \wedge e_{i_m}).$$

The integrand in (8.1) can thus be written

$$\kappa\,|\mathrm{d}X| = I^{-1}\underline{R}(\mathrm{d}X) = \mathrm{d}X\underline{R}(I^{-1}). \tag{8.3}$$

This is a differential form of degree m.

In Section 5-2 we defined the Gaussian curvature for a hypersurface $\mathscr{M}$ in $\mathscr{E}_{m+1}$ as the Jacobian of the Gauss mapping, the mapping which takes each point x in $\mathscr{M}$ into the unit normal $n = n(x)$; thus,

$$\underline{n}(I) = \kappa I. \tag{8.4}$$

This is perfectly consistent with (8.2) if m is even; indeed, considering (5-2.9b), we see that

$$\underline{n}(I) = R(e_1 \wedge e_2) \wedge \ldots \wedge R(e_{m-1} \wedge e_m) = \underline{R}(I), \tag{8.5}$$

with $I = e_1 \wedge e_2 \wedge \ldots \wedge e_m$. Thus, our definition of Gaussian curvature for an arbitrary even dimensional manifold is fully in accord with Gauss' original notion.

For hypersurfaces, the Gauss–Bonnet theorem can be proved by using the degree formula (5.10). Considering (8.2), we have

$$\oint_{\mathcal{M}} \kappa\,|\mathrm{d}X| = \oint_{\mathcal{M}} I^{-1}\underline{n}(\mathrm{d}S) = \deg(n)\odot_{m+1}, \tag{8.6}$$

where $\deg(n)$ is the degree of the Gauss mapping. Obviously (8.6) agrees with (8.1) if

$$\deg(n) = \tfrac{1}{2}\chi. \tag{8.7}$$

So by establishing (8.7) we confirm (8.1) for hypersurfaces. This approach is taken in [GP, p. 196]. We shall develop a proof which is not restricted to hypersurfaces. Then (8.7) is a simple consequence of the Gauss–Bonnet theorem.

Our proof of the Gauss–Bonnet theorem will appeal to some basic results in differential topology. We take it for granted that a smooth vector field $u = u(x)$ with a finite number of isolated zeros can be defined on $\mathcal{M}$. According to the *Poincaré–Hopf Index Theorem*, the sum $\Sigma\,\iota$ of indices at zeros of u is equal to the Euler characteristic of M, that is

$$\Sigma\,\iota = \chi. \tag{8.8}$$

(For a convenient proof of the theorem see [Mi, p. 35] or [GP, p. 134]. The Euler characteristic is defined and discussed in [GP] as well as in many other books on topology.)

This relation holds when $\mathcal{M}$ has a boundary if u points outward at boundary points. Moreover, it does not depend on the particular choice of a vector field, so it is a topological invariant of $\mathcal{M}$. Indeed (8.8) could be taken as a definition of the Euler characteristic.

Before considering the general case, let us prove the Gauss–Bonnet theorem for two-dimensional manifolds, since this case has some features of special interest. Let $u \equiv u(x)$ be a smooth vector field on $\mathcal{M}$ with isolated zeros. We may normalize u so that $u^2 = 1$ except at zeros. Following Eqn. (6-2.1), we can express the codifferential of u in terms of the angular velocity of u, which we denote here by $\omega(a)$ instead of ω_a; thus,

$$\delta_a u = \delta u(a) = \omega(a)\cdot u = \omega(a)u. \tag{8.9}$$

The last equality in (8.9) obtains because the trivector $\omega(a)\wedge u$ necessarily vanishes for a two-dimensional manifold. Solving (8.9) for the angular velocity, we get

$$\omega(a) = \delta u(a)u. \tag{8.10}$$

This quantity is well defined except at the zeros of u. Recall that the codifferential is the projection of a differential, that is,

$$\delta u(a) = a\cdot\nabla u = P(a\cdot\partial u) = P(\underline{u}(a)). \tag{8.11}$$

Hence,

$$I^{-1}\omega(a) = I^{-1}\ \delta u(a)u = \langle I^{-1}\underline{u}(a)u\rangle. \tag{8.12}$$

We recognize this as the 'index form' in Eqn. (7.13). So if we apply it to all the zeros of u simultaneously, we have

$$\oint_{\text{zeros}} I^{-1}\omega(\mathrm{d}x) = 2\pi\ \Sigma\ \iota. \tag{8.13}$$

Now consider the curvature tensor. From Eqn. (6-2.9) we see immediately that

$$R(I) = -\delta\omega(I), \tag{8.14}$$

since the commutator $\omega(a) \times \omega(b)$ necessarily vanishes, because all tangent bivectors are proportional. Using the fact that $\delta I = 0$, from (8.14) we get

$$I^{-1}R(I) = -I^{-1}\ \delta\omega(I) = -\delta[I^{-1}\omega] = -\mathrm{d}\langle I^{-1}\omega\rangle. \tag{8.15}$$

We recognize the right side of (8.15) as the exterior differential of the 1-form (8.12). Of course (8.14) and (8.15) are not defined at the zeros of u.

Now we are in a position to evaluate the integral of the Gaussian 2-form $I^{-1}R(\mathrm{d}X)$ over $\mathscr{M}$. Applying Stokes' theorem to (8.15) and using (8.13) to evaluate the integral at the zeros, we get

$$\int_{\mathscr{M}} I^{-1}R(\mathrm{d}X) = 2\pi\ \Sigma\ \iota - \oint_{\partial\mathscr{M}} \langle I^{-1}\omega(\mathrm{d}x)\rangle. \tag{8.16}$$

This supplies us with the desired proof of the Gauss–Bonnet theorem for $\partial\mathscr{M} = 0$ by appeal to the Poincaré–Hopf theorem.

For manifolds with boundary, the last integral in (8.16) can be put in standard form for ease of evaluation. Let $x = x(s)$ be a curve in $\mathscr{M}$ parametrized by arc length s. The velocity v of the curve is a unit vector, so it can be obtained from the unit vector $u = u(x)$ at any point x by a rotation described by the formula

$$v = \frac{\mathrm{d}x}{\mathrm{d}s} = \mathrm{e}^{I\theta}u = u\ \mathrm{e}^{-I\theta}. \tag{8.17}$$

The sign of the angle $\theta = \theta(s)$ depends on the orientation of the unit bivector I. The orientation of I is determined by Eqn. (2.2) relating the orientation of $\mathscr{M}$ to the orientation of $\partial\mathscr{M}$. We discussed this in connexion with complex variable theory, and we determined that I should be regarded as the generator of *clockwise* rotations. The sign of θ has been chosen in (8.17) so that positive θ determines a *counterclockwise* rotation, as is customary.

The codifferential of (8.17) gives us the expression for the 'coacceleration' of the curve compared to the vector field u;

$$v\cdot\nabla v = \delta_v v = (\delta_v uu + Iv\cdot\partial\theta)v. \tag{8.18a}$$

Using (8.10), this can be put in the form

$$k_g \equiv I^{-1} v\, \delta_v v = I^{-1}\omega(v) + \frac{d\theta}{dS}, \tag{8.18b}$$

where we have introduced the symbol k_g to identify the usual *geodesic curvature*. To see that the identification has been properly made, note that no special assumptions about the vector field u were made in deriving (8.18). If u is chosen to be 'parallel' on the curve, then $v \cdot \nabla u = 0$ and $\omega(v) = (v \cdot \nabla u)u = 0$. Whence $k_g = d\theta/dS$, which correctly describes the geodesic curvature as the rate of angular deflection of the curve from a geodesic.

Equation (8.18b) can be used to evaluate the boundary integral in (8.16). Suppose that $\partial\mathscr{M}$ consists of a single piecewise smooth curve e_k with exterior angles α_i at the vertices. Then from (8.18b) we have

$$\oint_{\partial\mathscr{M}} I^{-1}\omega(dx) = \sum_k \oint_e k_g\, dS - \sum_k \oint_{e_k} d\theta. \tag{8.19}$$

For conformity with the Poincaré–Hopf theorem, we assume that u points outward on each e_k. The angle θ, then, varies continuously on e_k between the values of 0 and π, except at the vertices where it suffers discontinuities α_i. It follows that

$$\oint_{e_k} d\theta = -\sum_i \alpha_i. \tag{8.20}$$

With (8.19) and (8.20) in (8.16), we finally get

$$\int_{\mathscr{M}} I^{-1}R(dX) + \oint_{\partial\mathscr{M}} k_g\, dS + \sum_i \alpha_i = 2\pi\chi, \tag{8.21}$$

where the summation is over exterior angles at all vertices on $\partial\mathscr{M}$. The result (8.21) is the famous *Gauss–Bonnet Formula*, in a form valid for a surface with any number of holes in it.

We prove the Gauss–Bonnet theorem for manifolds of arbitrary even dimension by a straightforward generalization of our proof for the two-dimensional case. Our proof is based on the original proof by Chern [Ch1, 2]. It incorporates some simplifications by Flanders [Fl]. Comparison with Chern's work displays the great economy of Geometric Calculus vis-à-vis conventional differential forms. The main problem is to express the Gaussian form as a form explicitly determined by a given vector field and its differentials. Then the theorems of Stokes and Poincaré–Hopf can be applied to complete the proof.

Our computations will be concerned with a sequence of multiforms of various grades and degrees. To simplify the formulas and computations, we introduce a number of abbreviations. To begin with, we define a k, k-form for $k = 1, 2, \ldots, m$ by

$$\begin{aligned}\underline{\delta u}_k = \underline{\delta u}_k(A) &\equiv \langle A e^{i_k} \wedge \ldots \wedge e^{i_2} \wedge e^{i_1}\rangle\, \delta u(e_{i_1}) \wedge \delta u(e_{i_2}) \wedge \ldots \wedge \delta u(e_{i_k})\\ &= k!\; P\underline{u}(\langle A_k\rangle).\end{aligned} \tag{8.22}$$

The last equality, expressing $\underline{\delta u_k}$ as the projection of the extended differential $\underline{u}$ into $\mathscr{G}^k(\mathscr{M})$, is an easy generalization of (8.11). For the k-blade argument $A = a_1 \wedge a_2 \wedge \ldots \wedge a_k$, definition (8.22) reduces to

$$\underline{\delta u_k}(a_1 \wedge \ldots \wedge a_k) = k!\, \delta u(a_1) \wedge \delta u(a_2) \wedge \ldots \wedge \delta u(a_k)$$
$$= k!\, P(\underline{u}(a_1 \wedge \ldots \wedge a_k)). \tag{8.23}$$

Note that $u^2 = 1$ implies $u \cdot \delta u = 0$, hence

$$u \cdot \underline{\delta u_k} = 0. \tag{8.24}$$

It follows that

$$\underline{\delta u_m} = 0. \tag{8.25}$$

From the curvature tensor $R = R(a \wedge b)$ we construct the $2k$, $2k$-form

$$R_k = R_k(A)$$
$$\equiv \langle A e^{i_{2k}} \wedge \ldots \wedge e^{i_2} \wedge e^{i_1} \rangle R(e_{i_1} \wedge e_{i_2}) \wedge \ldots \wedge R(e_{i_{2k-1}} \wedge e_{i_{2k}}). \tag{8.26}$$

Comparison with (8.2) shows that for $m = 2p$,

$$R^p(I) = m!\, \underline{R}(I) = m!\, I\kappa. \tag{8.27}$$

It is convenient to introduce pseudoscalar valued multiforms Y_k and Z_k of degrees $m - 1$ and m, defined for $k = 0, 1, 2, \ldots, p - 1$ by

$$Y_k = Y_k(a) = R_k \wedge \underline{\delta u}_{m-2k-1} \wedge u$$
$$\equiv \langle A e^{i_{m-1}} \wedge \ldots \wedge e^{i_1} \rangle R(e_{i_1} \wedge e_{i_2}) \wedge \ldots \wedge R(e_{i_{2k-1}} \wedge e_{i_{2k}}) \wedge$$
$$\wedge\, \delta u(e_{i_{2k+1}}) \wedge \ldots \wedge \delta u(e_{i_{m-1}}) \wedge u, \tag{8.28}$$

and

$$Z_k = Z_k(A) = R_{k+1} \wedge \underline{\delta u}_{m-2k-2}$$
$$\equiv \langle A e^{i_m} \wedge \ldots \wedge e^{i_1} \rangle R(e_{i_1} \wedge e_{i_2}) \wedge \ldots \wedge R(e_{i_{2k+1}} \wedge e_{i_{2k+2}}) \wedge$$
$$\wedge\, \delta u(e_{i_{2k+3}}) \wedge \ldots \wedge \delta u(e_{i_m}). \tag{8.29}$$

These multiforms are related by the exterior codifferential, specifically

$$\delta Y_k = Z_{k-1} - \frac{1}{2}\left(\frac{m - 2k - 1}{k + 1}\right) Z_k. \tag{8.30}$$

The proof is merely a matter of computation, but a little finesse is helpful.

To establish (8.30), we first prove that

$$Z_k \equiv R_{k+1} \wedge \underline{\delta u}_{m-2k-2} = -2(k+1) R_k \wedge \underline{\delta u}_{m-2k-2} \wedge \delta^2 u \wedge u. \tag{8.31}$$

Recall Eqn. (5-1.15) relating the codifferential to the curvature tensor:

$$\delta^2 u = u \cdot R. \tag{8.32}$$

Using this in connexion with the definition (8.26), we can write

$$R_1 = 2R = 2u \wedge \delta^2 u + h. \tag{8.33}$$

This may be regarded as a definition of h_1; it reduces to (8.32) if

$$u \cdot h = 0. \tag{8.34}$$

Using the binomial expansion and the asymmetry of the multiforms, we get

$$R_k = 2kh_{k-1} \wedge u \wedge \delta^2 u + h_k, \tag{8.35}$$

where h_k is defined in terms of h in the same way that R_k is defined in terms of R by (8.26). But (8.35) implies that $R_k \wedge u = h_k \wedge u$, hence

$$R_{k+1} = 2(k+1)h_k \wedge u \wedge \delta^2 u + u_{k+1}. \tag{8.36}$$

Now note that

$$h_{k+1} \wedge \underline{\delta u}_{m-2k-2} = 0, \tag{8.37}$$

because it has grade m but $u \cdot (h_{k+1} \wedge \underline{\delta u}_{m-2k-2}) = 0$ by virture of (8.24) and (8.34). So (8.31) is established by substituting (8.36) into the left side of (8.31) and using (8.37).

Now we are prepared to evaluate the codifferential of Y_k. We note that

$$\delta R_k = 0. \tag{8.38}$$

This is an immediate consequence of the Bianchi identity $\delta R = 0$ established in Section 5-1. Using (8.38) and (8.29) to evaluate the codifferential of (8.28), we get

$$\begin{aligned} \delta Y_k &= R_k \wedge \underline{\delta u}_{m-2k} + R_k \wedge \delta [\underline{\delta u}_{m-2k-1} \wedge u] \\ &= Z_{k-1} + (m-2k-1) R_k \wedge \underline{\delta u}_{m-2k-2} \wedge \delta^2 u \wedge u, \end{aligned}$$

which, when combined with (8.31), gives (8.30) as promised.

We can solve the set of Eqn. (8.30) for Z_k in terms of the δY_r; thus

$$Z_k = \frac{2(k+1)}{m-2k-1} \left\{ \delta Y_k + \frac{2k}{m-2k+1} \{\delta Y_{k-1} + \cdots\} \right\},$$

and finally,

$$Z_k = -\sum_{r=0}^{k} \frac{(k+1)k \ldots (r+1)2^{k-r+1}}{(m-2k-1)(m-2k+1)\ldots(m-2r-1)} \delta Y_r. \tag{8.39}$$

We will be concerned only with the special case of (8.39) with $m = 2p$ and $k = p - 1$. Considering (8.27) and (8.29), we have

$$(2p)!\, \underline{R}(I) = Z_{p-1} = -\sum_{r=0}^{p=1} \frac{p(p-1)\ldots(r+1)2^{p-r}}{1\cdot 3\ldots(2p-2r+1)}\,\delta Y_r$$

$$= \frac{-p!\,2^p}{1\cdot 3\ldots(2p-1)}\,\delta Y_0 + \cdots \tag{8.40}$$

This is the desired generalization of Eqn. (8.14).

We prove the Gauss–Bonnet theorem by applying Stokes' theorem to (8.40). First note that, by (8.28) and (8.23),

$$I^{-1} Y_0(\mathrm{d}S) = (2p-1)!\,\langle I^{-1}\underline{u}(\mathrm{d}S)\wedge u\rangle. \tag{8.41}$$

Moreover,

$$I^{-1}\,\delta Y_0 = \delta\,\langle I^{-1} Y_0\rangle = \mathrm{d}\langle I^{-1} Y_0\rangle, \tag{8.42}$$

so, by (7.12) and the theorems of Stokes and Poincaré–Hopf,

$$\int_{\mathcal{M}} I^{-1}\,\delta Y_0 = -(2p-1)!\,\odot_m \chi. \tag{8.43}$$

Therefore, from (8.40), we get

$$\int_{\mathcal{M}} I^{-1} R(\mathrm{d}X) = \frac{(p-1)!\,2^{p-1}}{1\cdot 3\ldots(2p-1)}\,\odot_m \chi. \tag{8.44}$$

All terms but the first from the right side of (8.40) make a vanishing contribution to (8.44) because of Stokes' theorem. Equation (8.44) agrees with (8.1), because

$$\odot_{m+1} = \frac{(p-1)!\,2^p}{1\cdot 3\cdot 5\ldots(2p-1)}\,\odot_m = \frac{2^{p+1}\pi^p}{1\cdot 3\cdot 5\ldots(2p-1)}. \tag{8.45}$$

This completes our proof of the generalized Gauss–Bonnet theorem. In developing the proof, we derived some general formulas, like (8.39), which undoubtedly have many other applications, but the matter will not be pursued here. Chern [Ch 2] applies them to evaluate integrals over hypersurfaces. His differential forms Φ_k and Ψ_k are related to our Y_k and Z_k by

$$\Phi_k = (-1)^{m-k-1}\langle I^{-1} Y_k\rangle \quad \text{and} \quad \Psi_k = (-1)^{k+1}\langle I^{-1} Z_k\rangle.$$

Chapter 8

Lie Groups and Lie Algebras

In this brief chapter we aim to lay the foundation for a formulation of the theory of Lie groups and Lie algebras in terms of Geometric Calculus. Such a reformulation of Lie theory appears to be desirable for several reasons. First, it is a step toward the unification of mathematics. Second, the coordinate-free methods of Geometric Calculus can be expected to simplify specific computations as well as the proofs of general results. Third, Geometric Algebra brings new methods and ideas to Lie theory which could simplify the theory and even lead to new results. Indeed, the structure of Geometric Algebra has so much in common with Lie algebra that we would be surprised if they could not be unified in a productive way.

Lie theory is so extensive, that even had we completed its reformulation, the results would certainly not fit in the remainder of this book. We will be content here to get the reformulation started.

Section 1 reformulates the basic concepts and results of Lie theory in sufficient detail to make application of Geometric Calculus to any problem in the theory of group manifolds a fairly straightforward matter. It will be noticed that little more is required than a few definitions. The proofs of Lie's fundamental theorems are quite trivial given the theory of vector manifolds developed in preceding chapters.

Section 2 shows how to make explicit coordinate-free computations on a group manifold by working out an important example.

Section 3 proposes a program for systematically classifying Lie algebras with methods of Geometric Algebra. This section presupposes familiarity with the problem of classifying Lie algebras, so well-known terms and results are freely employed and referred to.

8-1. General Theory

A Lie group $\mathscr{L}$ is a group which is also a manifold on which the group operations are continuous. Recall that a manifold $\mathscr{L}$ is a set of elements in one-to-one correspondence with the points of a vector manifold $\mathscr{M}$. Thus, there exists an invertible function T which maps each point x in $\mathscr{M}$ to a group element $X = T(x)$ in $\mathscr{L}$. This function enables us to apply to groups the powerful differential and integral calculus which we have developed for vector manifolds. On the other hand, the 'group structure' of $\mathscr{L}$ imposes stringent restrictions on the geometric structure of the *group manifold* $\mathscr{M}$. Let us see how this group structure can be expressed precisely.

A group is a set on which there is defined a binary product $XY = W$ which associates a unique element W in the group with each pair of elements X and Y. The group product is associative, so for any three elements,

$$(XY)Z = X(YZ). \tag{1.1a}$$

The group contains a unique identity element E with the property that

$$EX = X = XE \tag{1.1b}$$

for any element X in the group. Finally, to every element X in the group there corresponds a unique *inverse* element X^{-1} with the property

$$XX^{-1} = E = X^{-1}X. \tag{1.1c}$$

For the Lie group $\mathscr{L}$ we have a correspondence of elements with points (vectors) in $\mathscr{M}$ given by

$$X = T(x), \tag{1.2a}$$

$$X^{-1} = T(\overline{x}), \tag{1.2b}$$

$$E = T(e). \tag{1.2c}$$

Corresponding to the group product $XY = T(x)T(y) = T(w)$ defined on $\mathscr{L}$, we have a binary *product function* $\phi(x, y) = w$ defined on $\mathscr{M}$ which associates a unique vector w in $\mathscr{M}$ with each pair of vectors x and y. From (1.1) and (1.2) it follows that the product function has the group properties

$$\phi(\phi(x, y), z) = \phi(x, \phi(y, z)), \quad \text{(associativity)} \tag{1.3a}$$

$$\phi(e, x) = x = \phi(x, e), \quad \text{(identity)} \tag{1.3b}$$

$$\phi(x, \overline{x}) = e = \phi(\overline{x}, x). \quad \text{(inverse)} \tag{1.3c}$$

We have written $\phi(x, y)$ for the product function instead of the simpler notation xy, because we wish to retain the latter notation for the geometric product, and we will be using both kinds of product in our analysis of Lie groups. Indeed, we shall see that for a specific group the product function $\phi(x, y)$ is specified by expressing it in terms of the geometric product. To avoid confusion with the algebraic inverse $x^{-1} = x/x^2$, we have written $\overline{x}$ for the 'group inverse'. Of course, the group product XY should not be confused with the geometric product, although the two products are identical for the important case of vector groups and spinor groups discussed in Section 3-8.

We say that the group $\mathscr{M}$ is a *global parametrization* of the group $\mathscr{L}$. It should be realized that this parametrization is significantly different from the *local parametrization* of a group by coordinates so common in the literature. The two kinds of parametrization can be related by introducing a coordinate system on a *neighborhood* of the identity element in $\mathscr{M}$. This is best done by introducing a smooth invertible function

$$f: x \to x' = f(x), \tag{1.4a}$$

which maps points x in $\mathscr{M}$ to points x' in the tangent space $\mathscr{M}'$ at the identity element in $\mathscr{M}$ and satisfies the condition

$$f(e) = 0. \tag{1.4b}$$

The group product function $w = \phi(x, y)$ on $\mathscr{M}$ is mapped into

$$w' = f(\phi(f^{-1}(x'), f^{-1}(y')) = \phi'(x', y') \tag{1.5}$$

on $\mathscr{M}'$. The new product function $\phi'(x', y')$ obviously enjoys group properties with the same form as those for $\phi(x, y)$ given by (1.3a, b, c). Coordinates for the group are components $x_k' = x' \cdot e_k'$ of local group elements relative to a basis $\{e_k'\}$ in $\mathscr{M}'$. In terms of coordinates, the product function (1.5) takes the form

$$w_k' = \phi_k'(x_1', \ldots, x_n', y_1', \ldots, y_n'). \tag{1.6}$$

This is the form of the product function most used in the literature. Geometric Calculus enables us to work directly with the vector form of the product function (1.5) without decomposing it into the component form (1.6).

The neighborhood of the identity in the vector space $\mathscr{M}'$ which has been endowed with the group product function (1.5) is commonly called a *local Lie group* or *group germ*, because it does not describe the global properties of the group manifold $\mathscr{M}$, for the simple reason that the manifold $\mathscr{M}$ cannot in general be covered by a single coordinate system. In other words, the transformation (1.4a) usually has a singularity at some point of $\mathscr{M}$. The complications that result from this fact can be avoided by working directly with the manifold $\mathscr{M}$, and that is what we shall do. Geometric Calculus makes the transformation from global groups to local groups for computational purposes quite unnecessary, though easy to carry out if desired.

Once the product function $\phi(x, y)$ has been determined, the group structure of $\mathscr{L}$ has been transferred to $\mathscr{M}$ where it can be analyzed with the help of Geometric Calculus without further reference to $\mathscr{L}$. Since this can be done for any Lie group, we can develop an *abstract* theory of Lie groups by studying vector manifolds on which we can define a binary product $\phi(x, y)$ satisfying the group properties. We can get a classification of Lie groups by determining all such vector manifolds and all such product functions. We can get a good start on this program by determining how the geometry of group manifolds is restricted by the group structure.

The group properties (1.3) imply that a continuous group manifold $\mathcal{M}$ is necessarily a *smooth* manifold, which is to say that the unit pseudoscalar field $I = I(x)$ on $\mathcal{M}$ has finite differentials (derivatives) of all orders at every point x in $\mathcal{M}$. It also follows that $\phi(x, y)$ is a smooth function of each variable. For the sake of brevity, we simply assume these results in our analysis.

Our study of group geometry begins with the observation that each 'group element' y determines a transformation λ_y of the group manifold $\mathcal{M}$ onto itself defined by

$$\lambda_y : x \to x' = \lambda_y(x) \equiv \phi(y, x). \tag{1.7}$$

This transformation is called a *left translation* of the group by the element y. Its differential $\underline{\lambda}_y$ transforms each tangent vector $a = a(x)$ at a generic point x of the manifold to a tangent vector $a' = a'(x')$ at some point $x' = \lambda_y(x)$, as given by

$$\underline{\lambda}_y : a(x) \to a'(\lambda_y(x)) = \underline{\lambda}_y(a(x)) = a(x) \cdot \partial_x \lambda_y(x). \tag{1.8a}$$

We often suppress the argument, writing

$$\underline{\lambda}_y : a \to a' = \underline{\lambda}_y(a). \tag{1.8b}$$

We call the linear operator $\underline{\lambda}_y$ the *left differential* by the group element y. From our study of transformations in Section 4-5, we know that left differentials map vector fields into vector fields on the group manifold.

We can also define a *right translation* ρ_y of the group by

$$\rho_y : x \to x' = \rho_y(x) \equiv \phi(x, y),$$

with the corresponding *right differential*

$$\underline{\rho}_y(a(x)) = a(x) \cdot \partial_x \rho_y(x).$$

The properties of right translations are obviously so similar to those of left translations that we lose little by limiting our study to the latter.

Now let us establish the fundamental properties of group translations and differentials, namely, that the set of all left translations is a transformation group, while the set of all left differentials is a group of linear transformations of vector fields. Rewriting the associative rule (1.3a) with the notation of (1.7), we get

$$\lambda_x(\lambda_y(z)) = \lambda_{\phi(x,\, y)}(z). \tag{1.9a}$$

It will be convenient to write this in the operator notation

$$\lambda_x \lambda_y = \lambda_{\phi(x,\, y)}. \tag{1.9b}$$

This is the group product rule for left translations. Applying the rule twice, we have

$$(\lambda_x \lambda_y)\lambda_z = \lambda_{\phi(x,\, y)}\lambda_z = \lambda_{\phi(\phi(x,\, y),\, z)},$$

from which, by using (1.3a) once again, we get the associative rule

$$(\lambda_x \lambda_y)\lambda_z = \lambda_x(\lambda_y \lambda_z). \tag{1.10a}$$

Of course we could have written this down immediately, since the composition of functions is associative in general. Next, applying (1.3b) to (1.9b), we deduce

$$\lambda_x \lambda_e = \lambda_x = \lambda_e \lambda_x, \tag{1.10b}$$

thus, λ_e is the identity transformation. Finally, by applying (1.3c) to (1.9b), we get

$$\lambda_x \lambda_{\bar{x}} = \lambda_e = \lambda_{\bar{x}} \lambda_x, \tag{1.10c}$$

from which we conclude that the inverse of λ_x is $\lambda_x^{-1} = \lambda_{\bar{x}}$. This reformulation of the group properties in terms of left translations is mathematically trivial but conceptually powerful. Both mathematical and conceptual power are increased by proceding to the left differentials.

In an earlier chapter we established that application of the chain rule for differentiation to the composition of transformations yields the composition rule for differentials. Thus, by differentiating (1.9a) and using the notation of (1.8a), we get

$$\begin{aligned} \underline{\lambda}_{\phi(x,\,y)}(a(z)) &= a(z) \cdot \partial_z \lambda_{\phi(x,\,y)}(z) \\ &= [a(z) \cdot \partial_z \lambda_y(z)] \cdot \partial_w \lambda_x(w) \Big|_{w = \lambda_y(z)} \\ &= \underline{\lambda}_x(\underline{\lambda}_y(a(z))), \end{aligned} \tag{1.11a}$$

or, in operator notation,

$$\underline{\lambda}_x \underline{\lambda}_y = \underline{\lambda}_{\phi(x,\,y)}. \tag{1.11b}$$

Since differentials are linear operators, the associative property can be written down without further ado;

$$(\underline{\lambda}_x \underline{\lambda}_y)\underline{\lambda}_z = \underline{\lambda}_x(\underline{\lambda}_y \underline{\lambda}_z). \tag{1.12a}$$

Repeating the arguments used to get (1.10b) and (1.10c), we identify the identity differential

$$\underline{\lambda}_e(a) = a \tag{1.12b}$$

and the inverse

$$\underline{\lambda}_x^{-1} = \underline{\lambda}_{\bar{x}}. \tag{1.12c}$$

Equations (1.11) and (1.12) show that the group of left differentials is isomorphic to the abstract group $\mathscr{M}$. This is a group of linear operators, but it acts on the infinite dimensional space of vector fields on $\mathscr{M}$.

Besides transforming vector fields, the left differential can be used to *construct* vector fields on the group. A tangent vector $a' = a(e)$ at the identity can be extended to a smooth vector field $a = a(x)$ on the whole group by the definition

$$a(x) \equiv \underline{\lambda}_x(a') = a' \cdot \partial_e \phi(x, e). \tag{1.13a}$$

Since the x-dependence is obvious, it will often be convenient to reduce this to

$$a = \underline{\lambda}(a'). \tag{1.13b}$$

Throughout the rest of this section, we use primes exclusively to denote tangent vectors at the identity. The vector field defined by (1.13) is said to be left invariant, because it is transformed into itself by a *left translation* of the group. This follows from the transformation (1.8a) by using (1.13a) and (1.1b), thus

$$\begin{aligned} a'(\lambda_y(x)) &= \underline{\lambda}_y(a(x)) = \underline{\lambda}_y \underline{\lambda}_x(a') \\ &= \underline{\lambda}_{\phi(y,\, x)}(a') = a(\lambda_y(x)). \end{aligned}$$

The left invariance is best expressed by the equation

$$a(\lambda_y(x)) = \underline{\lambda}_y(a(x)). \tag{1.14}$$

We call each left invariant vector field on $\mathscr{M}$ a *generator* of the group $\mathscr{M}$. According to (1.13), every generator is uniquely determined by a tangent vector at the identity. It follows that the set of all generators is a linear space with the same dimension as the group manifold.

We are now in position to prove the *Fundamental Theorem* of Lie Group Theory: *The generators of a Lie group form a Lie algebra*. Multiplication in this algebra is taken to be the Lie bracket defined by Eqn. (4-3.17). With Eqn. (4-3.18) we proved that the Lie bracket $[a, b]$ of differentiable vector fields on any manifold is again a vector field. A set of vector fields $a, b, c, \ldots$ on a manifold form a Lie algebra if it closed under the Lie bracket and all fields satisfy the *Jacobi identity*

$$[[a, b], c] + [[b, c], a] + [[c, a], b] = 0. \tag{1.15}$$

The fundamental theorem was first formulated and proved by Sophus Lie using messy coordinate methods. He broke it into three theorems which have since become famous. Lie's first theorem is equivalent to Eqn. (1.11), and its importance is chiefly for the proof that group generators are left invariant, as we have already established. Lie's second theorem holds that the Lie bracket of generators is again a generator. This follows immediately by using the definition (1.13) in our general formula (4-5.37), to get

$$[a, b] = [\underline{\lambda}(a'), \underline{\lambda}(b')] = \underline{\lambda}([a', b']). \tag{1.16}$$

Lie's third theorem holds that generators satisfy the Jacobi identity. By applying (1.16) twice we see that

$$[[a, b], c] = \underline{\lambda}([[a', b'], c']),$$

so we need only establish the result for tangent vectors at the identity. To this end, we introduce the bilinear function

$$\underline{\phi}(a', b') \equiv a' \cdot \partial_x b' \cdot \partial_y \phi(x, y)\Big|_{x=y=e}. \tag{1.17}$$

This is related to the Lie bracket by applying the definition of the bracket (4-3.17) to (1.13a), whence

$$\underline{\phi}(a', b') - \underline{\phi}(b', a') = [a', b']. \tag{1.18}$$

By operating on the associativity relation (1.3a) with $a' \cdot \partial_x$, $b' \cdot \partial_y$ and $c' \cdot \partial_z$, we get

$$\underline{\phi}(a', \underline{\phi}(b', c')) = \underline{\phi}(\underline{\phi}(a', b'), c'). \tag{1.19}$$

This associativity relation differs from (1.3a) in that it is a linear function of each variable. Therefore,

$$[a', [b', c']] = \underline{\phi}(a', \underline{\phi}(b', c')) - \underline{\phi}(a, \underline{\phi}(b', c')) -$$
$$- \underline{\phi}(\underline{\phi}(b', c'), a') + \underline{\phi}(\underline{\phi}(c', b'), a').$$

A similar expansion of the other two terms in the Jacobi identity shows that all terms cancel when added up. This completes our proof of the Jacobi identity and of the fundamental theorem. We turn now to some other aspects of the general theory which can also be regarded as fundamental.

From left invariant vector fields we can get left invariant multivector fields by the general outermorphism property of Eqn. (4-5.16a); thus,

$$a \wedge b = \underline{\lambda}(a') \wedge \underline{\lambda}(b') = \underline{\lambda}(a' \wedge b') \tag{1.20}$$

is a left invariant bivector field. The left invariant pseudoscalar fields are of great importance for the theory of integration on Lie groups. Their properties can be read off at once from Eqn. (4-5.3), thus

$$M(x) \equiv \underline{\lambda}_x(I') = J(x) I(x), \tag{1.21}$$

where $I(x)$ is the unit pseudoscalar field and $J(x) = \det \underline{\lambda}_x = I^{-1} \underline{\lambda}_x(I')$ is the Jacobian of the left translation λ_x. Now let $\mathrm{d}X(x) = I|\mathrm{d}X|$ be a positively oriented pseudoscalar field on the group. Then

$$\mathrm{D}_\lambda(x) \equiv \frac{\mathrm{d}X(x)}{M(x)} = \frac{\underline{\lambda}_y(\mathrm{d}X)}{\underline{\lambda}_y(M)} = J^{-1}|\mathrm{d}X| \tag{1.22}$$

is obviously invariant under left translations. It is called the *Haar measure* for the group in the literature. With this observation it will not be difficult to apply our integration theory in Chapter 7 to standard problems in group integration theory.

The study of group structure is enhanced by the methods of differential geometry. Though we do not intend to pursue the subject here, we wish to point out that the necessary apparatus is available in this book. For example, all the results of Section 5-6 apply immediately to a group manifold if we take the extensor function h studied there to be

$$h \equiv \underline{\lambda}^{-1} = \underline{\lambda}_x^{-1} = \underline{\lambda}_{\bar{x}}, \tag{1.23a}$$

or, more explicitly,

$$h(a) = h(a(x), x) = \underline{\lambda}_x^{-1}(a(x)). \tag{1.23b}$$

Then Eqn. (5-6.14) provides us with a left invariant tensor on the group manifold

$$g = \bar{h}h = \bar{\lambda}_{\bar{x}}\lambda_{\bar{x}}, \tag{1.24}$$

and Section 5-6 provides us with a formulation of its general properties. Note that for left invariant vector fields $a = \underline{\lambda}(a')$ and $b = \underline{\lambda}(b')$,

$$a \cdot g(b) = a' \cdot b', \tag{1.25}$$

so g can be regarded as the extension to the whole manifold of a scalar product defined at the identity.

The geometrical methods of Chapter 6 are closer to conventional methods than those in Chapter 5. Let us examine briefly how they can be applied to Lie groups. Let $\{e_k'\}$ be a basis in the tangent space of the identity element in the group manifold. Then a frame of left invariant vector fields is defined on the manifold by

$$e_k(x) = \underline{\lambda}_x(e_k'). \tag{1.26}$$

Expanding the Lie bracket of basis vectors at the identity, we have

$$[e_i', e_j'] = c_{ij}^k e_k'.$$

This relation is immediately extended to the rest of the manifold by applying Lie's second theorem (1.16), thus,

$$[e_i, e_j] = \underline{\lambda}([e_i', e_j']) = \underline{\lambda}(c_{ij}^k e_k'),$$

so

$$[e_i, e_j] = c_{ij}^k e_k. \tag{1.27}$$

The coefficients c_{ij}^k are called *structure constants* of the Lie group. Introducing the reciprocal frame $\{e^k\}$ defined by the relations $e^k \cdot e_j = \delta_j^k$, we solve (1.27) to get

$$c_{ij}^k = e^k \cdot [e_i, e_j], \tag{1.28}$$

and from Eqn. (6-1.10), we get

$$\nabla \wedge e^k = \tfrac{1}{2} c_{ij}^k e^j \wedge e^k. \tag{1.29}$$

These equations are commonly called the ***Maurer–Cartan relations***. They can be immediately expressed in terms of differential forms with Eqn. (6-4.24). Of course, Eqn. (1.29) is completely equivalent to Eqn. (1.27). This should suffice to show how the apparatus of Chapter 6 can be effectively applied to Lie groups.

8-2. Computation

Computations in group theory are usually carried out with matrix algebra. We have established by general arguments in Chapter 3 that any computation with matrices can also be carried out with Geometric Algebra and in a coordinate-free manner, unless coordinates are essential to the given problem. Furthermore, we have demonstrated with many examples that computations with Geometric Algebra are usually simpler than those with matrices. Our extensive treatments of the orthogonal groups in Section 3-8 and the symplectic groups in Section 3-7 exhibit considerable advantages over matrix treatments. Here we wish to demonstrate by example how Geometric Algebra enables us to carry out coordinate-free computations on a group manifold.

We will study the group manifold of the spinor group Spin(3) = Spin$^+$(3, 0). This is the covering group of the three-dimensional rotation group and it is isomorphic to the special unitary group SU(2). The spinor groups were completely characterized in Section 3-8, and from that analysis we know that any element U in Spin(3) can be written in the form

$$U = \alpha + B, \tag{2.1}$$

where α is a scalar and B is a bivector in the three-dimensional space of bivectors $\mathcal{G}_3^2$. Thus, U is an element of the four-dimensional linear space of scalars and bivectors $\mathcal{G}_3^0 + \mathcal{G}_3^2$. However, the elements of Spin(3) are restricted by the condition

$$U^\dagger U = (\alpha - B)(\alpha + B) = \alpha^2 + |B|^2 = 1.$$

This suggests that the group can be parametrized by vectors of the unit sphere $\mathcal{M}$ in four-dimensional Euclidean space $\mathcal{E}_4$. To determine the parametrization, we select a unit vector e to represent the identity element and let x be the vector in $\mathcal{M}$ corresponding to U. From these two vectors we can form the scalar $\alpha = e \cdot x$, which will agree with (2.1) and reduce to $U = \alpha = 1$ when $x = e$ provided that B vanishes when $x = e$. The latter condition will be satisfied if B is a function of the bivector $e \wedge x$. We can satisfy the additional condition $e \cdot B = 0$ if $B = Ie \wedge x$, where I is the unit pseudoscalar of $\mathcal{E}_4$, with the properties

$$I^2 = 1, \tag{2.2a}$$

$$Ix = -xI. \tag{2.2b}$$

Thus, our parametrization takes the form

$$U = e \cdot x + Ie \wedge x, \tag{2.3}$$

and we observe that

$$U^\dagger U = (e \cdot x)^2 + |e \wedge x|^2 = x^2 = 1, \tag{2.4}$$

as required.

We can now find the group product function $z = \phi(x, y)$ on $\mathscr{M}$ from the simple group composition rule for spinors

$$U(z) = U(x)U(y). \tag{2.5}$$

Substituting (2.3) into (2.5) and expanding, we find, with the help of identity (1-1.69),

$$e \cdot z + Ie \wedge z = e \cdot xe \cdot y + (e \wedge x) \cdot (e \wedge y) + e(x \wedge e \wedge y) + I(e \cdot xe \wedge y + e \cdot ye \wedge x).$$

Separating this into scalar and bivector parts, we find

$$e \cdot z = e \cdot xe \cdot y + (e \wedge x) \cdot (e \wedge y) = 2e \cdot xe \cdot y - x \cdot y,$$

$$e \wedge z = e \cdot xe \wedge y + e \cdot ye \wedge x + Ie(x \wedge e \wedge y).$$

Adding these equations, we use the fact that $e \cdot z + e \wedge z = ez$ and solve for z, putting the result in the form

$$\phi(x, y) = \langle xey \rangle_1 + x \wedge e \wedge yI = \tfrac{1}{2}(xey + yex) + \tfrac{1}{2}(xey - yex)I. \tag{2.6}$$

Observe that this is a vector-valued bilinear function, although, of course, only its values for unit vectors apply to the group. Its symmetric part can be written in any of the equivalent forms

$$\begin{aligned} \langle xey \rangle_1 &= \tfrac{1}{2}(xey + yex) \\ &= x \cdot ey + (x \wedge e) \cdot y = e \cdot xy + e \cdot yx - x \cdot ye. \end{aligned} \tag{2.7a}$$

Its skewsymmetric part can be written in the equivalent forms

$$\begin{aligned} \langle xeyI \rangle_1 &= \tfrac{1}{2}(xey - yex)I = x \wedge e \wedge yI \\ &= (Ie) \cdot (x \wedge y) = (I \cdot e \wedge x) \cdot y. \end{aligned} \tag{2.7b}$$

As a check on (2.6), we note that

$$z^2 = \langle xey \rangle_1^2 + |\langle xey \rangle_3|^2 = |xey|^2 = y^2 e^2 x^2 = 1.$$

We should also verify that the group properties (1.3a, b, c) are satisfied. Verification of the identity property (1.36) is trivial, and we observe that (2.6) satisfies the inverse property (1.3c) if and only if the element inverse to x is $\bar{x} = -x$. Verification of the associative property requires a tiresome computation which we leave to the reader.

Computation of the left differential from (2.6) is trivial, with the result

$$\underline{\lambda}(a) = \underline{\lambda}_x(a) = \langle xea\rangle_1 + x \wedge e \wedge aI$$
$$= \tfrac{1}{2}(xea + aex) + \tfrac{1}{2}(xea - aex)I. \tag{2.8}$$

This function is defined for all tangent vectors on the group manifold. The condition that $a = a(y)$ is a tangent vector at some point y is simply $y \cdot a = 0$.

To clarify the significance of its functional form (2.8), we ascertain the effect of the left differential on products of vectors. The computation is easiest if we write (2.8) in the form

$$\underline{\lambda}(a) = \tfrac{1}{2}(A + B) + \tfrac{1}{2}(A - B)I = \tfrac{1}{2}(A + B) - I\tfrac{1}{2}(A - B),$$

where $A = xea$ and $B = aex$. Then

$$\underline{\lambda}(a_1)\underline{\lambda}(a_2) = \tfrac{1}{2}(A_1B_2 + B_1A_2) + \tfrac{1}{2}(B_1A_2 - A_1B_2)I,$$

hence

$$\underline{\lambda}(a_1)\underline{\lambda}(a_2) = \tfrac{1}{2}(a_1a_2 + xea_1a_2ex) + \tfrac{1}{2}(a_1a_2 - xea_1a_2ex)I. \tag{2.9}$$

From the scalar part of (2.9) we get

$$\underline{\lambda}(a_1) \cdot \underline{\lambda}(a_2) = a_1 \cdot a_2, \tag{2.10}$$

and from the bivector part we get the outermorphism

$$\underline{\lambda}(a_1 \wedge a_2) = \underline{\lambda}(a_1) \wedge \underline{\lambda}(a_2)$$
$$= \tfrac{1}{2}(a_1 \wedge a_2 + xea_1 \wedge a_2ex) + \tfrac{1}{2}(a_1 \wedge a_2 - xea_1 \wedge a_2ex)eI. \tag{2.11}$$

These results are much more difficult to obtain if computations are carried out with inner and outer products instead of the geometric product.

Equation (2.10) shows that $\underline{\lambda}$ is an orthogonal transformation, in fact, it is a rotation, because it is continuously connected to the identity. Equations (2.6) and (2.8) show that a left translation has the same functional form as its differential; hence the left translations are also rotations. This comes as no surprise, since the group manifold is a sphere in four dimensions. Clearly, the left translations provide us with a representation of SU(2) as a subgroup of the rotation group $O^+(4, 0)$. We proved in Section 3-5 that every such rotation can be written in the canonical form

$$\lambda(y) = \lambda_x(y) = S^\dagger yS. \tag{2.12}$$

In Section 3-8 we learned how to calculate the spinor $S = S(x)$ in (2.12) when the function λ is given. From Eqn. (2.6) we calculate the quantity

$$\partial\lambda = e^k\lambda(e_k) = \tfrac{1}{2}(e^k xee_k + e^k e_k ex) + \tfrac{1}{2}(e^k xee_k - e^k e_k ex)I.$$

From Eqns. (2-2.38a, c) we find $e^k e_k = 4$ and $e^k x \wedge ee_k = 0$. Hence,

$$\partial\lambda = 4[e \cdot x + e \wedge x \tfrac{1}{2}(1 - I)]. \tag{2.13}$$

Finally, from Eqn. (3-8.33) we get the explicit result

$$S = \pm \frac{\partial\lambda}{[\partial\lambda(\partial\lambda)^\dagger]^{1/2}}$$
$$= \pm \frac{e \cdot x + e \wedge x \tfrac{1}{2}(1 - I)}{[(e \cdot x)^2 + |e \wedge x|^2 \tfrac{1}{2}(1 - I)]^{1/2}}. \tag{2.14}$$

This should be compared with the spinor (2.3) which we started out with. Of course these spinors also have the group property $S(x)S(y) = S(\phi(x, y))$.

We can determine the left translation of pseudoscalars by evaluating $\underline{\lambda}(a_1 \wedge a_2 \wedge a_3)$ from $\underline{\lambda}(a_1 \wedge a_2)\underline{\lambda}(a_3)$ in the same manner that we evaluated $\underline{\lambda}(a_1 \wedge a_2)$. Writing $M = a_1 \wedge a_2 \wedge a_3$, we find that the result can be put in the form

$$\underline{\lambda}(M) = \tfrac{1}{2}(Mex + xeM) + \tfrac{1}{2}(xeM - Mex)I. \tag{2.15}$$

To get a left invariant pseudoscalar field we apply (2.15) to the pseudoscalar at the identity Ie, with the result

$$\underline{\lambda}(Ie) = Ix. \tag{2.16}$$

But Ix is just the pseudoscalar at a point x of the sphere. Thus, a left translation simply rotates the pseudoscalar at one point into the pseudoscalar at another with no change of scale. Of course, this is what we expected after we learned that the left differential is a rotation.

Now we are in a position to easily evaluate the Lie bracket of left invariant vector fields. According to (1.18), we can evaluate the bracket at the identity from the product function (2.6), and we find that

$$[a', b'] = Iea' \wedge b'.$$

According to (1.16), we can extend this to the entire group manifold by operating on it with the left differential. As we noted in writing Eqn. (3-5.38), since $\underline{\lambda}$ is a rotation, $\underline{\lambda}(Iea' \wedge b') = \underline{\lambda}(Ie)\underline{\lambda}(a' \wedge b')$, so we get

$$[a, b] = Ixa \wedge b. \tag{2.17}$$

This tells us that the bracket $[a, b]$ is simply the dual of the bivector field $a \wedge b$.

Now let us examine a *local* parametrization of the group. As an alternative to our global parametrization (2.3), we can employ the parametrization by angles determined by the equation

$$\begin{aligned} U = e^{IeX/2} &= \cos \tfrac{1}{2}X + Ie \sin \tfrac{1}{2}X \\ &= \cos \tfrac{1}{2}|X| + Ie\hat{X} \sin \tfrac{1}{2}|X| . \end{aligned} \tag{2.18}$$

As we saw in Section 3-5, the spinor U determines a rotation in three dimensions. The direction of the vector X specifies the axis of rotation, while its magnitude $|X|$ is the angle of rotation. The dual of X by the unit pseudoscalar Ie is the bivector angle of rotation IeX specified by (2.18). To relate the local and global parametrizations, we equate (2.18) to (2.3) and solve for x as a function of X, with the result

$$\begin{aligned} x = f^{-1}(X) &\equiv e \cos \tfrac{1}{2}X + \sin \tfrac{1}{2}X \\ &= e(\exp \tfrac{1}{2}eX), \end{aligned} \tag{2.19}$$

where $e \cdot X = 0$, because X is in the tangent space at the identity. Equation (2.19) is an example of our parametrization Eqn. (1.4) expressed in terms of the inverse function for simplicity.

We are most interested in the induced transformation of a tangent vector field $A = A(X)$ on 'angle space' into a vector field $a = a(x)$ on the sphere. So we calculate the differential of the function (2.19). Straightforward differentiation gives

$$\begin{aligned} \underline{f}^{-1}(A) &= A \cdot \partial_X f^{-1}(X) \\ &= (A \cdot \partial_X |X|) xe\hat{X} + (A \cdot \partial_X \hat{X}) \sin \tfrac{1}{2}|X| . \end{aligned}$$

From Section 2-1, we know how to evaluate the indicated derivatives, and we get the explicit result

$$\underline{f}^{-1}(A) = A \cdot \hat{X} xe\hat{X} + \frac{A \wedge \hat{X}\hat{X}}{|X|} \sin \tfrac{1}{2}|X| . \tag{2.20}$$

Since $\underline{f}^{-1}(A)$ is to be a tangent vector at x, we must have $x \cdot \underline{f}^{-1}(A) = 0$, and from (2.20) we get the requirement $x \cdot (X \wedge A) = 0$.

From (2.20) we compute the induced transformation of pseudoscalars. Many terms vanish, cancel or combine, leaving us with the result

$$\begin{aligned} \underline{f}^{-1}(A_1 \wedge A_2 \wedge A_3) &= \underline{f}^{-1}(A_1) \wedge \underline{f}^{-1}(A_2) \wedge \underline{f}^{-1}(A_3) \\ &= \left(\frac{\sin \tfrac{1}{2}|X|}{|X|} \right)^2 (xe\hat{X}) \wedge [\hat{X} \cdot (A_1 \wedge A_2 \wedge A_3)] . \end{aligned} \tag{2.21}$$

This can be further simplified to give us the transformation for the unit pseudoscalar:

$$\underline{f}^{-1}(Ie) = \left(\frac{\sin \frac{1}{2}X}{X}\right)^2 Ix. \tag{2.22}$$

In accordance with Eqn. (1.21) applied to angle space, from Eqn. (2.22) we conclude that $(\sin \frac{1}{2}X/X)^2$ is the Haar measure on angle space for SU(2) and the rotation group $O^+(3, 0)$.

Another important local parametrization is determined by stereographic projection. From the explicit Eqn. (4-6.31) for stereographic projection, our computations based on Eqn. (2.19) for parametrization by angles can be duplicated in straightforward fashion. The example we have worked out is sufficient to show how such computations can be carried out without coordinates.

Since the group manifold of Spin(3) is a sphere, it will not be necessary to discuss its geometry. We have already discussed the geometry of spheres in Section 5-2. However, we should take note of the fact that a sphere is a surface of constant curvature. There is a general connexion between Lie groups and surfaces of constant curvature which has been much explored in the literature. Our computation has some features which may be important for the general theory. In particular, we conjecture that any surface of constant curvature can be made into a Lie group manifold by endowing it with a bilinear product function $\phi(x, y)$ like (2.6). The linearity of $\phi(x, y)$ is of great importance, for, among other things, it leads directly to a spinor representation for the group, as we observed in writing Eqn. (2.12). This must be related to our observations about spinor representations in the next section, though much work will be required to work out the details. Note, finally, that the linearity of $\phi(x, y)$ is totally obscured in the product function for a local parametrization of the group, as is obvious from our example of parametrization by angles. So it should be no surprise that conventional approaches employing local parametrizations have failed to notice such an important fact.

8-3. Classification

Our analysis in Section 8-1 shows that to every Lie group there corresponds a Lie algebra of left invariant vector fields on the group manifold. Conversely, every simply connected Lie group is uniquely determined by its Lie algebra. It follows, then, that a classification of Lie groups will be achieved by classifying Lie algebras. The latter classification can be attacked with the methods of linear algebra, because every Lie algebra is a linear space. Traditionally, the computations required for detailed classification have been carried out with matrix algebra. But we have seen in Chapter 3 that Geometric Algebra supplies us with alternative ways to make computations. This leads us to ask how Geometric Algebra can best be employed to describe and classify Lie algebras.

To simplify the study of the Lie bracket $[a, b] = c$ on an abstract Lie algebra $\mathfrak{L}$, an *associative* algebra isomorphic to $\mathfrak{L}$ can be constructed in which the Lie bracket is represented as the commutator product

$$[A, B] = A \times B \equiv \tfrac{1}{2}(AB - BA) = C. \tag{3.1}$$

The possibility of identifying the elements $A, B, C, \ldots$ as linear transformations, or matrices, or differential operators has been studied at length in the literature. We shall consider a different possibility. Geometric Algebra is associative, so any subalgebra which is closed under the commutator product is a Lie algebra. We have previously noted that, as a consequence of Eqn. (1-1.67), the space of bivectors is closed under the commutator product. Let us call any linear subspace of bivectors which is closed under the commutator product (3.1) a *bivector algebra*. We now propose to investigate a *Most Interesting Conjecture* (MIC): *Every Lie algebra is isomorphic to a bivector algebra*. The bivector algebra isomorphic to a given Lie algebra $\mathfrak{L}$ is called the *bivector representation* of $\mathfrak{L}$.

To support our claim that MIC is interesting, we will:

(1) establish the existence of bivector representations for many of the most important Lie algebras, and
(2) show that bivector algebras provide us with a simple mechanism for analyzing and representing the structure of Lie algebras.

It follows that MIC is interesting even if it is false, for, at the very least, it should lead to a significant classification of Lie algebras into those with bivector representations and those without.

Now let us examine some specific bivector algebras. For reasons which will become apparent, we believe that the spinor groups characterized in Section 3-8 comprise the most important class of Lie groups. In our discussion here, we assume that the reader is familiar with the nomenclature and results established in Section 3-8. Recall that the spinor group $\text{Spin}^+(p, q)$ is the *covering group* of the special orthogonal group $\text{SO}(p, q) \equiv \text{O}^+(p, q)$ on a vector space $\mathscr{A}_{p,q}$ with signature (p, q). It follows that the Lie algebras of $\text{Spin}^+(p, q)$ and $\text{O}^+(p, q)$ are isomorphic. The Lie algebra of $\text{Spin}^+(p, q)$ is the linear space $\mathscr{B}(p, q) \equiv \mathscr{G}^2(\mathscr{A}_{p,q})$ of all bivectors in the Geometric Algebra of $\mathscr{A}_{p,q}$. If B is a bivector in $\mathscr{B}(p, q)$, then

$$\psi = e^B = \sum_{n=0}^{\infty} \frac{1}{n!} B^n \tag{3.2}$$

in a spinor in $\text{Spin}^+(p, q)$. This has a partial converse: Every element ψ in $\text{Spin}^+(p, q)$ which is continuously connected to the identity can be represented in the form (3.2) if and only if p or $q = 0$ or 1. The case $q = 0$ was treated in Section 3-5, and a complete proof is given in [Ri].

To describe the *structure* of the Lie algebra $\mathscr{B}(p, q)$, we choose a basis $e_1, e_2, \ldots, e_n$ of vectors in $\mathscr{A}_{p,q}$. Their inner products determine a 'metric tensor'

$$g_{ij} \equiv e_i \cdot e_j, \tag{3.3}$$

while their outer products determine a basis for $\mathscr{B}(p, q)$:

$$e_{ij} \equiv e_i \wedge e_j. \tag{3.4}$$

With the help of identity (1-1.68), from (3.3) and (3.4) we evaluate the commutator products

$$[e_{ij}, e_{mn}] = e_{ij} \times e_{mn} = g_{in}e_{jm} - g_{im}e_{jn} + g_{jm}e_{in} - g_{jn}e_{im}. \tag{3.5}$$

These are the well-known *structure relations* for the Lie algebra of the rotation group. But (3.5) is a consequence of the simpler relations (3.3) and (3.4), so these relations can, with equal merit, be referred to as *structure relations* for the bivector algebra $\mathscr{B}(p, q)$; thus, Eqn. (3.4) describes the *structure* of the bivector e_{ij} by giving its factorization into vectors.

Obviously, every bivector algebra is a subalgebra of some $\mathscr{B}(p, q)$. To show how structure relations for a subalgebra can be specified without commutation relations, we exhibit them for the bivector representation of the Lie algebra of the special unitary group SU(n) and its generalization SU(p, q). This algebra is a subalgebra of $\mathscr{B}(2p, 2q)$. To describe it, we select a basis in $\mathscr{A}_{2p,2q}$ of vectors $e_1, e_2, \ldots, e_n$, $f_1 = e_{n+1}, f_2 = e_{n+2}, \ldots, f_n = e_{2n}$ with the properties

$$e_i \cdot e_j = f_i \cdot f_j \equiv g_{ij}, \tag{3.6a}$$

$$e_i \cdot f_j = 0, \tag{3.6b}$$

where $i, j = 1, \ldots, n$. From these vectors we construct bivectors

$$E_{ij} = e_i \wedge e_j + f_i \wedge f_j, \tag{3.7a}$$

$$F_{ij} = e_i \wedge f_j - f_i \wedge e_j \quad (i \neq j), \tag{3.7b}$$

$$H_k = e_k \wedge f_k - e_{k+1} \wedge f_{k+1}, \tag{3.7c}$$

where $i, j = 1, \ldots, n$ and $k = 1, \ldots, n-1$. The bivectors E_{ij}, F_{ij}, H_k are linearly independent, so they form a basis for a space of bivectors of dimension $\frac{1}{2}n(n-1) + \frac{1}{2}n(n-1) + n - 1 = n^2 - 1$. The reader can verify that this space is closed under the commutator product, so it is a subalgebra of $\mathscr{B}(2p, 2q)$. The commutator products are easily evaluated with the help of (3.5) and seen to give the usual structure relations for the special unitary algebra when $g_{ij} = \delta_{ij}$. However, Eqns. (3.6) and (3.7) are not only simpler than those relations, they give us more insight into the structure of the algebra. In particular, the structural relations (3.7) show that *none of the bivectors in this algebra is simple*. Evidently, the distinction between simple and nonsimple bivectors is an important one in the classification of bivector algebras.

It is a well-known fact that every compact Lie group is isomorphic to a closed subgroup of some unitary group. Since we know how to represent the Lie algebras of the unitary groups, it follows that our MIC is true for the Lie algebras of all compact groups.

Every bivector algebra determines a unique spinor group by exponentiation, for the exponential (3.2) of any bivector is a spinor. Thus, from (3.7) we immediately obtain a spinor representation of $SU(p, q)$. More generally, our MIC obviously admits the following corollary: *Every simply connected Lie group is homomorphic to a spinor group*. Our corollary suggests a new approach to the theory of group representations. The conventional approach is based on homomorphisms with the general linear group. Our corollary implies that matrix representations can be replaced by spinor representations. For reasons which should be apparent from this section, we expect to find that the analysis and classification of spinor representations is generally simpler than that of matrix representations. Of course, we do not claim that the general linear group can be dispensed with just because it can be represented by spinors. Rather, we believe that spinor representations will give us a deeper understanding of all transformation groups, as they have most certainly done for the orthogonal groups already.

Let us return to the general problem of classifying Lie algebras. In Cartan's approach to this problem the so-called *Killing form* plays a key role. For the bivector algebra $\mathscr{B}(p, q)$, the Killing form $K(A, B)$ is given by

$$K(A, B) = \sum_{i<j} [A, [B, e_{ij}]] \cdot e^{ji},$$

where the sum is over a basis in $\mathscr{B}(p, q)$. Using (3.1) to reduce this to an expression in terms of the geometric product, we get

$$K(A, B) = \sum_{i<j} \tfrac{1}{2} \langle e^{ji} e_{ij} AB - e^{ji} A e_{ij} B \rangle.$$

From the identities (2-2.38a, c), we get

$$\sum_{i<j} e^{ji} e_{ij} = \binom{n}{2} = \tfrac{1}{2} n(n-1),$$

$$\sum_{i<j} e^{ji} A e_{ij} = \Gamma_2^2 A,$$

where $n = p + q \geqslant 2$, and

$$\Gamma_2^2 = \binom{2}{0}\binom{n-2}{2} - \binom{2}{1}\binom{n-2}{1} = \tfrac{1}{2}(n-2)(n-7).$$

Hence,

$$K(A, B) = \tfrac{1}{2}(4n-7)\langle AB \rangle.$$

The numerical coefficient is of no consequence, so we can regard the inner product $A \cdot B = \langle AB \rangle$ as the Killing form on any bivector algebra (either $\mathscr{B}(p, q)$ or any of its subalgebras). Clearly, both inner and outer products, in fact, the entire Geometric Algebra is needed for a full analysis of bivector algebras.

We are now in a position to examine some standard results translated into properties of bivector algebras. Consider *Cartan's criterion*: A Lie algebra is semisimple (i.e. has no nonzero abelian ideals) if and only if its Killing form is non-degenerate. This can be translated to assert that a bivector algebra is nondegenerate if it has a basis which does not include any null elements. A multivector A is said to be *null* if $\langle A^2\rangle = A \cdot A = 0$. Let us examine the structure of null bivectors. Consider, first, a simple bivector $A = a \wedge b$, for which

$$A^2 = A \cdot A = (a \wedge b) \cdot (a \wedge b) = (a \cdot b)^2 - a^2 b^2. \tag{3.8}$$

From this we can conclude that

$$A^2 = 0 \quad \text{iff } a \cdot b = 0 \text{ and } a^2 b^2 = 0, \tag{3.9a}$$

$$A^2 < 0 \quad \text{iff } a^2 b^2 > 0, \tag{3.9b}$$

$$A^2 > 0 \quad \text{iff } a^2 b^2 < 0. \tag{3.9c}$$

The vectors a and b are factors of A, and they are said to be *orthogonal factors* if $a \cdot b = 0$. From (3.9a) we conclude that a simple bivector is null iff it has a null vector as an orthogonal factor. Now consider a two-bladed bivector $B = B_1 + B_2$; obviously, its 'norm'

$$\langle B^2\rangle = B_1^2 + B_2^2$$

can vanish only if B_1^2 and B_2^2 have opposite sign. Equations (3.9b, c) show how this conditions depends on properties of vector factors; thus, $A^2 < 0$ if the factors of A have the same signature, and $A^2 > 0$ iff the factors have opposite signature. (The signature of a vector is the sign of its square.) These considerations are sufficient to explain how Cartan's criterion can be satisfied or violated by a bivector algebra. They also give us insight into Weyl's theorem that a semisimple connected Lie group is compact iff its Killing form is negative definite; for we have seen that all factors of a negative definite blade have the same signature. Thus, all questions of compactness can be reduced to questions of signature for factors of bivector representations.

It should not be difficult now to reproduce Cartan's classification of semisimple algebras for bivector algebras. However, to follow Cartan's method closely it would be necessary to 'complexify' the algebra. Formal algebraic 'complexification' without an underlying geometric basis is contrary to the philosophy of this book, so we believe it is desirable to look for an alternative method. The issue will be clarified by examining the simplest case.

The simplest Lie algebras are those for the three-parameter groups, and they can all be obtained directly from Eqn. (3.5). Let e_1, e_2 and e_3 be orthogonal vectors; for such vectors (3.5) reduces to

$$[e_{12}, e_{31}] = g_{11} e_{23}, \tag{3.10a}$$

$$[e_{23}, e_{12}] = g_{22} e_{31}, \tag{3.10b}$$

$$[e_{31}, e_{23}] = g_{33} e_{12}. \tag{3.10c}$$

By letting g_{11}, g_{22} and g_{33} have all combinations of the values 1, −1, 0, we get all the Lie algebras for the three-parameter groups. The choice $g_{11} = g_{22} = g_{33} = \pm 1$ gives us the Lie algebra of the isomorphic *compact* groups

$$\mathrm{Spin}^+(3) \sim \mathrm{SO}(3) \sim \mathrm{SU}(2) \sim \mathrm{Sp}(2), \tag{3.11}$$

where Sp(2) is the (real) symplectic group on two dimensions. The choice $g_{11} = g_{22} = -g_{33} = \pm 1$ gives us the Lie algebra for the isomorphic *noncompact* groups

$$\mathrm{Spin}^+(2,1) \sim \mathrm{SO}(2,1) \sim \mathrm{SU}(1,1) \sim \mathrm{SL}(2), \tag{3.12}$$

where SL(2) is the special linear group on two dimensions. The choices $g_{11} = \pm g_{22} = \pm 1$ and $g_{33} = 0$ give us the Lie algebras of the isometry groups on two dimensions, about which we shall have more to say later.

Cartans classification of Lie algebras is based on a systematic analysis of the 'eigenvalue' problem $[A, X] = \lambda X$. To find eigenbivectors of e_{12} in the algebra given by (3.10), we look for linear combinations satisfying

$$[e_{12}, \alpha e_{31} \pm \beta e_{23}] = \pm(\alpha e_{31} \pm \beta e_{23}).$$

Using the commutation relations (3.10), we find that this relation holds if and only if $g_{11}g_{12} = -1$, which implies that the algebra is noncompact. However, if we complexify the compact algebra by introducing 'Hermitian' generators $H_{ij} \equiv ie_{ij}$ where $i = (-1)^{1/2}$ is an imaginary scalar, then Eqn. (3.10a) becomes

$$[H_{12}, H_{31}] = ig_{11}H_{23},$$

and similar changes can be made to Eqns. (3.10b, c), and we find that the condition for H_{12} to have eigenbivectors is now $g_{11}g_{22} = 1$. Clearly, this artifice of complexification to accommodate the eigenvalue problem does not alter the structure relations, but it does introduce some danger of obscuring the crucial distribution between vectors of different signature in the g_{ij}, so we think it can and should be avoided. Besides, Geometric Algebra presents some attractive alternatives. For example, the analysis of 'root vector systems' which characterize the simple Lie algebras is known to be facilitated by employing finite reflexion groups, and our treatment in Sections 3-5 and 3-8 shows that Geometric Algebra is the ideal tool for describing reflexions. Also, our representation (3.7) for the Lie algebra of SU(p, q) suggests that a root vector system for a bivector algebra describes how elements of the algebra can be factored into vectors.

Now let us make a thorough analysis of the spinor representations for an important class of noncompact groups. Let $\mathscr{C}(p, q)$ be the *special conformal group* of the vector space $\mathscr{A}_{p,q}$. We found explicit expressions for the elements of this group in Section 5-5. Now we will prove by explicit construction that $\mathscr{C}(p, q)$ is isomorphic to SO(p + 1, q + 1), hence, Spin$^+$(p + 1, q + 1) is its covering group.

First, we extend $\mathscr{A}_{p,q}$ to $\mathscr{A}_{p+1,q+1}$ by introducing vectors e and $\bar{e}$ with the properties

$$e^2 = -\bar{e}^2 = 1, \qquad e \cdot \bar{e} = 0, \qquad e \cdot x = 0 = \bar{e} \cdot x, \tag{3.13}$$

for all x in $\mathscr{A}_{p,q}$. Next we introduce the function

$$F(x) = -(x-e)n(x-e) = -(x-e)e(x-e) + (x-e)^2\bar{e}, \tag{3.14}$$

where $n = e + \bar{e}$. This function can be regarded as a representation of the stereographic projection of $\mathscr{A}_{p,q}$ onto the unit sphere in $\mathscr{A}_{p+1,q}$ in terms of 'homogeneous coordinates'; for,

$$f(x) = \frac{\bar{e}e \wedge F(x)}{e \cdot F(x)} = -(x-e)^{-1}e(x-e) \tag{3.15}$$

is exactly our formula (4-6.31) for a stereographic projection, and it is obviously invariant under an arbitrary scale transformation

$$F(x) \to \lambda(x)F(x).$$

We will show that rotations of the point $x' = F(x)$ in $\mathscr{A}_{p+1,q+1}$ correspond exactly to conformal transformations of the point x in $\mathscr{A}_{p,q}$. In our computations we will use the relations

$$\begin{aligned} \bar{n} \equiv e - \bar{e} = ene, \qquad & n^2 = 0 = \bar{n}^2, \qquad nen = 2n \\ n\bar{n} = 2ne \qquad \bar{e}en = n, \qquad & xe = -ex, \qquad x\bar{e} = -\bar{e}x, \end{aligned} \tag{3.16}$$

which follow from the relations (3.13).

Rotations on $\mathscr{A}_{p,q}$ are obviously rotations on $\mathscr{A}_{p+1,q+1}$, so if U is an element of $\text{Spin}^+(p, q)$, then

$$UF(x)U^\dagger = F(UxU^\dagger). \tag{3.17}$$

Explicit proof of this relation from (3.14) uses only the properties that U commutes with e and $\bar{e}$ and $UU^\dagger = 1$.

The correspondence between rotations and translations is given by the equation

$$T_aF(x)T_a^\dagger = F(x+a), \tag{3.18}$$

where T_a is a spinor defined by

$$T_a = e^{na/2} = 1 + \tfrac{1}{2}na = 1 - \tfrac{1}{2}an. \tag{3.19}$$

Equation (3.18) is derived by operating with T_a on (3.14); thus,

$$T_a(x - e)n = (1 - \tfrac{1}{2}an)(x - e)n$$

$$= (x + a - e)n,$$

with a similar result from the action of $T_a^\dagger$.

The special conformal transformation (5-5.51a) corresponds to the rotation

$$K_a F(x) K_a^\dagger = \sigma(x) F(x(1 - ax)^{-1}), \tag{3.20}$$

where

$$K_a = e^{\bar{n}a/2} = 1 + \tfrac{1}{2}\bar{n}a, \tag{3.21}$$

and

$$\sigma(x) = (1 - xa)(1 - ax).$$

We have already noted that the scale factor $\sigma(x)$ in (3.20) does not affect the correspondence established via the stereographic projection (3.15). Equation (3.20) follows by operating with K_a on (3.14), for

$$K_a(x - e)n(x - e) = (1 - \tfrac{1}{2}a\bar{n})(x - e)n(x - e)$$

$$= (x - e + aex)n(x - e)$$

$$= [x(1 - ax)^{-1} - e](1 - ax)n(x - e).$$

In a similar way, by operating on this expression from the right with $K_a^\dagger$ we get (3.20) as desired.

The correspondence between dilations and rotations is given by

$$D_\alpha F(x) D_\alpha^\dagger = e^{-\alpha} F(e^\alpha x), \tag{3.22}$$

where

$$D_\alpha = e^{\alpha e\bar{e}/2}, \tag{3.23}$$

as the reader can readily verify.

To sum up, we have shown that the spinors K_a, U, D_α, T_a are respectively, in two-to-one correspondence with the special conformal transformation, the rotation, the dilation and the translation specified by Eqns. (5-5.51a, b, c, d). This group of spinors is therefore a representation of the conformal group. The generators of this spinor group are bivectors

$$(e - \overline{e}) \wedge e_i \quad \text{for } K_a, \tag{3.24a}$$

$$e_i \wedge e_j \quad \text{for } U, \tag{3.24b}$$

$$\overline{e} \wedge e \quad \text{for } D_\alpha, \tag{3.24c}$$

$$(e + \overline{e}) \wedge e_i \quad \text{for } T_a, \tag{3.24d}$$

where $\{e_i\}$ is a basis for $\mathscr{A}_{p,q}$. But these bivectors form a complete basis for $\mathscr{B}(p + 1, q + 1)$, which is the Lie algebra for $\text{Spin}^+(p, q)$. Therefore, the special conformal group $e(p, q)$ is isomorphic to the complete orthogonal group $\text{SO}(p + 1, q + 1)$ as claimed.

References

[Ar] Artin, E., *Geometric Algebra*, Interscience, New York (1957).

[Be] Bellman, R., *Introduction to Matrix Analysis*, McGraw-Hill, New York (1960).

[Ber] Bers, L., *Riemann Surfaces*, Courant Institute Lecture Notes, New York University, N.Y. (1958).

[Bo] Bourbaki, *Éléments de Mathématique VII*, partie 1, livre 2, *Algèbre*, chapitre 3, 'Algébre multilinèaire', Hermann, Paris (1958).

[Bot] Bott, R., *Lectures on K(X)*, Benjamin, N.Y. (1969).

[Ca] Carson, A., 'Analogue of Green's theorem for multiple integral problems in the calculus of variations', in *Contributions to the Calculus of Variations (1938–1941)*, University Chicago Press, Chicago (1942).

[Ch1] Chern, S., 'A simple intrinsic proof of the Gauss–Bonnet formula for closed Riemannian manifolds', *Ann. Math.* **45**, 747–52 (1944).

[Ch2] Chern, S., 'On the curvatura integral in a Riemannian manifold', *Ann. Math.* **46**, 674–684 (1945).

[Ch] Chevalley, C., *The Algebraic Theory of Spinors*, Columbia University Press, New York (1954).

[Do] Doublet, P., Rota, G.-C., and Stein, J., 'On the foundations of combinatorial theory. IX, Combinatorial methods in invariant theory', *Studies Appl. Math.* **53**, 71 (1974).

[Fl] Flanders, H., 'Development of an extended differential calculus', *Trans. Am. Math. Soc.* **75**, 311–26 (1953).

[GS] Gelfand, I. and Shilov, G., *Generalized Functions*, Vol. I, Academic Press, New York (1964).

[GP] Guillemin, V. and Pollack, A., *Differential Topology*, Prentice-Hall, New Jersey (1974).

[Gr] Greub, W., *Multilinear Algebra*, Springer-Verlag, New York (1967), especially Chapters 5 and 6.

[Ha] Halmos, P., *Finite Dimensional Vector Spaces*, 2nd edn., Van Nostrand, New York (1958).

[H1] Hestenes, D., *Space-Time Algebra*, Gordon & Breach, New York (1966).

[H2] Hestenes, D., 'Real spinor fields', *J. Math. Phys.* **8**, 798 (1967).

[H3] Hestenes, D., 'Spin and isospin', *J. Math. Phys.* **8**, 809 (1967).

[H4] Hestenes, D., 'Vectors, spinors and complex numbers in classical and quantum physics', *Am. J. Phys.* **39**, 1013 (1971).

[H5] Hestenes, D. and Gurtler, R., 'Local observables in quantum theory', *Am. J. Phys.* **39**, 1028 (1971).

[H6] Hestenes, D., 'Local observables in the Dirac theory', *J. Math. Phys.* **14**, 893 (1973).

[H7] Hestenes, D., 'Proper particle mechanics', *J. Math. Phys.* **15**, 1786 (1974).

[H8] Hestenes, D., 'Proper dynamics of a rigid point particle', *J. Math. Phys.* **15**, 1778 (1974).

[H9] Hestenes, D., 'Observables, operators and complex numbers in the Dirac theroy', *J. Math. Phys.* **16**, 556 (1975).

[H10] Hestenes, D. and Gurtler, R., 'Consistency in the formulation of the Dirac, Pauli and Schroedinger theories', *J. Math. Phys.* **16**, 573 (1975).

[H11] Hestenes, D., 'Spin and uncertainty in the interpretation of quantum mechanics', *Am. J. Phys.* **47**, 399 (1979).

[H12] Hestenes, D., 'Space-time structure of weak and electromagnetic interactions', *Found. Phys.* **12**, 153 (1982).

[H13] Hestenes, D., 'Multivector calculus', *J. Math. Anal. Appl.* **24**, 313 (1968).

[H14] Hestenes, D., 'Multivector functions', *J. Math. Anal. Appl.* **24**, 467 (1968).

[He] Hestenes, M. 'Analogue of Green's theorem in the calculus of variations', *Duke Math. J.* **8**, 300 (1941).

[Hi] Hicks, N., *Notes on Differential Geometry*, Van Nostrand, New York (1965).

[Hil] Hille, E., *Analytic Function Theory*, Vol. I, Ginn and Company, New York (1959).

[Hu] Husemoller, D., *Fibre Bundles*, McGraw-Hill, New York (1966).

[La] Lang, S., *Differentiable Manifolds*, Addison-Wesley, Reading, Mass. (1972).

[Mc] McShane, E., 'A unified theory of integration', *Am. Math. Monthly* **80**, 349 (1973).

[Mi] Milnor, J., *Topology from the Differentiable Viewpoint*, University of Virginia Press, Charlottesville (1965).

[MTW] Misner, C., Thorne, K., and Wheeler, J., *Gravitation*, Freeman, San Francisco (1973).

[Po] Porteous, I. R., *Topological Geometry*, Van Nostrand, London (1969).

[Ri] Riesz, M., *Clifford Numbers and Spinors*, Lecture Series No. 38, The Institute for Fluid Dynamics and Applied Mathematics, University of Maryland (1958).

[Ro] Rota, G.-C. and Stein, J., 'Applications of Cayley algebras', *Teorie Combinatoire*, Tomo II, Accademia Nazionale Dei Lincei, Roma (1976), p. 71.

[S1] Sobczyk, G., *Mappings of Surfaces in Euclidean Space Using Geometric Algebra*, Thesis, Arizona State University (1971).

[S2] Sobczyk, G., 'Spacetime algebra approach to curvature', *J. Math. Phys.* **22**, 333 (1981).

[S3] Sobczyk, 'Plebanski Classification of the Tensor of Matter', *Acta Physica Pol.* **B11**, 579 (1980).

[S4] Sobczyk, G., 'Conjugations and Hermitian Operators in Space Time', *Acta Physica Polonica* **B12**, 509 (1981).

[S5] Sobczyk, G., 'Algebraic Classification of Important Tensors in Physics', *Phys. Letters* **84A**, 49 (1981).

[St] Struik, D., *Lectures on Classical Differential Geometry*, Addison-Wesley, Reading, Mass., (1950).

[Sy] Synge, J., *The Petrov Classification of Gravitational Fields*, Communications of the Dublin Institute for Advanced Studies, Series A, No. 15 (1964).

[Tu] Turnbull, H., *The Theory of Determinants, Matrices and Invariants*, Dover, N.Y. (1960).

[Wh] Whitney, H., *Geometric Integration Theory*, Princeton University Press, Princeton, New Jersey (1957).

[We] Westwick, R., 'Linear transformations of Grassmann spaces', *Pacif. J. Math.*, 1123–1127 (1964).

Additional References

Since this manuscript was completed in 1975 there has been increasingly widespread research on Clifford Algebras. The following articles are classified by topic. Unfortunately, we learned about them too late to integrate them with results in the text.

The Universal Property of Clifford Algebra

Karoubi, M. (1978). *K-Theory*, Springer-Verlag, New York.

Micali, A. and Revoy, Ph. (1977). 'Modules quadratique', *Cah. Mathe.* **10**, Montpellier.

Graded Structure of Clifford Algebras

Bass, H. (1967). *Lectures on topics in algebraic K-theory*. Tata Institute of Fundamental Research, Lectures on Math. No. 41, Bombay.

Bass, H. (1967). 'Clifford algebras and spinor norms over a commutative ring', *Am. J. Math.* **96**, 156–206.

Wall, C. T. C. (1964). 'Graded Brauer groups', *J. Reine Angew. Math.* **213**, 187–199.

Wall, C. T. C. (1968). 'Graded algebras, anti-involutions, simple groups and symmetric spaces', *Bull. Am. Math. Soc.* **74**, 198–202.

Classification of Clifford Algebras

Edwards, B. H. (1978). 'On classifying Clifford algebras', *J. Indian Math. Soc.* **42**, 339–344.

Lounesto, P. (1981). 'Scalar products of spinors and an extension of Brauer-Wall groups', *Found. Phys.* **11**, 721–740.

Function Theory on Clifford Algebra

Battle, G. A. (1981). 'Generalized analytic functions and the two-dimensional Euclidean Dirac operator', *Comm. Partial Differential Equations* **6**, 121–151.

Delange, R. (1970). 'On regular-analytic functions with values in a Clifford algebra', *Math. Ann.* **185**, 91–119.

Delange, R. (1972). 'On the singularities of functions with values in a Clifford algebra', *Math. Ann.* **196**, 293–319.

Delange, R. and Brackx, F. (1978). 'Hypercomplex function theory and Hilbert modules with reproducing kernel', *Proc. Lond. Math. Soc.* **37**, 545–576.

Gilbert, R. P. and Hile, G. (1976). 'Hypercomplex function theory in the sense of L. Bers', *Math. Nach.* **72**, 187–200.

Goldschmidt, B. (1981). 'Regularity properties of generalized analytic vectors in R^n', *Math. Nachr.* **103**, 245–254.

Habetha, K. (1976). *Eine Bermerkung zur Functionentheorie in Algebren. Function theoretic methods for partial differential equations*, Proc. Internat. Sympos., Darmstadt, 1976, pp. 502–509. Lecture Notes in Math., Vol. 561, Springer, Berlin.

Lounesto, P. (1979). *Spinor Valued Regular Functions in Hypercomplex Analysis*, Helsinki University of Technology, Mathematical Report A154.

Sprossig, W. (1979). 'Taylor-und Laurententwicklungen raumlich veralgemeinerter analytischer Functionen', *Wiss. Z. Tech. Hochsch. Karl-Marx-Stadt* **21**, 889–894.

Functional Analysis on Infinite Dimensional Clifford Algebras

de la Harpe, P. (1972). 'The Clifford algebra and spinor group of a Hilbert space', *Compositio Math.* **25**, 245–261.

Plymen, R. J. (1976). 'Spinors in Hilbert space', *Math. Proc. Camb. Phil. Soc.* **80**, 337–347.

Plymen, R. J. (1980). 'The Laplacian and the Dirac operator in infinitely many variables', *Composito Math.* **41**, 137–152.

Shale, D. and Stinesprung, W. (1964). 'States of the Clifford algebras', *Ann. Math.* **80**, 365–381.

Differential Geometry with Clifford Algebra

Crumeyrolle, A. (1971). 'Groupes de spinorialité', *Ann. Inst. H. Poincaré* **A14**, 309–323.

Crumeyrolle, A. (1980). 'Algèbres de Clifford dégénérées et revêlements des groupes conformes affines orthogonaux et symplectiques', *Ann. Inst. H. Poincaré* **A33**, 235–249.

Lawson, H. B. and Gronow, M. (1980). 'Spin and scalar curvature in the presence of a fundamental group', *Ann. Math.* 111, 209–230 and 423–434.
Lounesto, P. (1980). 'Conformal transformations and Clifford algebras', *Proc. Am. Math. Soc.* 79, 533–538.
Rzewuski, J., Z. Oziewicz, and R. Abłamowicz (1982). 'Clifford Algebra Approach to Twistors', *J. Math. Phys.* 23, 231–242.

Index